Introduction to Biomechatronics, Second Edition

Other related titles:

You may also like

- PBHE022 | Nazarpour | Control of Prosthetic Hands: Challenges and emerging avenues | 2020
- PBCE105 | Chappell | Mechatronic Hands: Prosthetic and robotic design | 2016
- PBCE099 | Yamamoto | Practical Robotics and Mechatronics: Marine, space and medical applications | 2016
- PBHE014 | Xie, Zhang, and Meng | Soft Robots for Healthcare Applications: Design, modelling, and control | 2017
- PBCE109 | Chin | Embedded Mechatronics System Design for Uncertain Environments: Linux®-based, Rasbpian®, ARDUINO® and MATLAB® xPC Target Approach | 2018
- PBCE104 | Levent Güvenç, Aksun Güvenç, Demirel, and Emirler | Control of Mechatronic Systems | 2017

We also publish a wide range of books on the following topics:
Computing and Networks
Control, Robotics and Sensors
Electrical Regulations
Electromagnetics and Radar
Energy Engineering
Healthcare Technologies
History and Management of Technology
IET Codes and Guidance
Materials, Circuits and Devices
Model Forms
Nanomaterials and Nanotechnologies
Optics, Photonics and Lasers
Production, Design and Manufacturing
Security
Telecommunications
Transportation

All books are available in print via https://shop.theiet.org or as eBooks via our Digital Library https://digital-library.theiet.org.

IET Healthcare Technologies 067

Introduction to Biomechatronics, Second Edition

Volume 1: Mechatronic considerations

Graham M. Brooker

The Institution of Engineering and Technology

About the IET

This book is published by the Institution of Engineering and Technology (The IET).

We inspire, inform and influence the global engineering community to engineer a better world. As a diverse home across engineering and technology, we share knowledge that helps make better sense of the world, to accelerate innovation and solve the global challenges that matter.

The IET is a not-for-profit organisation. The surplus we make from our books is used to support activities and products for the engineering community and promote the positive role of science, engineering and technology in the world. This includes education resources and outreach, scholarships and awards, events and courses, publications, professional development and mentoring, and advocacy to governments.

To discover more about the IET, please visit https://www.theiet.org/.

About IET books

The IET publishes books across many engineering and technology disciplines. Our authors and editors offer fresh perspectives from universities and industry. Within our subject areas, we have several book series steered by editorial boards made up of leading subject experts.

We peer review each book at the proposal stage to ensure the quality and relevance of our publications.

Get involved

If you are interested in becoming an author, editor, series advisor, or peer reviewer, please visit https://www.theiet.org/publishing/publishing-with-iet-books/ or contact author_support@theiet.org.

Discovering our electronic content

All of our books are available online via the IET's Digital Library. Our Digital Library is the home of technical documents, eBooks, conference publications, real-life case studies and journal articles. To find out more, please visit https://digital-library.theiet.org.

In collaboration with the United Nations and the International Publishers Association, the IET is a Signatory member of the SDG Publishers Compact. The Compact aims to accelerate progress to achieve the Sustainable Development Goals (SDGs) by 2030. Signatories aspire to develop sustainable practices and act as champions of the SDGs during the Decade of Action (2020–30), publishing books and journals that will help inform, develop, and inspire action in that direction.

In line with our sustainable goals, our UK printing partner has FSC accreditation, which is reducing our environmental impact on the planet. We use a print-on-demand model to further reduce our carbon footprint.

Published by The Institution of Engineering and Technology, London, United Kingdom

The Institution of Engineering and Technology (the "**Publisher**") is registered as a Charity in England & Wales (no. 211014) and Scotland (no. SC038698).

First published: 2012

Second edition published: 2025

The Institution of Engineering and Technology
Futures Place,
Kings Way, Stevenage,
Herts, SG1 2UA
United Kingdom

www.theiet.org

British Library Cataloguing in Publication Data
A catalogue record for this product is available from the British Library

ISBN 978-1-83724-103-3 (Volume 1 hardback)
ISBN 978-1-83724-104-0 (Volume 1 PDF)

ISBN 978-1-83724-105-7 (Volume 2 hardback)
ISBN 978-1-83724-106-4 (Volume 2 PDF)

Typeset in India by MPS Limited

Cover image credit: MTStock Studio/E+ via Getty Images

Contents

About the author

Graham M. Brooker is a senior lecturer at the School of Aerospace, Mechanical and Mechatronic Engineering in the Engineering Department of the University of Sydney, Australia. He has published more than 120 journal and conference papers as well as numerous books and book chapters. His research interests include environmental radar imaging, biomechatronics and rehabilitation engineering.

Chapter 1

Introduction to biomechatronics

1.1 Introduction

The second edition of the Introduction to Biomechatronics comes in two volumes, each of which is stand-alone. The first volume provides a comprehensive background to most of the disciplines required to understand mechatronics in general, as summarised in Figure 1.1 and would be a good introductory text in its own right. Volume 2 introduces a number of specific biomechatronic topics that would be of interest to both Mechatronics and Biomedical students.

Mechatronic engineering can be defined as the synergistic combination of mechanical, electronic, computer and control systems, along with a dash of systems engineering, as illustrated in Figure 1.1. This interdisciplinary combination brings together the requisite technology and skills to design new and improve existing electromechanical systems.

Biomechatronics is the application of mechatronic engineering to human biology and as such, it forms an important subset of the overall biomedical engineering discipline. As with mechatronics, which is often considered to be synonymous with robotics, biomechatronics is often considered to be restricted to the development of prosthetic limbs. However, in reality, biomechatronics covers a much wider genre than this, and along with prosthetic limbs, Volume 2 examines some of the more interesting applications including those that relate to hearing, respiration, visual and neural interfaces, the cardiovascular system and exoskeletons.

1.2 Biomechatronic systems

Ultimately, biomechatronics can be thought of similarly to any other engineering system, with one of the elements, generally the most complex one, being the human being. Unfortunately, the human element is not only the most complex and least understood, but also the most difficult to interface with. Attempts to measure and stimulate the human body are not completely deterministic, and repeated application of a set of inputs will not always produce the same response. In fact, even when under conscious control, responses (or actions) are seldom identical. Consider, for example, the best sportsmen in the world, with practice and talent they are able to produce fairly repeatable performances, but subtle changes in initial conditions, within and external to their bodies, result in some variations. This uncertainty is

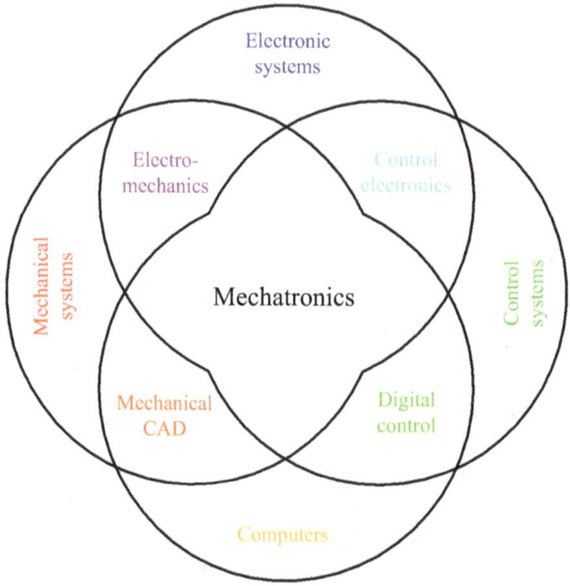

Figure 1.1 Mechatronic engineering as a combination of mechanical, electronic, computer and control systems

manifest across the complete range of physiological responses, from slight variations in the resting heart rate through the apparently chaotic nature of firing neurons.

In a typical biomechatronic system, a number of components can be identified. These include the following:

- The human (or animal) subject,
- Stimulus or actuation,
- Transducers and sensors,
- Signal conditioning elements,
- Recording and display and
- Feedback elements.

These are clearly illustrated in Figure 1.2.

1.2.1 The human subject

The human subject adds the 'bio' to this mechatronic control and monitoring process. What makes biomechatronics particularly interesting compared to other mechatronic systems is the diversity and complexity of human physiology. Unlike the usual engineering systems, the behaviour of which can be more-or-less predicted, each human being is unique and ever-changing. This is not a book on human physiology, but to give the reader some essential background, its various aspects are considered in more detail in Volume 2 when a specific type of biomechatronic system is introduced.

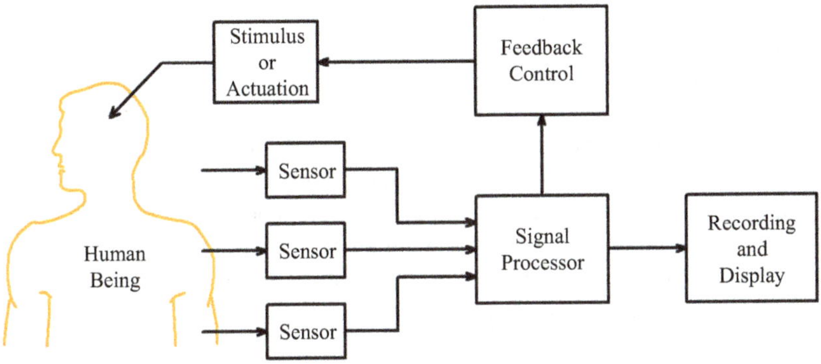

Figure 1.2 Block diagram showing the elements of a biomechatronic system

1.2.2 Stimulus or actuation

The process of stimulus can be introduced as a feedback element, as shown in Figure 1.2, or as a naturally occurring input. Sources of stimulus can include any modality that has an effect on the human element. This can include electrical stimuli, an audio tone, control of air or blood flow, a source of light, a tactile stimulus or even the physical actuation of a limb. Volume 1 includes an entire chapter on actuators, their implementation and analysis in general terms, as well as considering them as part of a number of individual biomechatronic systems.

1.2.3 Transducers and sensors

Transducers and sensors are the devices that convert physiological outputs into signals that can be utilised. In most cases, these are sensors that amplify electrical signals or convert them from chemical concentration, temperature, pressure or flow into electrical signals that can be further processed. Interfacing with the human body is not a trivial task, as embedded sensors must be biocompatible, flexible and extremely robust to survive in the aggressive internal environment. While surface sensors, particularly electrodes, must be able to form a compatible and relatively stable conductive interface across the skin. As with actuators, sensors and sensing of physiological processes are considered to be sufficiently important to devote a full chapter of Volume 1 to the physical mechanisms that underpin their operation.

1.2.4 Sensory substitution extension and addition

In the second edition of the book, the section on sensory substitution has been expanded in scope and given its own chapter in Volume 1.

1.2.4.1 Substitution

Sensory substitution refers to the use of one sensory modality to supply information typically provided by another. This technique allows individuals, particularly those with sensory impairments, to perceive the world in alternative ways. For instance, a common device used in sensory substitution is the tactile-vision substitution system where visual information is converted into tactile feedback on the skin, enabling visually impaired users to 'see' through touch. Another example is auditory-tactile substitution, where sounds are converted into vibrations or tactile sensations that can be felt, aiding those who are deaf or hard of hearing. These devices and systems work by capturing information from the environment using sensors and then translating this data into a form that can be understood through another sense, effectively replacing one sensory input with another. Sensory substitution explores and utilises the brain's plasticity, demonstrating its ability to adapt and process sensory information in new and varied forms.

1.2.4.2 Extension

Sensory extension refers to the enhancement or augmentation of human sensory capabilities through technology or biological means. This concept involves extending the natural range, resolution or sensitivity of our senses, enabling individuals to perceive beyond the normal boundaries of human sensation. Examples include night vision goggles that extend our ability to see in low-light conditions.

1.2.4.3 Addition

Sensory addition involves augmenting the human sensory system with new capabilities that provide additional types of information beyond the traditional five senses. This concept is explored primarily in the field of human enhancement technology, where devices or implants are designed to give individuals new sensory experiences or perceptions that they could not naturally achieve. Examples of sensory addition include the creation of devices that allow humans to detect electromagnetic fields, sense ultraviolet or infrared light, or perceive real-time geospatial information. These augmentations can be achieved through wearable technology or more invasive methods such as implants that interface directly with the body's nervous system.

1.2.5 Signal processing elements

Signal processing involves modification of the electrical signal to some form that is more useful. This generally involves amplification and filtering to extract salient features. However, it often involves the conversion of the analogue signal to a digital equivalent that allows for the application of complex algorithms to obtain more subtle characteristics. Volume 1 includes a chapter that concentrates on the classical aspects of signal processing before introducing the reader to modern machine learning algorithms. The latter are becoming essential tools in the quest to better understand complex physiological processes.

1.2.6 Recording and display

In many cases, the biomechatronic device functions to monitor a physiological process or response. In these cases, it may be important to display the information in a form that is easy to interpret or to store it for later analysis. Common examples of such devices are the now ubiquitous 12-lead electrocardiograph, pneumotachographs and sphygmomanometers. In the past, many of these devices were mechanical and recorded onto paper tape or photographic film, but with the advent of modern electronics, most have been replaced by their electronic equivalents – random access memory and liquid crystal displays.

1.2.7 Feedback elements

In a closed-loop control application, any stimulus or excitation signal is conditioned by the processed outputs of one or a number of sensors monitoring the physiological process. The link that connects the sensing output back to the stimulus includes further processing through control elements. This feedback can be used to close an external loop, or one that operates through the human being. An example of the latter would be a vibrotactile stimulus that is triggered in response to some changes in the direction of the Earth's magnetic field to help a blind person to navigate. The chapter on classical control in Volume 1 investigates the use of feedback elements in terms of their effect on loop performance.

1.3 Physiological systems

To interact effectively with the human part of the structure shown in Figure 1.2, it is essential to have some understanding of the subject on which the measurements are being made or to which the stimulus is applied. The major functional systems of the body include the cardiovascular, respiratory and musculoskeletal systems along with those that interpret taste, sight, hearing and touch. These are illustrated in Figure 1.3.

Most of Volume 2 considers the application of biomechatronic systems to these physiological components.

1.3.1 Biochemical system

The human body is controlled and powered by a complex system of chemical processes. Biochemical processes convert the food we eat into amino acids that are

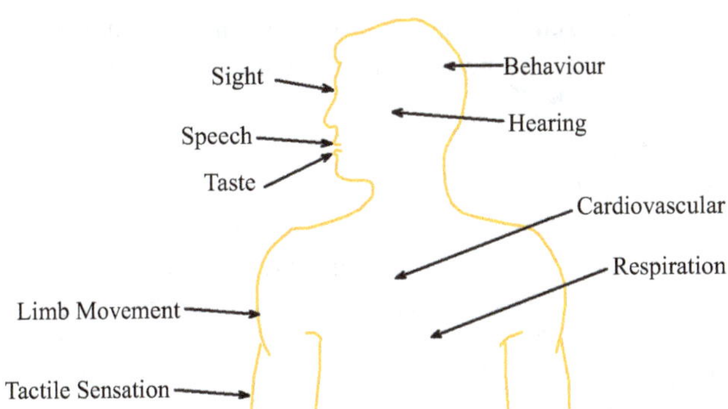

Figure 1.3 Major physiological components that can be incorporated into biomechatronic systems [adapted from Cromwell et al. (1973)]

used for building and repair; they break down sugars and fats and store them for later use as sources of energy. In addition, our blood and tissues are awash with hormones and other organic molecules that control, signal and regulate the amazingly interconnected functions that keep the system alive and healthy. Later sections of Volume 2 examine some of the sensors and actuators that interact with these biochemical processes for monitoring and control purposes.

1.3.2 Nervous system

Whereas the biochemical system is mostly concerned with long-term signalling and control, the nervous system is the high-speed communications network for the body. At its centre is the brain, which performs the processing and memory storage tasks, while to it are connected myriad input/output nerve channels. These nerves convey sensory and status information from specialised sensory organs like the eyes, ears and skin as well as from the various internal organs to the brain. Nerve outputs from the brain provide feedback to control some of the internal processes and to adjust the tension in the hundreds of muscles in the body that allow us to interact with the environment. Later chapters of Volume 2 discuss some of these interactions in more detail, particularly with regard to biomechatronic prostheses and interfaces that can restore function when sensory organs have been damaged.

1.3.3 Respiratory system

The respiratory system is pneumatic and consists of an air pump (the diaphragm and ribcage) which alternately produces negative and positive pressures in a sealed chamber (the thoracic cavity) that draws air into and then expels it from a pair of balloon-like organs (the lungs) that are linked to the atmosphere. The lungs are designed to provide hundreds of square metres of highly vascularised membranes (the alveoli) that allow for the free exchange of gas between the regularly replenished air and the blood to ensure that essential oxygen levels remain high and that waste carbon dioxide is flushed from the system. The pump regulatory mechanism is automatic, but it is provided with a conscious override that can be used to accelerate or inhibit flow as the circumstances dictate. One chapter of Volume 2 deals with some of the biomechatronic mechanisms that can augment or replace the regulatory process should its performance become degraded or even if it fails altogether.

1.3.4 Cardiovascular system

From an engineering perspective, the cardiovascular system is a closed-circuit hydraulic system comprising a dual-action multi-chambered electrically controlled pump (the heart) interconnected through hundreds of kilometres of flexible tubing (arteries, veins and capillaries). Pressure and flow regulation are achieved by changing the pump stroke and rate, as well as altering the diameters of the arteries. The hydraulic fluid (blood plasma) contains organic molecules for regulation as well as larger particles to aid with puncture repair (blood platelets), defence against intruders (white cells), and the transport of oxygen to the tissues and waste products back to the lungs (red cells). One chapter of Volume 2 is dedicated to analysing the performance of the heart and its mechatronic replacement.

1.3.5 Musculoskeletal system

The musculoskeletal system performs two major functions. Firstly, it maintains the integrity of the body by providing a firm structure to both support and protect the internal organs. Secondly, it provides a means for the organism to interact with the outside world by means of locomotion, and by manipulation. For human beings, locomotion is primarily provided by the legs and feet, and our capability for manipulation by the arms, the hands and particularly, the fingers. Two chapters of Volume 2 are concerned with replacing lost limbs and exoskeletons for the augmentation of limb function in cases of diminished performance.

1.4 Summary of contents

Most biomechatronic systems operate by sensing some particular aspect of the environment or the human body, processing the information and then responding in some way. It is therefore important that the student is aware of sensor technology, processing methods and actuation systems. These are all covered in Volume 1 of the two-volume set. Chapter 2 introduces the concept of biometrics, the science of measuring physiological signals, and myriad sensors that can be used to measure these. Such measurements include bioelectric signals in the form of the electrocardiogram, the electroencephalogram and the electromyogram, amongst others. Microphones can be used to measure heart sounds, while pressure and flow sensors measure the characteristics of the cardiovascular and respiratory systems. Other sensors measure the position and speed of limb elements, real and prosthetic. Chapter 3 considers various forms of actuation that use electrics, hydraulics or pneumatics. Chapter 4 discusses techniques to replace or enhance existing human senses. Chapter 5 introduces concepts of both analogue and digital signal processing that are applicable to biomechatronic systems. These include filtering in the analogue and digital domains, rectification and detection of signals, as well as sampling and digitisation. It also introduces the reader to machine learning algorithms that are becoming important in improving our ability to interpret complex physiological processes. Chapter 6 is the last of the introductory chapters and closes the loop around the measurement and actuation process. It examines various feedback types and methods of analysis.

Volume 2 is focussed on specific aspects of human physiology, with respect to biomechatronic possibilities for their improvement or rehabilitation. Chapter 1 is concerned with hearing and it considers conventional hearing aids, bone-anchored hearing aid devices, middle ear implantable hearing devices (MEIHDs) as well as cochlear and brainstem implants. Chapter 2 examines respiration with a focus on negative pressure ventilators (iron lungs) and positive pressure respiration devices that became particularly important during the initial phases of the COVID-19 pandemic in 2020. Chapter 3 examines one of the less mature fields – that of visual prosthetics based on neural and cortical implants, as well as the expanding field of neural interfaces in general. The following three chapters address the rehabilitation or replacement of organs or systems that move, with Chapter 4 focussing on the cardiovascular system, particularly with regard to the total artificial heart and ventricular assist devices. Chapter 5 examines powered and passive limb prostheses in detail, as these are some of the most important biomechatronic devices. Finally, Chapter 6 examines the design and use of exoskeletons to both replace and extend the physical capabilities of human subjects.

1.5 The future of biomechatronic systems

An introductory text cannot hope to cover even a small fraction of the biomechatronic devices and systems that exist or will be available in the foreseeable future. Both volumes of this book provide some insight into the technology and applications that have been developed over the last 70 years or so. But, as with all technology, the rate of advance in the field is accelerating, so it is expected that some amazing new biomechatronic devices will become available within the next decade.

1.5.1 Neural interfaces

Improvements in our ability to interface directly with the human neural system will provide the greatest advances in a range of applications. Already, neural reinnervation that offers both actuation and feedback is providing better interfaces to prosthetic limbs (MacIsaac and Englehart 2006; Kuiken *et al.* 2007; Hebert and Marasco 2021). This is just an interim measure, and direct interface to neurons in the brain allowing seamless integration of increasingly sophisticated limb prostheses to human amputees. Also known as brain–computer interfaces (BCIs), these have made significant advancements in both medical and technological domains in the last decade. These devices interface directly with the brain's neural architecture to restore or enhance cognitive or sensory-motor functions.

Devices such as those developed by companies like Neuralink have shown promise in helping patients with paralysis regain control over their limbs through direct neural interfaces. Early trials have demonstrated the ability of patients to control robotic arms or regain some motor control (Neuralink 2023).

Advances in signal processing and the reliable and safe electrical stimulation of neurons have made the cochlear implant the most common prosthetic in the world. This success has fostered research into other sensory prosthetics focussed mostly on restoring sight to the blind. Electrode arrays are now routinely inserted into the visual cortex and onto the retina to provide rudimentary vision. Progress has been slower than expected and whereas researchers had envisaged that these implants would allow blind people to navigate through their environments with confidence and even to read again by 2020 (Lovell *et al.* 2007), this proved premature. However, research continues to refine these technologies to enhance their performance and integration (Wang *et al.* 2022)

Cortical implants are being used for more precise monitoring and modulation of brain activity in patients with epilepsy or major depressive disorder, enabling more tailored and effective treatments (Ghosh *et al.* 2023). Modern cortical implants increasingly feature wireless communication capabilities, reducing the need for invasive surgeries and the risks associated with wired connections (Parker 2023).

The use of sophisticated machine learning algorithms has significantly improved the interpretative capabilities of current BCIs, enhancing the accuracy and responsiveness of the devices to user intentions (Chenna 2023). There is ongoing research into integrating AI more seamlessly with neural interfaces to predict and adapt to patient needs in real time, potentially opening up new treatment paradigms for complex neurological conditions (Guo *et al.* 2023).

1.5.2 Exoskeletons and nanomachines

Where the body is intact, but dysfunctional, new lightweight materials, improved batteries and small but powerful actuators will be used to provide full-body powered exoskeletons that will allow the wheelchair bound to walk again (Pons 2008; Nicolelis 2011). Already, exoskeletons have been developed as walking aids or assistive limbs for the aged and as mechanisms to provide users with superhuman strength (Stevens 2010; Babič *et al.* 2021; Qui *et al.* 2023).

Other exciting biomechatronic applications that have been promised for some time but are not yet available are nano-machines that could be injected into the body to perform microsurgery in inaccessible areas. To date, improvements in micro-electromechanical systems (MEMS) technology have facilitated the development of small devices that are capable of locomotion through liquids (Sanchez *et al.* 2011). Advances in nanotechnology and materials science have enabled the development of smaller, more biocompatible implants that reduce the risk of rejection and minimise damage to brain tissue.

Micro actuators have been developed that are capable of stimulating the ossicles within the human ear to restore hearing to the deaf. This technology is advancing quickly, and may replace audio amplification-based hearing aids in the near future (Shohet 2008; Tisch 2017).

In future, it may be possible to provide other artificial organs, including lungs (Downs 2002), improved kidneys and maybe even livers.

References

Babič, J., M. Laffranchi, F. Tessari, *et al.* (2021). "Challenges and for application and wider adoption of wearable robots." *Wearable Technologies* 2021(2): e14

Chenna, S. (2023). "Exploring the use of artificial intelligence in brain–computer interfaces for improved control and communication." SSRN: http://dx.doi.org/10.2139/ssrn.4447601.

Cromwell, L., F. Weibell, E. Pfeiffer and L. Usselman (1973). *Biomedical Instrumentation and Measurements*. Englewood Cliffs, New Jersey: Prentice-Hall, Inc.

Downs, M. (2002). "Artifical lung closer to clinical trial." *WebMD*. Retrieved March 2011 from http://www.webmd.com/lung/features/artificial-lung-closer-to-clinical-trial.

Ghosh, S., J.K. Sinha, S. Ghosh, H. Sharma, R. Bhaskar and K.B. Narayanan (2023). "A comprehensive review of emerging trends and innovative therapies in epilepsy management." *Brain Science* 13(9): 1305.

Guo, Y., Y. Li, H. Wei and Y. Zhao (2023). "Editorial: new theories, models, and AI methods of brain dynamics, brain decoding and neuromodulation." *Frontiers of Neuroscience* 17: 1302505.

Hebert, J. and P. Marasco (2021). "Chapter 7 – Targeted reinnervation for somatosensory feedback." *Somatosensory Feedback for Neuroprosthetics*. B. Güçlü, Academic Press: 245–263.

Kuiken, T., L. Miller, R. Lipschutz, B. Lock and K. Stubblefield (2007). "Targeted reinnervation for enhanced prosthetic function in a woman with a proximal amputation: a case study." *The Lancet* 369(9559): 371–380.

Lovell, N., L. Hallum, S. Chen, *et al.* (2007). "Chapter 20: Advances in retinal neuroprosthetics." In *Handbook of Neural Engineering*, M. Akay (ed.). Wiley-IEEE Press, pp. 337–356.

MacIsaac, D. and K. Englehart (2006) "The science fiction's artificial men." *CrossTalk – The Journal of Defense Software Engineering* 19(10): 4–8.

Neuralink (2023). "Neuralink's first-in-human clinical trial is open for recruitment." https://neuralink.com/blog/first-clinical-trial-open-for-recruitment/.

Nicolelis, M. (2011). "Mind out of body: Controlling machines with thought." *Scientific American*. 61–63.

Parker, M. (2023). "Bridging spinal cord injuries wirelessly." *Nature Electronics* 6(6): 400.

Pons, J. (ed.), (2008). *Wearable robots – biomechatronic exoskeletons*. Chichester, England: John Wiley & Sons.

Qui, S., Z. Pei, C. Wang and Z. Tang (2023). "Systematic review on wearable lower extremity robotic exoskeletons for assisted locomotion." *Journal of Bionic Engineering* 20: 436–469.

Sanchez, S., A. Soloved, S. Harazim and O. Schmidt (2011). "Microbots swimming in the flowing streams of microfluidic channels." *Journal of the American Chemical Society* 133(4): 701–703.

Shohet, J. (2008). "*Implantable Hearing Devices*." Retrieved July 2008 from http://www.emedicine.com/ent/TOPIC479.HTM.

Stevens, T. (2010). "HULC exo-skeleton ready for testing, set to hit the ground running next year." Retrieved March 2010 from http://www.engadget.com/2010/07/21/hulc-exo-skeleton-ready-for-testing-set-to-hit-the-ground-runni.

Tisch, M. (2017). "Implantable hearing devices." *GMS Current Topics in Otohrinolaryngology – Head and Neck Surgery* 16: 1–22.

Wang, B.-Y., Z. C. Chen, M. Bhuckory, T. Huang, A. Shin, V. Zuckerman, E. Ho, E. Rosenfeld, L. Galambos, T. Kamins, K. Mathieson and D. Palanker (2022). "Electronic photoreceptors enable prosthetic visual acuity matching the natural resolution in rats." *Nature Communications* 13(1): 6627.

Chapter 2
Sensors and transducers

2.1 Introduction

Much of biomechatronic engineering involves the measurement of physical processes either stemming from the biological organism or from an associated prosthesis. These measurements can include voltage, chemical concentration, pressure, position, displacement and rate amongst many. In general, they are processed in some way and then used to apply some form of actuation such as the motion of a prosthetic arm, the contraction of an artificial heart or the electrical stimulation of the cochlea.

In most cases, the measured parameter is converted to an electrical signal for analysis or processing, though a number of devices such as the sphygmomanometer, used to measure blood pressure, often still use pneumatics. However, since the introduction of low-cost lightweight and reliable electronics in the last seventy years, and particularly since the invention of the microprocessor, far more complex analysis and processing can be undertaken.

After introducing switches and relays as well as a range of power-generating mechanisms, this chapter focuses on the more common sensors and transducers that are used in mechatronic and biomechatronic engineering. Examples of some panel switches are shown in Figure 2.1.

2.2 Switches

The fundamental electrical building block is the switch. These are devices that control the flow of current in a binary (on–off) manner, either using a mechanical contact or an electronic device such as a transistor. Switches are commonly used to isolate thc supply of current to a device, as selectors or, often in mechatronic devices, to identify the limits of travel of some mechanical structure.

2.2.1 Toggle switches

Toggle switches are available in a number of different configurations depending on the application, they are characterised by the number of switching contacts (or poles) and the number of terminals. For example, a single on–off switch shown in Figure 2.2(a) is referred to as a single pole single throw (SPST) device. In the case where the switch can connect to a pair of terminals as shown in Figure 2.2(b), it is referred to as a single pole double throw (SPDT) device. Multi-pole switches are also common in

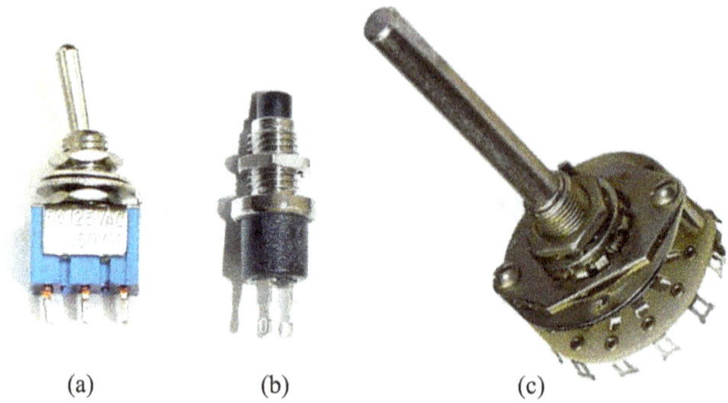

Figure 2.1 Examples of some panel switches (a) toggle, (b) push-button and (c) rotary

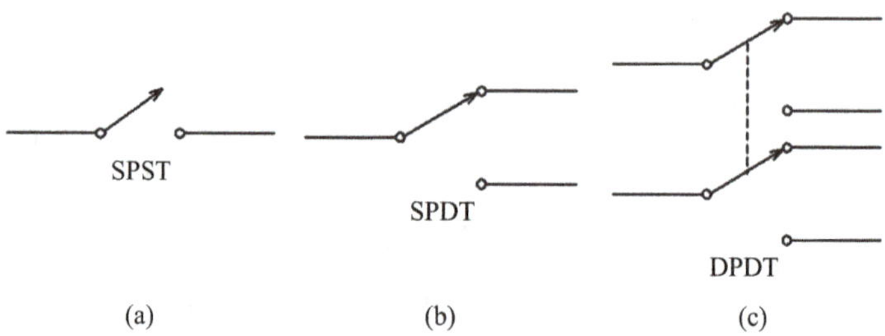

Figure 2.2 Fundamental switching configurations

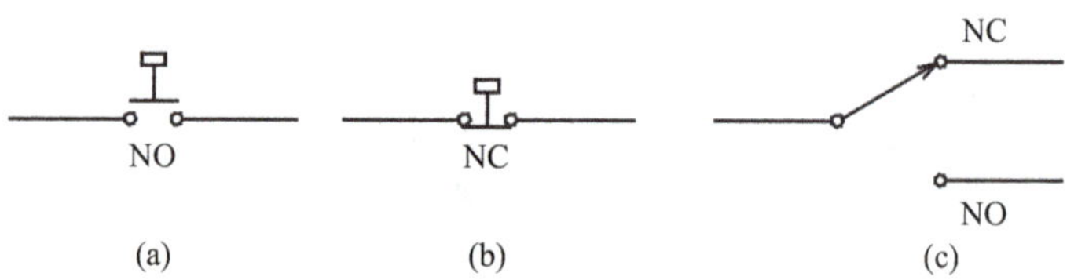

Figure 2.3 Push-button switch configurations

which individual switching elements are ganged together to form double, triple and even quadruple devices. The nomenclature follows that of the previous cases so that a double pole double throw switch would be referred to as DPDT (Figure 2.2(c)).

All double throw switches are break-before-make in that the pole never contacts both of the terminals simultaneously. This is important to ensure proper isolation and to avoid short circuits.

2.2.2 Push-button switches

Push-button switches are momentary contact devices because they only operate when force is applied. They come in two varieties, with a normally open (NO) or a normally closed (NC) contact for single throw devices but are often also configured as double throw devices as shown in Figure 2.3.

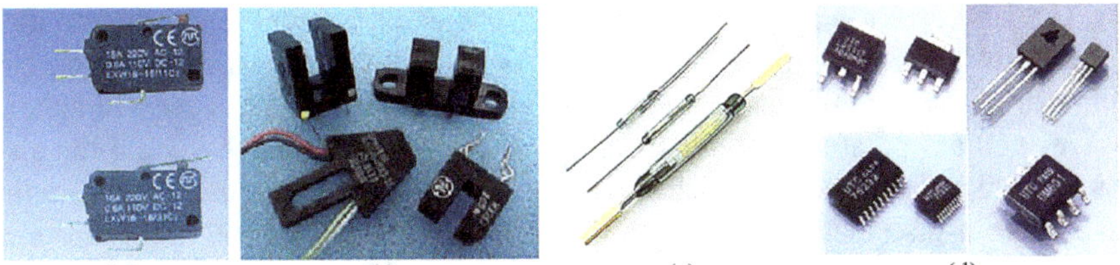

(a) (b) (c) (d)

Figure 2.4 Examples of limit switches (a) micro-switch, (b) optical (c) reed and (d) hall device

2.2.3 Limit switches

Mechanical limit switches are a variation of the standard push-button configuration fitted with a lever arm. These are often known as micro-switches because of their small size. In the interests of reliability, mechanical limit switches have mostly been superseded by optical or hall switches (discussed later) which have no moving parts and are generally easier to interface. Figure 2.4 shows photos of some examples of micro-switches and their electronic equivalents.

2.2.4 Rotary switches

Rotary switches consist of a number of circular wafers containing many poles operated by a spindle that passes through the centre of each wafer. There are many types including both shorting (make before break) and non-shorting (break before make). However, as with most other mechanical switches, these are not particularly reliable for long-term operation and are being replaced by their electronic counterparts.

2.2.5 Optical switches

Optical switches consist of an infrared light-emitting diode (IR LED) and a phototransistor sensitive to a similar wavelength, mounted into the same plastic holder. These are mostly configured so that the IR LED is directed towards the phototransistor across a slot in a transmission configuration, but some are mounted at an angle to operate in a reflective configuration. In the transmission configuration, the phototransistor operates as a switch that is conductive (on) if it is illuminated by the IR LED. In the transmission configuration, if an opaque object is introduced into the slot, then the light path is blocked and the phototransistor turns off, while in the reflective case, the switch is off until a reflective material is introduced to reflect the light from the IR LED back into the phototransistor.

2.2.6 Other switches

Reed switches generally consist of a normally open contact pair mounted on ferromagnetic reeds in a hermetically sealed glass tube so that, in the presence of a strong magnetic field, the contacts are drawn together to close the switch.

The Hall effect generates a potential difference orthogonal to the direction of current flow in the presence of a magnetic field. Hall switches are integrated circuits that exploit this effect to produce the electronic equivalent of reed switches.

As well as operating as limit switches, these electronic switches are often also used to detect rotary motion. In the optical case, a slotted disc is used, while in the other cases a disc that has been appropriately magnetised can be used. Up until recently these devices had a large sensitive region and were only effective for low count/revolution applications. However, magnetic technology driven by increased storage density in hard drives has resulted in much higher resolution capability, as discussed later in this chapter.

It is important to note that all switches, whether they are mechanical or electronic, have a limited current and voltage rating that must be adhered to if reliable operation is required. This specification is particularly important if inductive loads such as motors or relays are switched. The high voltage induced during turn-off can result in arcing across the contacts of mechanical switches and will destroy electronic switches.

Switches of all varieties are common in most mechatronic devices. These range from the conventional toggle on–off switch that controls the power supplied to the device, through a range of limit switches that ensure that actuators remain within the required physical limits. One common application for micro switches is in the socket of conventional powered upper limb prostheses as discussed in Volume 2, Chapter 5. These can be controlled by remaining muscles above the amputation, or in the case of thalidomide deformity, by using the nub digits that sometimes remain.

2.2.7 Relays

Relays are electrically controlled switches. In their mechanical embodiment, they consist of an electromagnet that operates a mechanical lever arm to operate a switching mechanism, as illustrated in Figure 2.5. As with toggle switches, many switching configurations are possible.

Electronic switches have replaced their mechanical counterparts in many applications, both for switching high-voltage mains and low-voltage control signals.

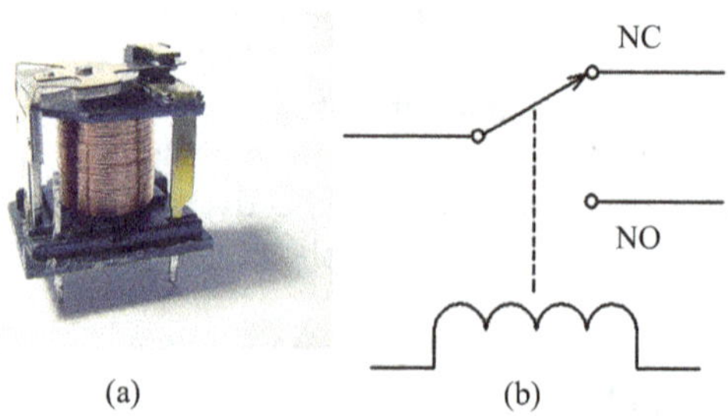

Figure 2.5 Mechanical relay (a) photograph and (b) SPDT switching configuration

(a) (b)

Figure 2.6 Examples of electronic switches (a) solid-state relays and (b) analogue switch

Solid-state relays (SSRs) generally have an opto-isolated low-voltage input that controls high current-handling transistors, usually MOSFETs or IGBTs and sometimes, thyristors. When designed for lower current applications, these devices are usually called analogue switches. Some examples of these devices are shown in Figure 2.6.

In biomechatronic applications, relays are often used to provide the mandatory electrical isolation between the various parts of a system to ensure patient safety. SSRs are also quite common where simple on–off control of electric motors is required in primitive prostheses.

2.3 Power supplies

Medical and biomechatronic devices generally run from mains power if they are static and from batteries if they need to be portable. In most cases, this supply voltage will need to be converted to one or a number of other voltages to power the various modules within any device. For example, a prosthetic arm may require 12 V to power the motors, ± 5 V for the analogue electronics and 3.3 V for the signal processor.

It is now common to provide power to embedded systems such as cochlear devices and artificial hearts using electromagnetic induction through the unbroken skin, as this eliminates a common source of infection. This technology, referred to as transcutaneous energy transfer (TET), is discussed in more detail in Volume 2, Chapter 4. An alternative that is becoming feasible as the efficiency of implants improves it to scavenge power from flexing muscles or changes in pressure driven by the heartbeat.

2.3.1 Linear power supplies

Until recently, most mains-powered devices consisted of a step-down transformer followed by a full-wave rectifier to produce rectified AC. This was followed by large capacitors and chokes to supply the smoothed DC voltage required by the system. If different voltages were required, the transformer would include a number of secondary windings as shown in Figure 2.7. As transformers are heavy and expensive, modern power supplies often rectify the mains directly and then use switch-mode power supplies to provide the various output voltages.

Irrespective of the method used, good isolation must be maintained between the input and output voltages, and the raw DC must be further regulated and filtered to remove mains hum or switching noise and to maintain a constant and clean output voltage.

Linear regulators are active control systems that compare the output voltage to a fixed reference and use the error to adjust a series transistor to keep the output constant irrespective of changes in the input voltage or the load as shown schematically in Figure 2.8. In this example, a Zener diode is used to generate the reference voltage which is compared to the output sense voltage using an operational amplifier (op amp), the output of which drives the NPN transistor.

It is possible to construct voltage regulators from discrete components but modern voltage regulator integrated circuits are cost effective and extremely sophisticated. They offer excellent regulation, low dropout voltage (the difference between V_{in} and V_{out}) and good input noise rejection. Fixed three-terminal regulators like the LM7812 (+12 V) and LM7912 (−12 V) and their relatives operating at other voltages are easy to use, require few external components and provide

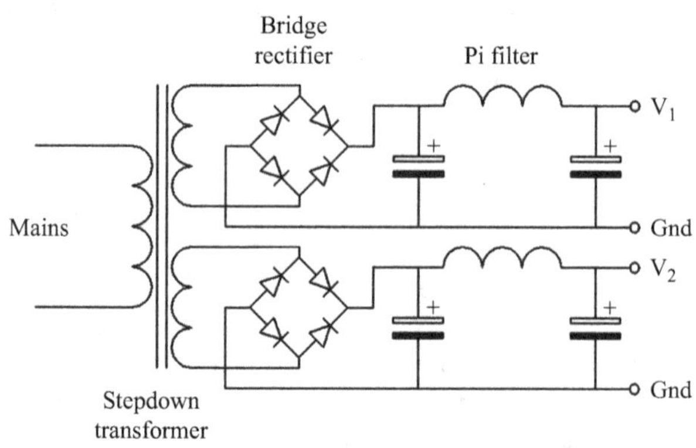

Figure 2.7 Schematic diagram for a dual rail mains power supply

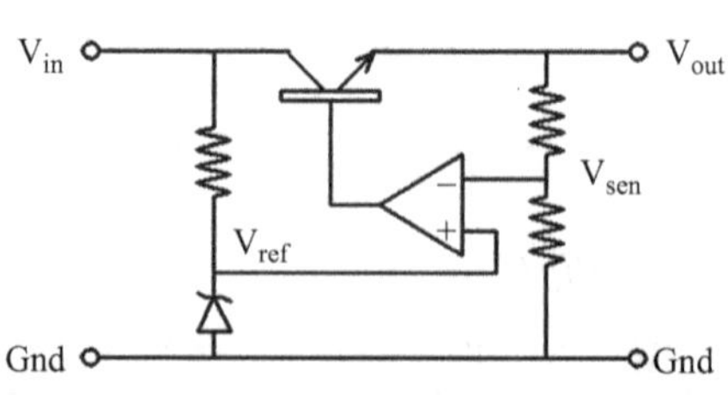

Figure 2.8 Schematic diagram of a generic voltage regulator

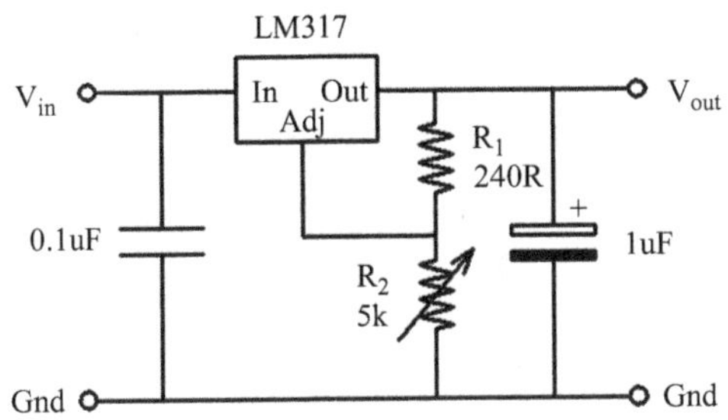

Figure 2.9 Schematic diagram of a voltage regulator based on the LM317

Table 2.1 Specifications of a range of linear voltage regulators

Component	Feature	Output voltage (V)	Dropout voltage	Current (max)	Line regulation	Load regulation	Ripple reject
LM317L	Adj	1.2 to 37		100 mA	0.01%/V	0.1%	80 dB
LM337L	Adj	−1.2 to −37		100 mA	0.01%/V	0.1%	80 dB
LM317	Adj	1.2 to 37		1.5 A	0.01%/V	0.3%	80 dB
LM337	Adj	−1.2 to −37		1.5 A	0.01%/V	0.3%	77 dB
LM2941	Adj LDO	5 to 20	0.5 V @ 1 A	1 A			0.005%/V
LM2991	Adj LDO	−3 to −24	0.6 V @ 1 A	1 A			60 dB
LM340T-xx	Fixed	5, 12, 15 (2%)		1 A			70–80 dB
LM2940-xx	Fixed LDO	5, 12, 15	0.5 V @ 1 A	1 A	50 mV	50 mV	54–60 dB
LM2990-xx	Fixed LDO	−5, −12, −15	0.6 V @ 1 A	1 A	40 mV	40 mV	58 dB

good performance for loads of up to about 1A (depending on the voltage drop). Adjustable, three-terminal regulators like the LM317 (for +ve voltages) and LM337 (for –ve voltages) have become the regulators of choice for many designs. They are low cost, require few peripheral components as can be seen from the schematic in Figure 2.9 and are capable of adjusting the output voltage between 1.2 V and 37 V for a 40 V input voltage.

Because all linear regulators produce output voltages that are lower than the input, some power, P_{disp}, is dissipated. This is equal to the product of the average current, I_{out}, and the difference between the input and the output voltage

$$P_{disp} = I_{out}(V_{in} - V_{out}) \tag{2.1}$$

A wide range of linear regulators is available commercially for most applications as listed in Table 2.1. For more information on these and other devices, examine the catalogues of electronics mail-order companies such as DigiKey, Element14 or RS-Electronics.

2.3.2 Switch-mode power supplies

Another method of generating a regulated DC voltage is to use a switching regulator. A transistor operating as a switch periodically applies the unregulated DC input voltage across an inductor. The inductor's current builds up during the on period, storing energy in its magnetic field, and this energy is transferred to a filter capacitor at the output, where it is smoothed. As with linear regulators, feedback is used to control the output, but in this case, it adjusts the oscillator's duty cycle or frequency (Horowitz and Hill 1989). A generic example of a step-up switching regulator is shown in Figure 2.10.

Switching regulators have some unusual properties that have made them particularly useful in portable applications. Because the control element is either on or off, there is very little power dissipation within the device, which makes the conversion efficiency very good even if there is a large voltage drop between input and output. They are also capable of generating output voltages higher than the input as well as outputs with the opposite polarity.

Switching power supplies can be designed with no DC path from the input to the output and are therefore useful for isolation. This allows them to be run directly from the rectified mains with no step-down transformer to produce the small, lightweight and efficient DC power supplies that are now ubiquitous for powering modern electronics.

The main problems with switch-mode power supplies are residual switching noise on the output and noise fed back onto the input lines. This requires filtering and results in a decreased overall efficiency and a higher component count.

RMS output noise of a typical switcher can be 50 mV, which may be acceptable for digital applications but poses a major problem for sensitive analogue electronics. Passive LC filtering can reduce this somewhat, but it is often practical to install a linear regulator with good noise rejection at the output. Many DC–DC converters include these considerations and are also housed within a shielded enclosure to minimise noise radiation. In spite of this, circuits including these power supplies must be carefully designed to minimise residual noise problems that may swamp sensitive biological signals.

Switching power supplies are also available with multiple outputs. However, these are generally provided from multiple windings on the same transformer, with feedback regulation applied only to one of the rails. This results in cross-coupling between the sections and poorer regulation of the uncontrolled outputs.

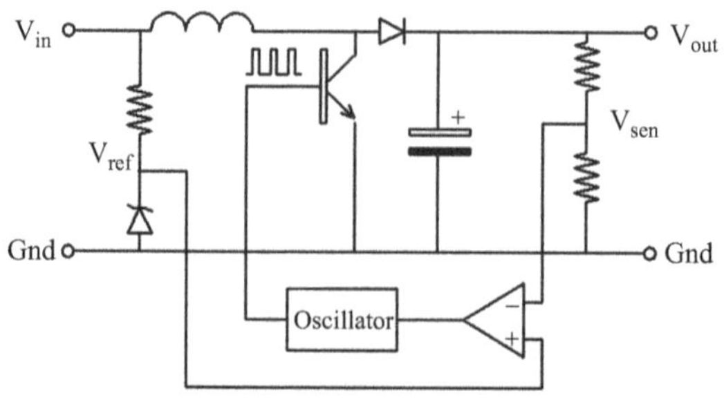

Figure 2.10 Schematic diagram of a generic switching regulator

Table 2.2 Specifications of a small range of switch-mode power supply modules that could be useful in biomechatronic applications

Manufacturer	Model	Power rating	Input (V)	Output (V)	Efficiency	Noise
C&D Tech.	LME0515SC	250 mW	5 V±10%	15 V	<75%	50 mV pp
C&D Tech.	NME0515SC	1 W	5 V±10%	15 V	<80%	110 mV pp
C&D Tech	NYD2405C	3 W	18–36 V	5 V	<83%	
ASTEC	AA05E-024L-150D	5 W	18–36 V	15 V @ 160 mA–15 V @ 160 mA		
ASTEC	AEE02A24	10 W	18–36 V	5 V @ 2 A	78–85%	
Power One	HAS030YJ-A	25–30 W	18–36 V	15 V @ 2 A		150 mV pp
C&D tech	WPA60R48D0515	60 W	36–75 V	5 V @12 A1.5 V @12 A	<90%	90 mV pp
SunPower	SDS-100B05	100 W	18–36 V	5 V @ 20 A	72%	
Mascot	9970 24/12V	276 W	20–30 V	13.8 V @ 20 A	High	

A final consideration is that switching power supplies often have a minimum load current below which their operation can be erratic, or they may not even start. Care should be taken during the design and implementation phases to ensure that this does not occur.

The following advice (Horowitz and Hill 1989) should be followed when designing power supplies:

- For digital systems which usually need +5 V or +3.3 V at a high current, use a commercial line-powered switcher.
- For analogue circuits with low-level signals (small-signal amplifiers with input levels of less than 100 μV), use a linear regulator unless efficient operation from a battery supply is required. Switchers will cause problems!
- For high-power applications, use switchers as they are lighter and more efficient than transformer-based power supplies
- For high-voltage low-power applications (photomultipliers, defibrillators etc) use low-power step-up (boost) converters.

In general, low-power DC–DC converters are easy to design and require few external components thanks to the available integrated circuit building blocks. However, wherever possible it is preferable to use complete off-the-shelf modules as there are many types ranging from less than 1 W to hundreds of watts, as listed in Table 2.2

2.3.3 Batteries

Many of the biomechatronic systems discussed in this book are portable and therefore need to operate using battery power. Where the power consumption is significant and the battery pack is integral to the device for reliability and sealing issues, or where access to the device is difficult, rechargeable batteries (secondary batteries) are used. Such devices include prosthetic limbs. In low-power applications where battery life can be months and even years, it is more practical to use primary batteries.

The oldest form of rechargeable battery still in use is the lead-acid battery. These are "wet cell" devices and mostly need to be kept upright and placed in well-ventilated areas as they generate hydrogen if overcharged. One convenient alternative form is the gel cell that contains a semi-solid electrolyte that prevents spillage.

Most portable rechargeable batteries are "dry cell" types that are hermetically sealed. These include nickel-cadmium, nickel-metal hydride and lithium types. From a practical perspective, the following rules of thumb can be used when selecting batteries for a specific application:

- Nickel-cadmium (NiCd): These are good for devices that include motors and other high-discharge requirements. They can accommodate heavy drain rates, but their mAh rating is lower than more modern rechargeables and they also have a strong memory effect.
- Nickel-metal hydride (NiMH): These batteries have a high mAh rating and can sustain moderate to high current drain.
- Lithium-ion (Li-Ion): These have a very long shelf life and are excellent for moderate to low power applications.
- Lithium-polymer (Li-Po): These have similar chemistry to Li-Ion but, because they are manufactured in flat sheets rather than cylinders, have a higher power density.

A large number of primary battery types are available for different applications:

- Zinc-carbon: This is a low-cost battery good for light current drain applications.
- Zinc-chloride: Similar to the zinc-carbon but with a slightly higher power density.
- Alkaline: These are long-life batteries and are suitable for low- and high-current drain applications. Their energy density is significantly higher than that of zinc-carbon types.
- Silver-oxide: Commonly used for some older hearing aids and watches where the current drain is low.
- Mercury: Formerly used in a wide range of devices but seldom used today because of toxicity issues.
- Zinc-air: used previously in hearing aids.

The main advantages of primary batteries are their high energy density, long storage lives and ease of disposal (most primary batteries contain little toxic material). A regular alkaline battery provides 50% more power than a Li-Ion, and a primary Lithium battery has three times the energy of a similar sized Li-Ion battery.

Because primary batteries have relatively high internal resistances, the maximum current rating is limited. The energy density ratings shown in Figure 2.11 are determined at the optimum discharge rate for that particular cell type. However, if the comparison is made for a high-current mode application such as a prosthetic limb, then the useable energy for each battery type is as shown in Figure 2.12.

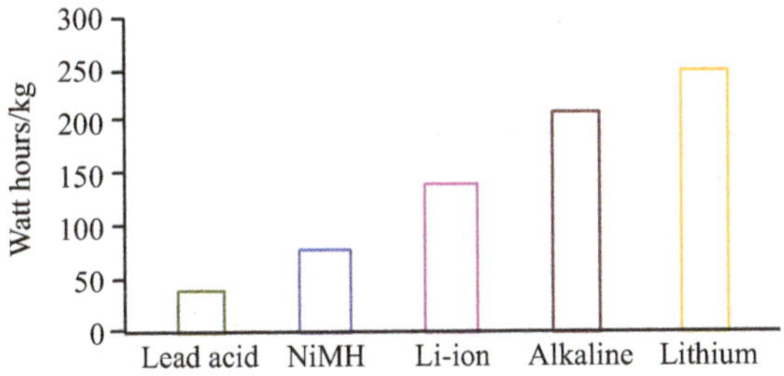

Figure 2.11 *Energy density comparison of primary and secondary batteries [adapted from Buchmann (2005)]*

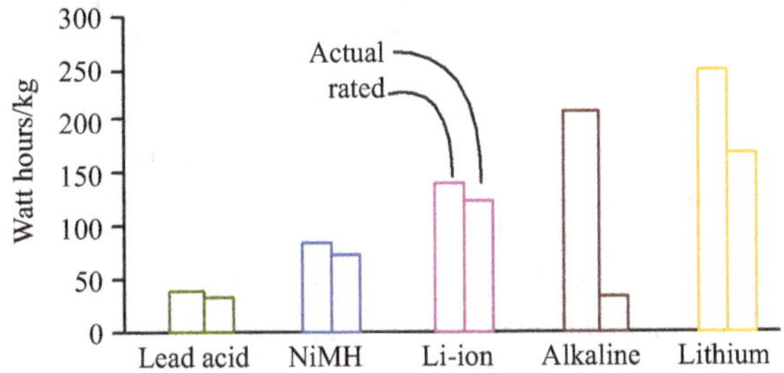

Figure 2.12 *Energy comparison under load [adapted from Buchmann (2005)]*

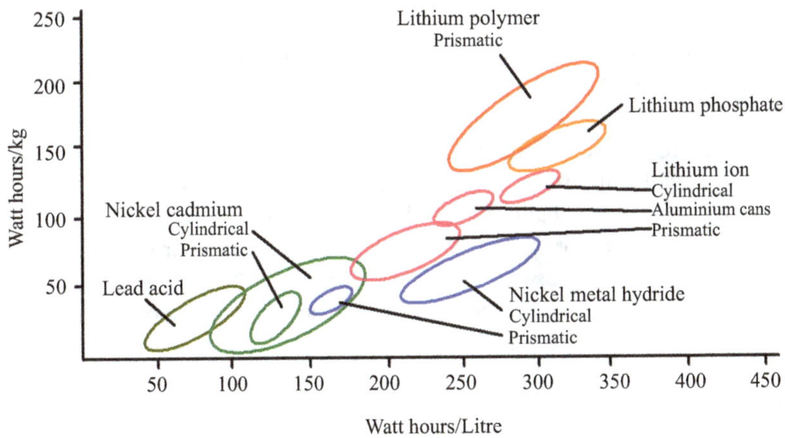

Figure 2.13 *Rechargeable battery energy density [adapted from wikipedia]*

It can be seen in Figure 2.12 that though the alkaline battery is good for low discharge rates, its performance in this case is poor. It can also be seen that the performance of the Li-Ion secondary battery is approaching that of the primary Lithium type. A comparison of the energy densities of a number of rechargeable batteries is shown in Figure 2.13.

Table 2.3 Comparison of battery technologies and performance (Martin-Garcia et al. 2020)

Parameter/technology	Lead acid	Ni–Cd	Ni–MH	Li-ion	Li–Po
Cell voltage (V)	2 V	1.2 V	1.2 V	3.3 V	3.7 V
Self-discharge (%/month)	>20%	10%	30%	8%	5%
Cycles	500–800	1500–2000	300–500	>2000	>1000
Lifetime (year)	5–15	10–20	10–15	2–5	2–7
Specific energy (Wh/kg)	30–40	40–60	30–80	100–250	130–200
Efficiency (%)	50%–90%	70%–90%	66%	80%–95%	90%–95%
Overload tolerance	–	Very good	Good	Poor	Poor
Impact robustness	Good	Very good	Good	Poor	Poor
High temperature operation	Average	Very good	Average	Poor	Poor

Rechargeable batteries have a very low internal resistance. This allows high current to be drawn on demand, which is essential for many digital devices, and to drive switch-mode power supplies that have high "inrush" currents.

Their main disadvantages include a limited shelf life and high self-discharge. Whereas a primary battery may have a shelf life of 10 years, lithium-based batteries are good for 2–3 years in normal use, though cool storage at moderate charge prolongs longevity. Nickel-based batteries have a shelf life of up to 5 years but require careful priming to regain performance after long storage.

Nickel-based batteries exhibit a 10–20% self-discharge per month compared to a 5–10% value for lithium and lead-based batteries. This rate increases at high temperatures.

Finally, secondary batteries provide a limited number of charge–discharge cycles. The number that can be achieved is determined by the depth of discharge, the environmental conditions, charge methods and maintenance. Nickel-based batteries need to be deep-discharged periodically to reverse crystal growth (memory effects), while lithium- and lead-based batteries have no memory and can therefore be operated in shallow-cycle applications, with only an occasional deep discharge to verify performance. Table 2.3 provides a summary of the characteristics and performance of a wide range of secondary batteries.

In the future, batteries could be made with materials like iron or plastic, and they might use water instead of organic solvents as charge carriers. This addresses some of the concerns about the safety of lithium-ion battery installations. A promising new battery technology based on Sodium rather than Lithium may also replace Li-ion cells in many applications. These batteries are cheaper, offer stability against extreme temperatures and pose no risk of overheating. With a new high-capacity electrode made from nanostructured hard carbon, their power density is reaching that of Li-ion phosphate (LFP) batteries, as shown in Figure 2.14 (Crownhart 2023).

2.3.4 Energy scavenging

Energy scavenging from a biomechatronic perspective can include various sources of power including gross body motion, vibration, changes in shape or volume and

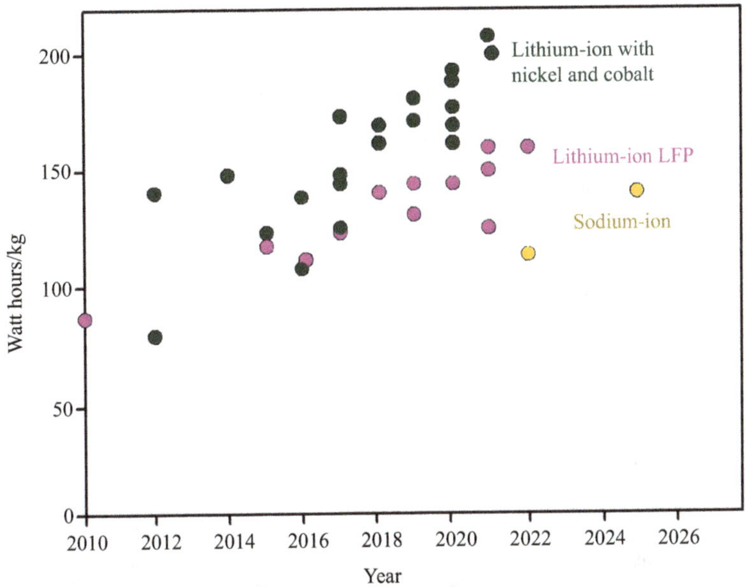

Figure 2.14 Improvements in metal-ion battery energy density since 2010 [adapted from Crownhart (2023)]

pressure, temperature gradient and even fuel cells that oxidise blood glucose. These can range in size from external devices weighing 1 kg or more that provide sufficient energy to power devices like wireless ECG monitors, to minute implantable devices to monitor medical conditions like hypoglycaemia (Campbell 2010).

2.3.4.1 External devices

Compared to the amount of power used by the human body, as listed in Table 2.4, the proportion that can be usefully extracted is rather insignificant. A naked human being radiates about 150 W of infrared heat into the surroundings if the ambient temperature is 17 °C (290 K), but with a Carnot efficiency of only 5.5%, a maximum of 8 W could be harvested from their complete surface area. Even this is optimistic as the best thermoelectric generators are only about 1% efficient for temperature gradients of between 5 °C and 20 °C and it is unlikely that the whole body could be used. In a typical application where the exposed neck comprising 1% of the total surface area of the body is covered by thermoelectric generators with an efficiency of 1%, only about 15 mW of power could be harvested.

Capturing all the work of exhalation based on the differential pressure that can be exerted without interfering with normal breathing is another potential source of power. It is shown in Volume 2, Chapter 2 that the average breathing rate at rest is about 15 breaths per minute for a tidal volume of 400 ml, making the total flow,

Table 2.4 Human energy use for a range of activities [adapted from Starner and Paradiso (2004)]

Activity	Power (W)
Sleeping	80
Laying quietly	95
Sitting	120
Standing at ease	130
Talking	130
Eating a meal	130
Strolling	165
Driving a car	165
Playing violin or piano	165
Housekeeping	175
Carpentry	270
Hiking (6.5km/h)	410
Swimming	580
Mountain climbing	700
Long-distance running	1050
Sprinting	1630

Q = 6 l/min (0.1 l/s). For a pressure differential, ΔP, of 10 cmH_2O (980 Pa) the power is

$$\begin{aligned} P &= Q\Delta P \\ &= 0.1 \times 10^{-3} \times 980 \\ &= 98 \text{ mW} \end{aligned}$$

A facemask housing a small turbine-driven generator with a good overall efficiency would be required to capture this energy.

A tight band around the chest can also be used to capture energy from breathing. For normal breathing an increase in circumference, Δx of about 10 mm occurs during inspiration. For a restraining force, F, of 100 N, the total energy per breath is

$$\begin{aligned} E &= F\Delta x \\ &= 100 \times 10 \times 10^{-3} \\ &= 1 \text{ J} \end{aligned}$$

For a breathing rate of 15 breaths per minute, or 0.25 breaths per second, the total power output is only 250 mW. Once again, a harvesting efficiency of about 50% for ratchet and flywheel or a piezoelectric material could be expected. The use of internal piezoelectric materials is discussed in the next section.

The power required for arm motion can easily be calculated from its mass and the change in height of the centre of gravity per unit time. For example, to lift an arm weighing 3 kg through a height of 0.6 m once per second requires about 18 W.

This is obviously unnatural and would be unsustainable over long periods. However, normal arm movement could still provide a reasonable amount of energy without loading the limb or joint significantly.

Activities specifically designed to generate electricity such as shake-driven torches or wind-up radios and smartphone chargers can be reasonably efficient. For example, the power supply of the Freeplay radio stores about 500 J of energy from 60 s of winding (Starner and Paradiso 2004).

Of course, running uses the most energy of any human activity, so capturing even a fraction of that could produce significant amounts of power. A good example of one of the larger devices is a knee brace designed and built by researchers at Simon Fraser University in British Columbia lead by Max Donelan. The 1.5 kg device consists of a flexible joint that converts knee flexion into rotary motion that drives a generator. When used at end swing (when the leg is decelerating), it is capable of producing 5 W of electrical energy for a measured expenditure of only 5 W. This should be considered in comparison to the 6 W of energy required to produce each watt of electrical power for a hand cranked generator.

Unfortunately, just carrying the prototype device, even when it is not generating any energy, uses about 60 W. This is a significant percentage of the 300 W that is used for walking. The researchers suggest that this overhead could be reduced to about 15 W by moving the device higher up the leg and reducing its mass. When these overheads are taken into account it is not any more efficient than other techniques such as shoes with power generation mechanisms in the soles, or oscillating backpacks (Johnston 2008).

Piezoelectric or rotary generators that convert some of the heel-strike energy into electrical power for storage and later use are the most common energy scavenging devices. Consider a device that can capture the energy over a distance $x = 30$ mm during heel-strike of a typical 70 kg person. The power per stride per leg is

$$\begin{aligned} P &= mgx \\ &= 70 \times 9.8 \times 30 \times 10^{-3} \\ &= 20.6 \text{ W} \end{aligned}$$

At two strides per second, the total power available is 41 W.

Rotary generators need to spin rapidly to achieve good efficiencies, so pure mechanical coupling involves high gear ratios and a fairly complex mechanism that is prone to failure. Alternatives include miniature hydraulic pump/turbine combinations or pneumatic systems that store power in compressed air.

Piezoelectric materials produce an electrical charge when mechanically stressed. It is interesting to note that human skin and bone both exhibit this property, albeit with very low coupling efficiencies. Common alternatives are polyvinylidene fluoride (PVDF) and lead zirconate titanate (PZT) whose characteristics are shown in Table 2.5.

Because a piezoelectric ceramic is anisotropic, physical constants relate to both the direction of the applied mechanical or electric force and the directions

Table 2.5 Piezoelectric characteristics of PVDF and PZT

Material property	PVDF	PZT
Density ρ (g/cm^3)	1.78	7.6
Relative permittivity ε	12	1700
Young's modulus E (N/m^2)	3×10^9	83×10^9
Piezoelectric constant (C/N)	$d_{31} = 20 \times 10^{-12}$	$d_{31} = 180 \times 10^{-12}$
	$d_{33} = 30 \times 10^{-12}$	$d_{33} = 360 \times 10^{-12}$
Coupling constant (CV/Nm)	0.11	$k_{31} = 0.35$
		$k_{33} = 0.69$

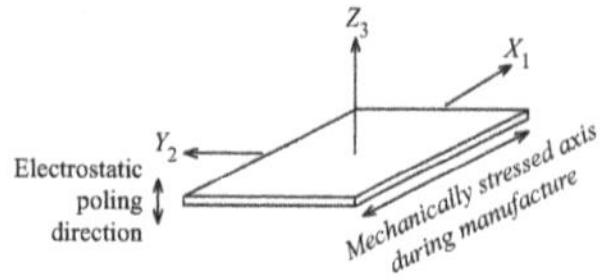

perpendicular to the applied force. Consequently, each constant generally has two subscripts that indicate the directions of the two related quantities, such as stress (force on the ceramic element/surface area) and strain (change in length of element/original length) for elasticity. The direction of positive polarisation usually coincides with the Z-axis of a rectangular system of X, Y and Z axes. Direction X, Y or Z is represented by the subscripts 1, 2 or 3, respectively.

The piezoelectric charge constant, d, is the polarisation generated per unit of mechanical stress applied to a piezoelectric material. The first subscript to d indicates the direction of polarisation generated in the material when the electric field is zero or, alternatively, is the direction of the applied field strength. The second subscript is the direction of the applied stress or the induced strain, respectively. Because the strain induced in a piezoelectric material by an applied electric field is the product of the value for the electric field and the value for d, d is an important indicator of a material's suitability for strain-dependent (actuator) applications.

The definitions and values for the piezoelectric charge constant and elasticity for PZT are summarised below:

- $d_{33} = 360 \times 10^{-12}$ C/N. induced strain in direction 3 per unit electric field applied in direction 3,
- $d_{31} = 180 \times 10^{-12}$ C/N. induced strain in direction 1 per unit electric field applied in direction 3,
- Young's Modulus $E = 83 \times 10^9$ N/m^2.

It is not feasible to extract power by compressing the piezoelectric material as the Young's modulus is too high. However, it is possible to bend the material to take advantage of the 31 mode. In the case of a PZT cantilever beam, the maximum allowed deflection, x (m), at the tip is determined by the yield stress, $S = 50$ MPa (Fett *et al.* 1999) using the following relationship

$$x = \frac{2l^2}{3Et}S \tag{2.2}$$

where l (m) is the length of the beam, t (m) is the thickness of the material and E (N/m^2) is the Young's modulus for PZT.

Consider mounting a 150 mm cantilever beam made of a 0.5 mm thick slab of PZT within the sole of a shoe. The maximum deflection before the material fractures will be

$$\begin{aligned} x &= \frac{2l^2 S}{3Et} \\ &= \frac{2 \times \left(250 \times 10^{-3}\right)^2 \times 50 \times 10^6}{3 \times 83 \times 10^9 \times 0.5 \times 10^{-3}} \\ &= 50 \text{ mm} \end{aligned}$$

It is therefore feasible to develop the mechanics to deflect the beam by 30 mm while maintaining a reasonable safety margin. This consideration notwithstanding, the superior flexibility of PVDF film makes it a better candidate for power extraction.

Because the total energy output is proportional to the volume of the film stressed, thicker films should be used. However, because they generate higher voltages but form smaller capacitors, it is often preferable to use laminates made of many layers of thinner films separated by a compatible material such as polyester.

A 16-layer bimorph PVDF insole developed at the MIT Media Lab produced peak powers of about 15 mW at heel-up into a matched resistive load, with an average of about 1.3 mW. Improved efficiency can be achieved by "tuning" the device so that its electrical resonance is matched to the mechanical excitation frequency (Starner and Paradiso 2004).

Newer research has focused on using PZT or PVDF nanofibres to produce smaller devices as shown in Figure 2.15. The ultimate goal is to weave these fibres into normal fabrics so that they can be used to produce power directly from a vest or shirt (Chen *et al.* 2010).

Triboelectric nanogenerators (TENGs) combine the effects of contact electrification and electrostatic induction. They can effectively convert mechanical energy into electric power by moving two different materials, initially in contact, to separate the charges. There are four common mechanisms that can be used to achieve this, as illustrated in Figure 2.16.

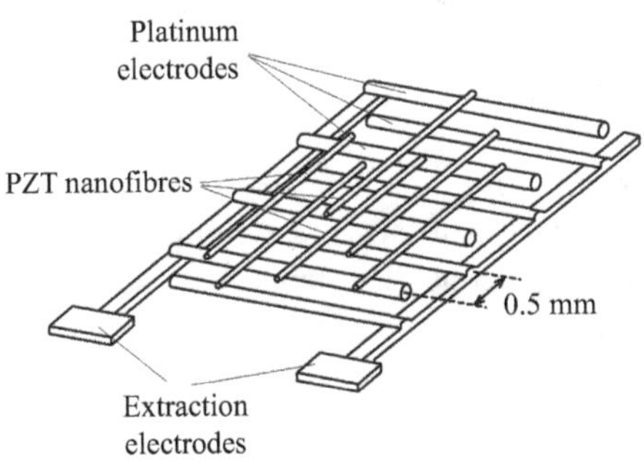

Figure 2.15 Nanofibre-based energy scavenging module [adapted from Chen et al. (2010)]

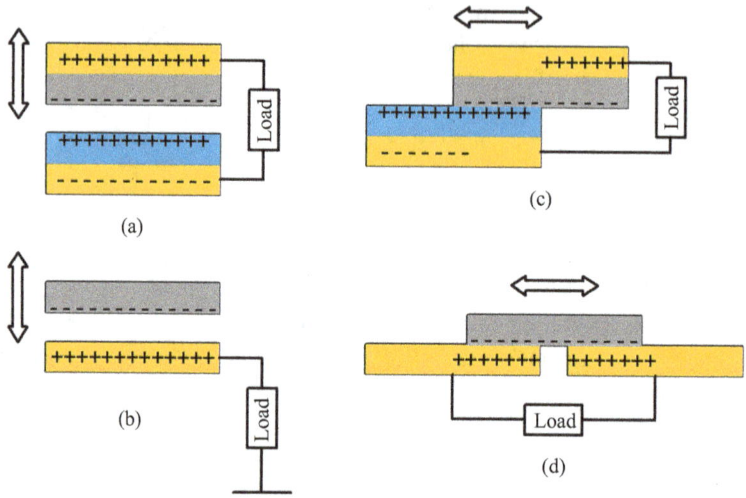

Figure 2.16 *Four basic models of triboelectric nanogenerators (TENGs): (a) vertical contact separation model, (b) single-electrode model, (c) lateral sliding model, (d) freestanding model. Note that the dielectric layer material is represented by the light blue cuboid and the metal electrode by the golden colour.*

TENGs are most effective for converting irregular, low-frequency and low-amplitude mechanical energy into electric power, making them ideal for harvesting energy from the living environment. Since the mechanism was discovered in 2012, TENG development has progressed rapidly, from fundamental scientific understanding to advanced technologies and applications (Cheng *et al.* 2023). As an example, Gao from the California Institute of Technology has been developing self-powered electronic skins (e-skins) containing a number of sensors including those to measure heart rate, body temperature, and blood sugar level along with a self-contained TENG power source. This power source consists of layers Teflon, copper and polyimide and can generate a maximum of 0.94 mW when attached to a person (Pepitone 2023).

The use of metamaterials in energy scavenging is also becoming more common. For example, researchers at ETH in Zurich have developed metamaterial-based sensors that are sensitive to a specific frequency or combination of frequencies. The sensor consists of dozens of similarly structured plates linked together and embedded in silicone. The links form tiny springs that are tuned to provide the required response for energy scavenging or vocal identification (Dubček *et al.* 2024).

2.3.4.2 Internal devices

Zhong Lin Wang and his colleagues from the Georgia Institute of Technology have developed a "nanogenerator" that is able to scavenge power from involuntary movements such as breathing or a beating heart. Zinc-oxide nanofibres, which exhibit piezoelectric properties, are bonded onto a 2 × 5 mm flexible polymer substrate before the complete unit is coated in silicone to protect it from corrosion. When bonded to a rat's diaphragm using a tissue adhesive, the device generated a current of 4 pA at 2 mV. The output increased to 30 pA and 3 mV when a similar device was attached to a rat's heart.

Other researchers have implanted larger piezoceramic into muscles to obtain power from normal body movement (Campbell 2010).

2.4 Sensors and transducers

Sensors in the natural world include those that equip us with our five senses – sight, hearing, smell, taste and touch. These convert the various and diverse inputs to electrochemical messages that can be used to inform or control the living organism.

In a similar way, in fabricated devices, sensors are also used to measure various stimuli. However, because of the broad range of potential inputs and outputs, the accepted definition of a sensor is refined. In this definition, all devices that convert input energy into output energy are referred to as *transducers*, and sensors form a small subset of the group as defined below:

> "*A sensor is a transducer that receives an input signal or stimulus and responds with an electrical signal bearing a known relationship to the input*" (Fraden 2003)

2.4.1 Resistive displacement sensors

2.4.1.1 Strain gauges

Small displacements are generally measured using a strain gauge bridge configuration. A strain gauge consists of a long, narrow metal conductor such as a piece of metal foil (usually constantan) mounted on a polyimide film, or fine gauge wire stretched over a frame. If it is stretched within its elastic limits, it will increase in length, L (m), and decrease in cross-sectional area, A (m^2). Because the resistance between the two ends of this foil or wire can be given by

$$R = \rho \frac{L}{A} \tag{2.3}$$

where ρ is the electrical resistivity of the material, this stretching will result in an increase in resistance. Because the change in length can only be small for the foil or wire to remain within its elastic limit, the resultant change in resistance will also be small.

To determine how this resistance changes under deformation, the derivative of equation (2.3) must be obtained (Alciatore and Histand 2003). First, take the natural log of both sides

$$\ln R = \ln \rho + \ln L - \ln A \tag{2.4}$$

Taking the derivative yields the following expression for the change in resistance as a function of the changes in the geometry of the conductor and the material property

$$\frac{dR}{R} = \frac{d\rho}{\rho} + \frac{dL}{L} - \frac{dA}{A} \tag{2.5}$$

Since the cross-sectional area of the conductor is the product of the width, w (m), and the height, h (m), the area derivative is

$$\frac{dA}{A} = \frac{w.dh + h.dw}{wh} = \frac{dh}{h} + \frac{dw}{w} \tag{2.6}$$

Poisson's ratio, v, is defined as the ratio of the transverse $\varepsilon_{transverse}$, and axial ε_{axial}, strain for a cylinder with diameter, D (m), and length, L (m)

$$v = \frac{\varepsilon_{transverse}}{\varepsilon_{axial}} = -\frac{\Delta D/D}{\Delta L/L} \tag{2.7}$$

Therefore

$$\frac{dh}{h} = -v\frac{dL}{L} \text{ and } \frac{dw}{w} = -v\frac{dL}{L} \tag{2.8}$$

which makes

$$\frac{dA}{A} = -2v\frac{dL}{L} = -2v\varepsilon_{axial} \tag{2.9}$$

Substituting equation (2.9) back into equation (2.5)

$$\frac{dR}{R} = \varepsilon_{axial}(1 + 2v) + \frac{d\rho}{\rho} \tag{2.10}$$

Dividing through by ε_{axial} gives

$$\frac{dR}{R}\frac{1}{\varepsilon_{axial}} = 1 + 2v + \frac{d\rho}{\rho}\frac{1}{\varepsilon_{axial}} \tag{2.11}$$

Note that the first two terms on the RHS, 1 and $2v$, represent the change in resistance due to the increased length and decreased diameter of the conductor. The last term represents the piezoresistive effect of the material, that is, the change in resistivity of the material with strain.

The relative sensitivity of such a device is given by the gauge factor, γ, which is defined as

$$\gamma = \frac{\Delta R/R}{\varepsilon_{axial}} \tag{2.12}$$

where ΔR is the change in resistance when the structure is stretched by an amount ΔL. This factor is usually specified on the data sheet of the strain gauge, and it can be used to determine the axial strain directly from the change in resistance.

$$\varepsilon_{\text{axial}} = \frac{\Delta R/R}{\gamma} \tag{2.13}$$

Typical strain gauges have $R \approx 120\ \Omega$ and $\gamma \approx 2$

Example: Strain gauge

If a 120 Ω strain gauge with a gauge factor of 2.0 measures a strain of 50 με (50×10^{-6}) read as "fifty microstrain." What is the change in resistance from the unloaded to the loaded state?

From equation (2.13)

$$\begin{aligned}\Delta R &= R\varepsilon_{\text{axial}}\gamma \\ &= 120 \times 50 \times 10^{-6} \times 2 \\ &= 0.012\ \Omega\end{aligned}$$

2.4.1.2 Measuring with a Wheatstone bridge

Because the change in resistance is small, measurement can be quite challenging and a Wheatstone bridge configuration shown in Figure 2.17 is generally used. It

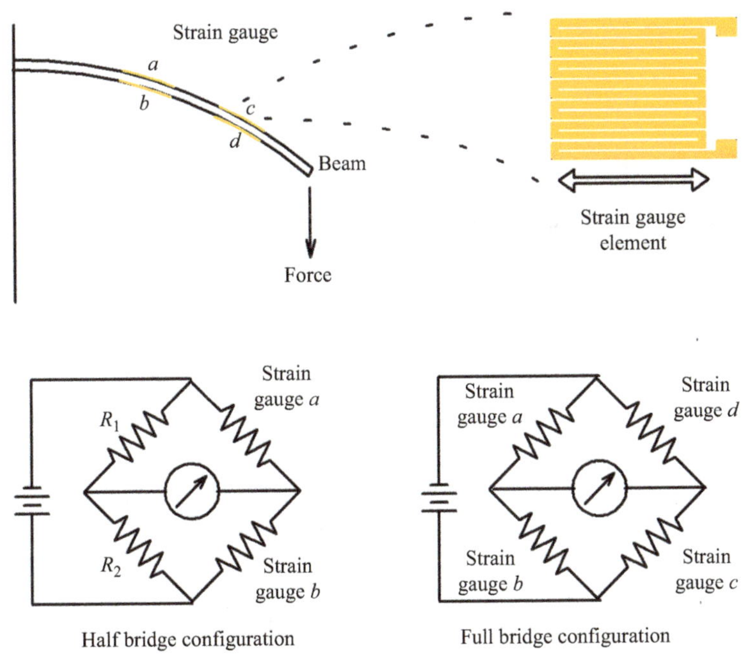

Figure 2.17 Strain gauges used to measure the deflection of a cantilever beam

should be noted that temperature changes can result in resistance changes of the same order as those caused by strain and so temperature compensation is essential if good accuracy is required.

A typical example of strain gauge-based displacement measurement is shown in Figure 2.17. In this application, the strain gauges are attached to the upper and lower faces of a cantilever beam. When the beam deflection is downward, the length of the strain gauges on the upper surface increases and those on the lower decrease.

It is possible to use a single strain gauge element and three fixed resistors to make up the bridge, but it is more common to use two or even four elements which provide increased sensitivity and also compensate for temperature variations (if all the elements on the beam remain at the same temperature).

Two different measurement modes are possible; they are the static balanced mode and the dynamic unbalanced mode (Alciatore and Histand 2003).

In the static balanced mode, R_1 and R_2 are precision resistors while R_3 is a precision potentiometer with an accurate scale displaying its resistance scale. R_s is the resistance of the strain gauge in the circuit as seen in Figure 2.18.

To balance the bridge, the potentiometer is adjusted until the voltage between nodes A and B is zero. The voltage at A must therefore equal the voltage at B and

$$i_s R_s = i_1 R_1 \tag{2.14}$$

Assuming that a high impedance voltmeter that draws no current is used to measure the voltage, then

$$i_s = i_3 = \frac{V}{R_s + R_3} \tag{2.15}$$

and

$$i_1 = i_2 = \frac{V}{R_1 + R_2} \tag{2.16}$$

where V is the DC excitation voltage applied to the bridge.

Substituting into equation (2.14) and simplifying gives

$$\frac{R_s}{R_3} = \frac{R_1}{R_2} \tag{2.17}$$

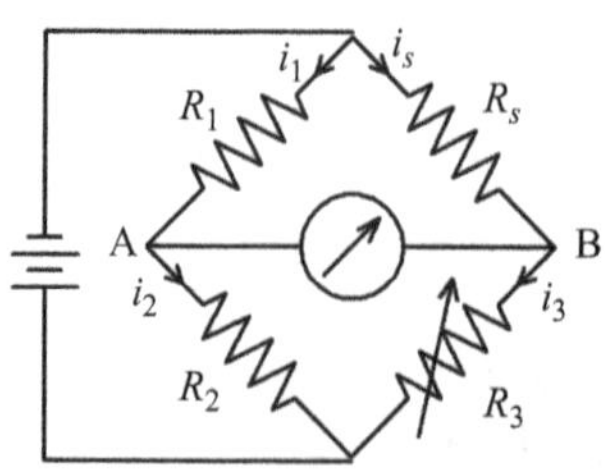

Figure 2.18 Static balanced bridge circuit

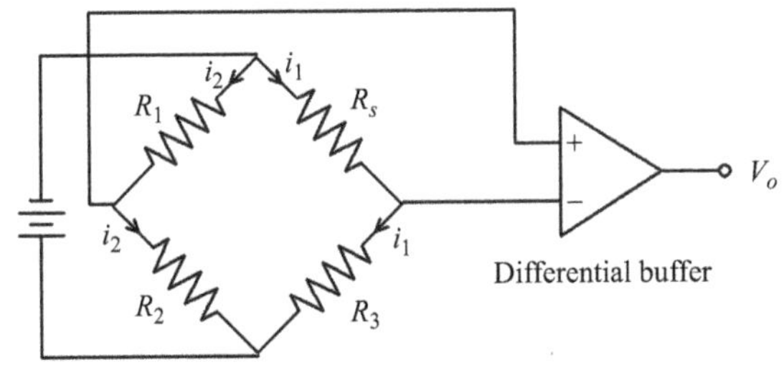

Figure 2.19 Dynamic unbalanced bridge circuit

If R_1 and R_2 are known, and R_3 can be determined accurately from the angular displacement of the potentiometer, then R_s can be determined

$$R_s = \frac{R_1 R_3}{R_2} \tag{2.18}$$

Note that this result is independent of the excitation voltage.

In the dynamic deflection mode shown in Figure 2.19, the bridge is first balanced under no-load conditions and then the measured output voltage can be used to determine the strain gauge resistance as a load is applied.

The output voltage can be written in terms of the volt drop from the +ve terminal and also in terms of the volt drop from the negative terminal

$$V_o = i_1 R_s - i_2 R_1 = i_2 R_2 - i_1 R_3 \tag{2.19}$$

and the applied excitation voltage can also be written in terms of the current and resistance in each arm

$$V = i_1(R_s + R_3) = i_2(R_1 + R_2) \tag{2.20}$$

Solving for i_1 and i_2 in terms of V and substituting into equation (2.19) gives

$$V_o = V\left[\frac{R_s}{R_s + R_3} - \frac{R_1}{R_1 + R_2}\right] \tag{2.21}$$

If the bridge is balanced under no-load conditions so that $V_o = 0$ in this case, then as the strain gauge is loaded, the voltage change ΔV_o will be related to the new resistance $R_s + \Delta R_s$ according to

$$\frac{\Delta V_o}{V} = \left[\frac{R_s + \Delta R_s}{R_s + \Delta R_s + R_3} - \frac{R_1}{R_1 + R_2}\right] \tag{2.22}$$

This equation can be rearranged to give the relationship between the relative change in resistance and the measured output voltage.

Let

$$A = \left[\frac{\Delta V_o}{V} + \frac{R_1}{R_1 + R_2}\right] \tag{2.23}$$

then by simplifying equation (2.22) it can easily be shown that

$$\frac{\Delta R_s}{R_s} = \frac{AR_3}{(1 - A)R_s} - 1 \tag{2.24}$$

Therefore, by measuring the change in output voltage ΔV_o the change in strain gauge resistance ΔR_s can be determined using equations (2.23) and (2.24). Finally, the actual strain is determined using equation (2.13).

A similar analysis can be performed for the bridge with two or four strain gauge elements either mounted on the flexible beam as shown in Figure 2.17 or as dummy elements to be maintained at the same temperature as the active element.

A more compliant structure that has found applications in biomedical instrumentation is the liquid metal strain gauge. Instead of using a solid electrical conductor, a mercury-filled compliant silicone rubber tube is used. As the tube is stretched, the length increases and the cross-sectional area decreases as with a conventional strain gauge (Bronzino 2006). Due to the toxicity of mercury, Gallium-based liquid alloys are now considered superior for stretchable strain sensors because they retain good electrical conductivity even under extreme stretch conditions. They also exhibit low viscosity and fluidic nature, which minimises the problem of hysteresis. (Wu *et al.* 2021).

2.4.1.3 Potentiometers

Larger displacements can be measured using precision potentiometers. These are available in rotary, linear-motion and string pot forms as shown in Figure 2.20. String pots are also known as cable pots, yo-yo pots or draw-wire transducers, and measure the extended length of a spring-loaded cable. Rotary pots are single- or multi-turn (commonly 3, 5 or 10 turns), while linear-motion pots are available with maximum strokes ranging from 5 mm to over 4 m (Webster 1999).

Potentiometers may be characterised as wire-wound or non-wire-wound. The former are made from tightly wound resistive wire that quantises measurements, while the latter are made from a sheet of resistive material that is, in theory, capable of unlimited resolution. Wire wound pots offer good temperature stability and high power handling capability, but have a limited life. However, as most measurement applications are low power, potentiometers made from conductive plastic film are the

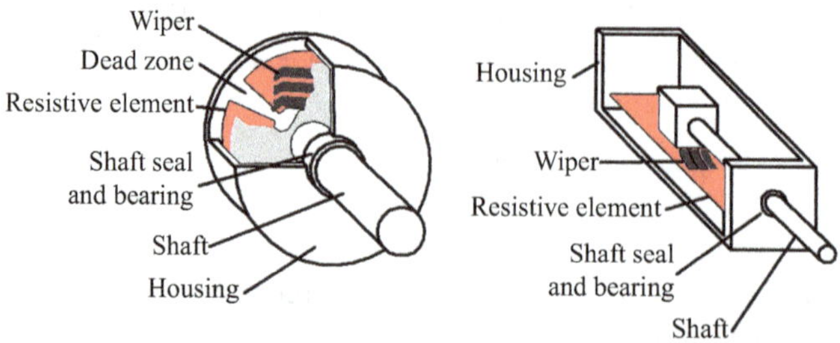

Figure 2.20 Schematic cutaways of rotary and linear-motion potentiometers [adapted from Webster (1999)]

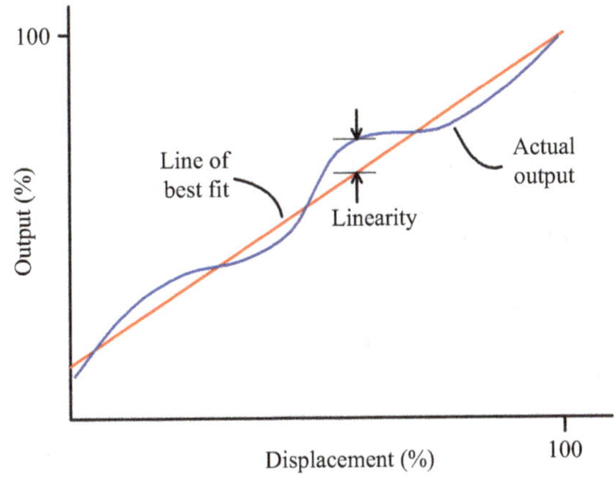

Figure 2.21 Definition of potentiometer linearity is the percentage deviation of the output from a best fit straight line

components of choice. They are low friction and low noise and have a long operational life. Unfortunately, they are sensitive to temperature and other environmental factors.

For displacement measurements, most potentiometers have a linear taper in that the output varies linearly with wiper motion. The linearity is defined as the maximum deviation of the output function from a straight line and is illustrated in Figure 2.21. With laser trimming, overall linearities of better than 0.1% of full-scale output are routinely achieved.

To avoid electrical loading, the pot-wiper should drive into a high impedance. This is generally achieved by making the potentiometer two arms of a Wheatstone bridge and measuring the difference voltage using an operational amplifier that has a high input impedance as shown in the circuit in Figure 2.19.

From a mechanical perspective, potentiometers add inertia and friction to the moving parts of the system they are measuring. As a result, they increase the force required to move these parts. Rotary pot manufacturers commonly list the equivalent mass moment of inertia of the rotating part, the dynamic torque required to maintain rotation at a constant angular rate and the starting torque required to initiate motion.

Despite constant mechanical wear, a conductive plastic potentiometer may last hundreds of millions of cycles if the conditions are good and abrasive dirt is kept off the film. Wire wound pots do not usually last as long, but they can be expected to survive over one million cycles.

2.4.2 Inductive displacement sensors

Mutual inductance between two coils can be used to provide a sample displacement sensor. If one coil is driven by an AC signal, then the induced voltage in the other will be proportional to the distance between coils. A variation of this would be to have two fixed coils coupled together by a movable core in which the position of the core controlled the mutual inductance, and hence the induced voltage (Bronzino 2006).

2.4.2.1 Linear variable differential transformer

Probably the most common inductive device for measuring linear displacement is the linear variable differential transformer (LVDT). It consists of a primary and two secondary cores mounted axially over a movable iron core as shown in Figure 2.22. The two secondary coils are generally connected in a series-opposing configuration and their combined output describes the magnitude and direction of the core motion when the primary coil is excited by an AC signal.

When the core is positioned midpoint between the symmetrical windings, each of the secondary windings provides signals with the same amplitude, but phase shifted by 180° with the result that the output voltage, Vo, is zero. As the core is displaced away from the balance point, the induced voltage in one of the secondary windings will increase, and that in the other will decrease, with the result that the magnitude of the output signal will grow in a fairly linear fashion as shown in Figure 2.23.

Determining the direction of core displacement requires that the phase of the output signal be accounted for. This can be found using a phase-sensitive detector, a synchronous detector, or the demodulation scheme shown in Figure 2.24. The bridge rectifiers produce a positive or negative rectified sine wave depending on

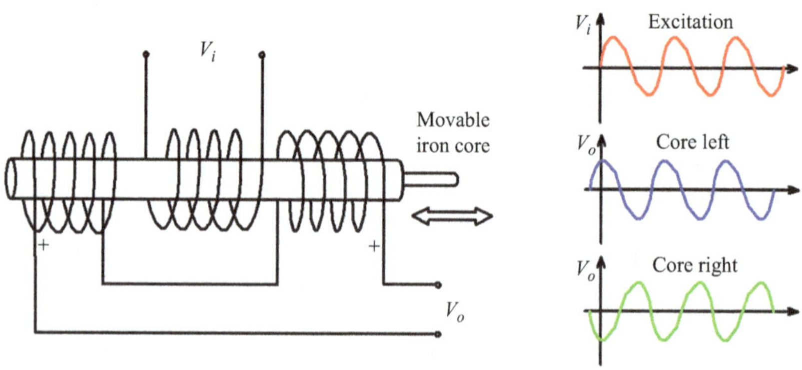

Figure 2.22 Linear variable differential transformer

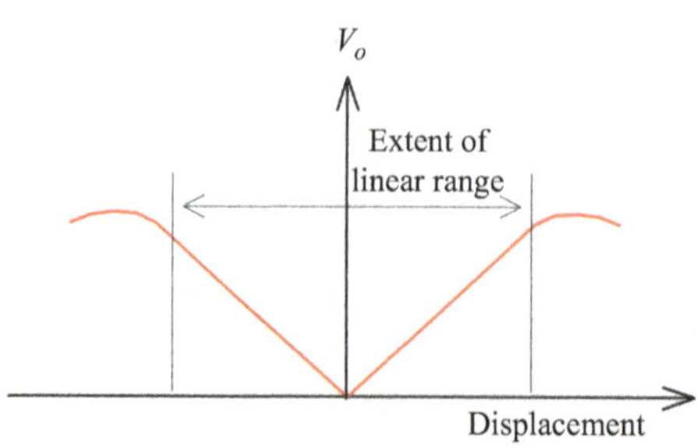

Figure 2.23 RMS voltage output from an LVDT as a function of core displacement

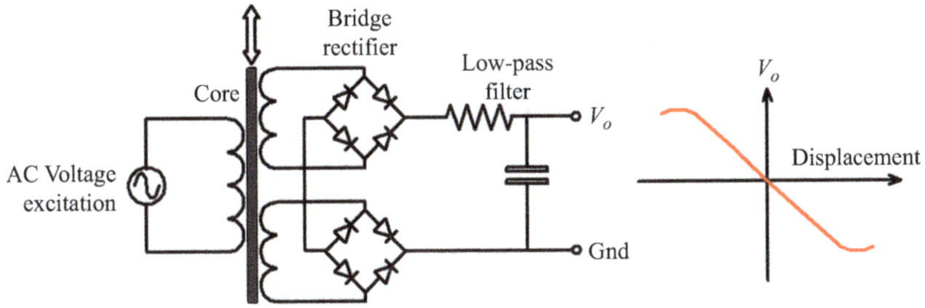

Figure 2.24 Demodulation of LVDT output

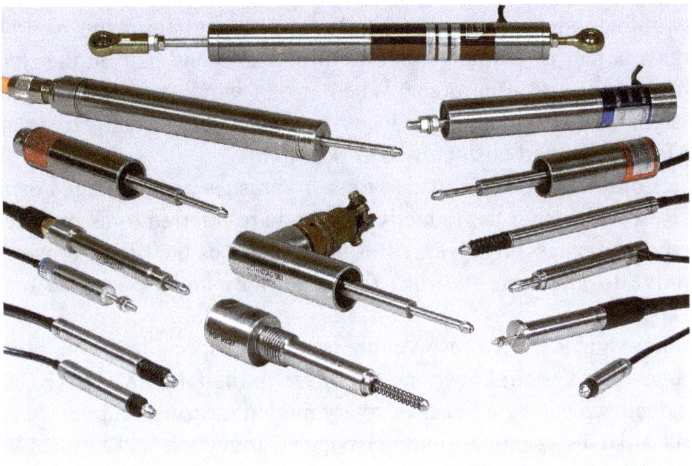

Figure 2.25 Photograph of a range of LVDTs, reproduced courtesy RDP Electronics with permission

which side of the null the core is located, and a low-pass filter then smooths the output to provide a DC voltage proportional to the displacement.

Commercial LVDTs are available with different diameters, lengths and strokes. They mostly include internal electronics to provide the AC signal and perform the demodulation, so that the output is a DC voltage proportional to the displacement. Good accuracy over the linear range is provided, and the analogue output generally does not require amplification. LVDT nonlinearities are typically 0.25% of full scale with precision units being as low as 0.05%. The units shown in Figure 2.25, manufactured by RDP Electronics in the UK, have a specified accuracy of 0.1%, infinite resolution and strokes from 0.5 mm up to 470 mm.

LVDTs are also less sensitive to variations in temperature than the other linear encoders discussed so far, but their main disadvantage is a limited frequency response. This is determined by the inertial effects associated with the core's mass, the choice of excitation frequency and the bandwidth of the low-pass filter at the output (Alciatore and Histand 2003).

Measurement of angles using the same principle is by means of a rotary variable differential transformer (RVDT). In this case, a rotary ferromagnetic core is used with linear ranges of ±40° possible, with a nonlinearity error of about 1% (Fraden 1996).

2.4.2.2 Inductosyns and resolvers

Other inductive angle measurement systems include resolvers and inductosyns. Resolvers resemble small motors and are essentially rotary transformers designed so that the coefficient of coupling between rotor and stator varies with the shaft angle. Fixed windings are placed on a laminated iron stack to form the stator, and movable windings are placed on a similar structure to form the rotor. Usually, resolvers have a pair of windings on a rotor and a second pair on the stator, positioned at right angles to each other. When a rotor winding is excited with an AC reference signal, stator windings produce AC voltage outputs that vary in amplitude according to the sine and cosine of shaft position.

Connection to the rotor is made either by brushes and slip rings or inductive coupling. Resolvers using the inductive method are referred to as brushless types. The inductive (brushless) resolvers offer up to 10 times the life of brush types and are insensitive to vibration and dirt; therefore, they are used in the majority of industrial applications.

The stator signals from a resolver are routed to a specialised type of analogue-to-digital converter system known as a resolver-to-digital converter (RDC).

The inductosyn can be a linear or rotary motion measurement device consisting of two parts: a fixed serpentine winding having a 2 mm pitch, and a movable winding known as a *rotor or slider* depending on whether rotary or linear motion is being measured. The rotor/slider has a pair of windings having the same pitch as the fixed winding that are mutually offset by a quarter pitch so both sine and cosine waves are produced by movement. One slider winding is adequate for counting pulses but provides no direction information. The 2-phase windings provide direction information in the phasing of the sine and cosine waves. Movement by one pitch produces a cycle of sine and cosine waves; multiple pitches produce a train of waves.

2.4.3 Magnetic displacement sensors

Magnetic field strength is most commonly measured using the Hall effect and magnetoresistance.

In the Hall effect, a voltage, V_H, appears across a conductor when a magnetic field is applied at right angles to the current flow. The direction is perpendicular to both the current and the magnetic field. The magnitude of the Hall voltage, V_H (V), is proportional to the magnetic flux density, β (Weber/m^2) and the current, I (A).

$$V_H = \frac{K_H \beta I}{z} \tag{2.25}$$

where K_H is the hall constant and z (m) is the thickness of the conductor (Webster 1999).

Magnetoresistance (MR) is a property of most magnetic materials. A decrease in electrical resistance occurs when a magnetic field is applied perpendicular to the direction of current flow. The resistance decreases as the magnetic flux density increases until the material reaches magnetic saturation. The total change in resistance is about 0.3% for iron and 2% for nickel. Giant magnetoresistance (GMR) was discovered in 1988. It is the phenomenon where the resistance of certain layered materials drops dramatically as a magnetic field is applied. It is described as giant since it is a much larger effect than had ever been previously seen in metals. With GMR, changes in resistance of up to 10% have been measured.

As with other sensors that use changes in resistance, MR and GMR sensors are often incorporated into a Wheatstone bridge configuration to reduce temperature dependence and to increase sensitivity. In these configurations, two of the elements are shielded from the applied magnetic field and the opposite pair exposed to it. Commercial sensors are available with outputs that vary by more than 5% of the applied voltage.

For short distances, linear displacement measurement can be achieved using a pair of opposing magnets and a single magnetic sensor as illustrated in Figure 2.26. This configuration partially solves two of the more annoying problems commonly encountered in magnetic distance sensing schemes, that is lack of linearity and the lack of a well-defined reference point. The pair of opposing magnets mounted on a movable, nonmagnetic yoke creates a magnetic null point exactly halfway between the two magnets. This is convenient as a stable reference point. Another benefit offered by opposed magnets is that the sensed field vs. displacement is a nearly linear function over a significant range of travel.

A rotary sensor can be made using the same principles, but with a plastic yoke formed into an annulus as shown in Figure 2.27. In this case, when the magnets are rotated around the sensor, it sees a sinusoidally varying field with good linearity available over a span of ±30° of rotation. Multiple sensors and linearisation processes can be used to extend this technique to operate over the full 360°.

Most magnetic linear or angular displacement encoders use magnets with multiple poles to extend the maximum displacement that can be measured or to improve the resolution of angular encoders. The principles involved are similar to those used by the more common optical encoders that are described later in this chapter.

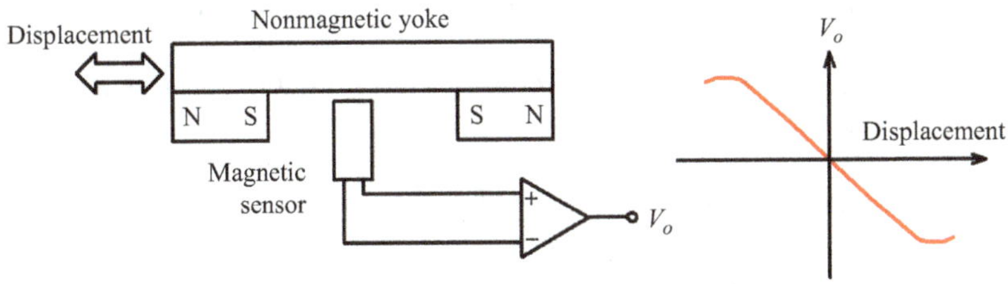

Figure 2.26 Magnetic linear position sensor

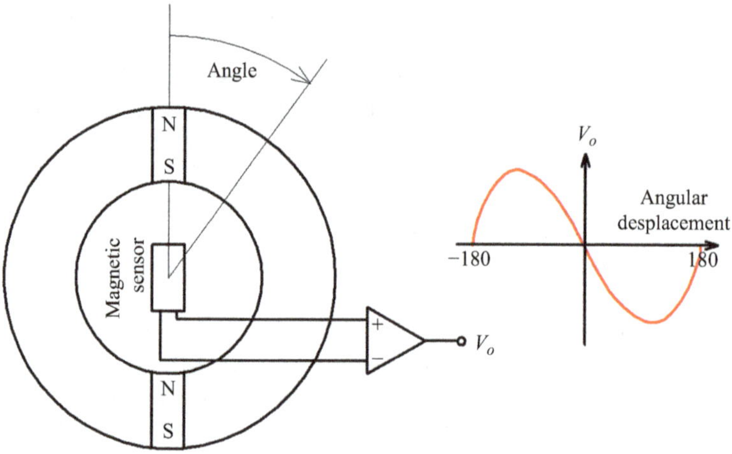

Figure 2.27 Magnetic rotary position sensor

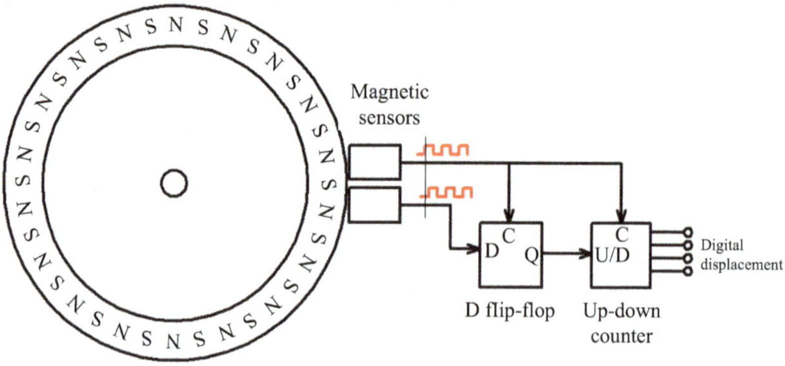

Figure 2.28 Magnetic incremental encoder for rotation

Encoders can be absolute or incremental. At each measurement position, absolute encoders include a binary word made up of a number of magnetic elements that encode for the complete displacement or angle, while incremental encoders include only a single bit, and rely on the associated electronics to maintain a count from some datum point. Incremental encoders are simpler and cheaper, but if power is lost then position information is lost, and it is necessary to return to the datum point.

To increase accuracy and to enable incremental encoders to determine the direction of travel requires that two magnetic sensors are used. The one is displaced from the other along the axis of travel by 90° so that the outputs are in quadrature. This allows a simple D type flip-flop to determine the direction of travel and control an up-down counter to produce a quantised measure of the displacement as shown in Figure 2.28.

By "unwinding" the ring magnet used in the previous application, a device that can measure linear motion is produced. While a rigid rod magnet could be used to provide the required alternating pole pattern, flexible magnetic materials have made this sensor much easier to implement, especially for long linear runs. For low-resolution applications where the sensor can be very close to the magnetic strip, flexible ferrite materials offer good resolution. These materials can be purchased in rolls of varying thickness and width, often with pressure-sensitive adhesive backing for easy installation. For more demanding applications, either in terms of the number of poles per mm or the working distance from sensor to magnet strip, rare-earth materials (neodymium-iron-boron or samarium-cobalt) mixed with plastic binders can be used.

An example of a linear incremental encoder is the LM10 manufactured by RLS/Renishaw shown in Figure 2.29. This is a solid-state, non-contact encoder that features a compact IP68 read head that rides between 0.1 and 1.5 mm above a self-adhesive magnetic strip. Devices can operate with a maximum velocity of 25 m/s for sinusoidal outputs, while digital models provide resolutions of 100, 50 and 10 μm at 25 m/s, improving to 5 μm at 20 m/s, and 1 μm at 4 m/s. Resistant to shock, vibration, and pressure, they operate over a temperature range of −20 °C to +85 °C.

2.4.4 Capacitive displacement sensors

The capacitance of a parallel plate capacitor is directly proportional to the area of the two plates and inversely proportional to the distance between them. Capacitance is also directly proportional to the dielectric constant of the material between the plates. A simple, albeit very short-range sensor could be made by having one fixed plate and one plate that moves. However, more sophisticated capacitance-based sensors use a pair of coaxial capacitors with a movable dielectric sleeve that fits between the outer and inner electrodes of one capacitor. The proportion of the area between the two electrodes that is filled with dielectric determines the capacitance (Fraden 1996).

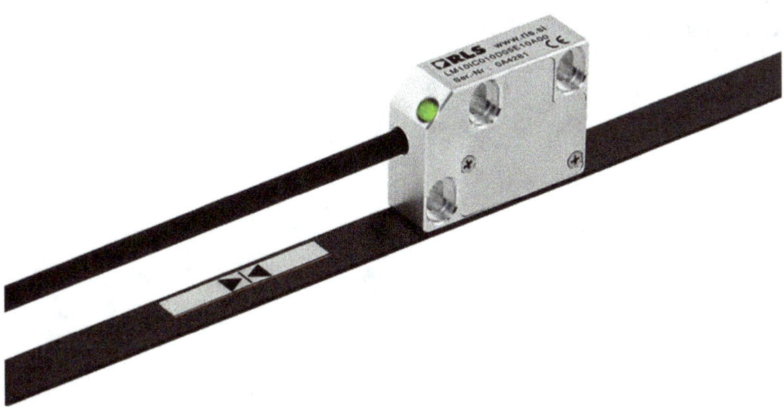

Figure 2.29 Photograph of the LM10 linear magnetic encoder, reproduced courtesy RLS/Renishaw with permission

These sensors are ratiometric because they measure the ratio of two capacitors maintained under the same conditions of temperature and humidity to minimise drift. This ratio can be determined using an AC equivalent of the Wheatstone bridge described earlier, or by incorporating the capacitors into a pair of radio frequency LC oscillators and measuring the frequency difference between the two.

2.4.5 Optical displacement sensors

2.4.5.1 Digital optical encoders

Digital optical encoders are probably the most common of all the sensors used for angle and angular rate measurement. They consist of a light source and a number of phototransistors separated by a rotating mask made from alternating opaque and transparent radial bands. As with magnetic encoders, rotary optical encoders come in two types: absolute encoders, in which a unique digital word corresponds to every rotational position (within the quantisation level), and incremental encoders which produce a sequence of digital pulses as the shaft rotates, allowing measurement of the relative displacement of the shaft.

Rotary encoder discs can be made from glass or plastic onto which has been deposited a radial pattern organised into tracks or, for more robust applications, a metal disc through which holes have been cut. Astride this disc is placed the LED source on one side and a number carefully aligned phototransistors on the other. As the disc rotates, it will periodically interrupt the beam between the emitter–receiver pair to produce a change in the digital level as shown in Figure 2.30.

The optical disk of an absolute encoder is designed to produce a digital word that distinguishes N distinct positions on the shaft. For example, if there are eight tracks, then the encoder is capable of measuring $2^8 = 256$ distinct angles each corresponding to an angular resolution of 360°/256 = 1.406°.

Most absolute encoder discs are encoded with either binary or Grey codes. Figure 2.31 shows the encoder pattern for a 4-bit binary disc that is the obvious method to encode for angle. Unfortunately, if the level transitions are not simultaneous, large errors will occur in the digital output. For example, in the transition from 360° to 0°, all four bits must change simultaneously, otherwise any output between 1 and 14 could occur for a brief period during the transition.

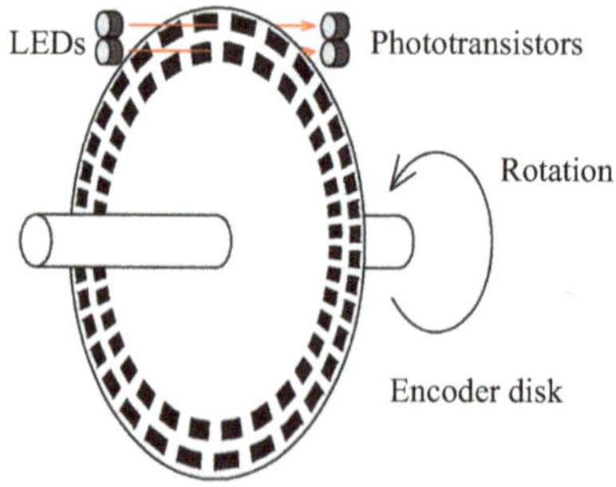

Figure 2.30 Schematic diagram of a generic rotary optical encoder

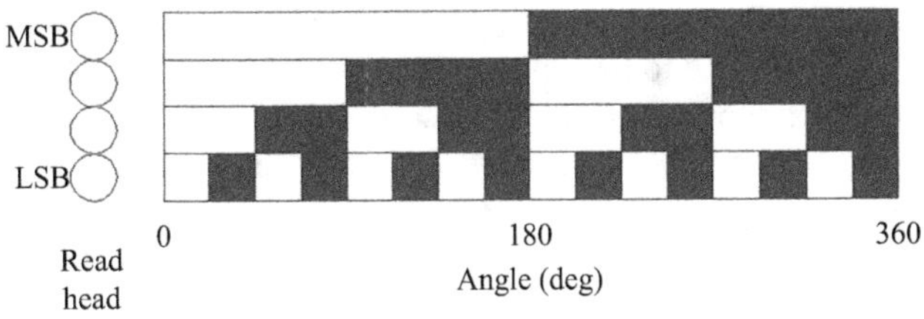

Figure 2.31 Encoder pattern for a 4-bit binary disc

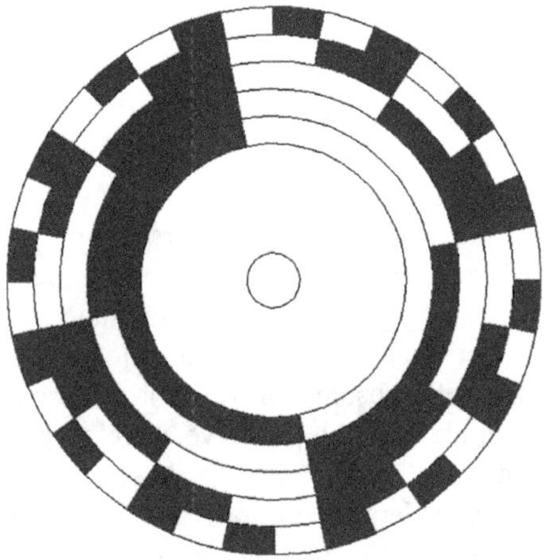

Figure 2.32 Encoder disc pattern for a 5-bit binary encoder

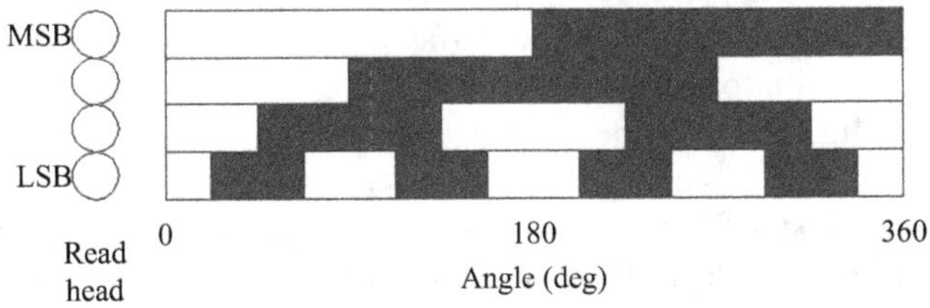

Figure 2.33 Encoder pattern for a 4-bit Grey code disc

The actual encoder disc will obviously be circular, with the inner rings that encode for the more significant bits being smaller as can be seen from Figure 2.32. However, because the transitions are aligned along the radial, and the individual read heads are smaller than the size of a bit on the innermost ring, the readout should be accurate. However, as the number of bits increases, the size of the outermost track becomes smaller and the devices become more expensive and less robust.

The Grey code is designed so that only one bit changes state for each count transition; therefore, the largest error can be a single count as can be seen if the transitions shown in Figure 2.33 are considered.

Digital electronics uses the binary system; therefore to be useful the Grey code outputs of the encoder need to be converted to binary. This is easy to achieve using a number of exclusive-or gates, $\oplus$, because the Boolean expression that relates binary bits to Grey code bits is as follows for a four bit word:

$$B_3 = G_3$$
$$B_2 = B_3 \oplus G_2$$
$$B_1 = B_2 \oplus G_1$$
$$B_0 = B_1 \oplus G_0$$

This sequence can be extended to any number of bits starting with the premise that $B_n = G_n$.

As the number of bits increases in an absolute encoder, the number of tracks increases. For example, a 12-bit encoder would have 12 individual tracks. Because each position has a unique signal, the absolute encoder will not lose positional information if power is interrupted.

A common application for absolute encoders is on a prosthetic arm, where accurate positional information is paramount to the operation's accuracy and safety. Located at the joints or pivot points of an articulated prosthesis, the absolute encoder can monitor exact angular position, direction and speed of arm travel. Another critical application for absolute encoders is in aircraft, where they can be mounted to the plane's rudder and aileron drive shafts to provide exact positional and absolute data during flight.

Absolute multi-turn encoders incorporate standard absolute technology but rely upon an additional internal counting process to monitor and track the number of rotations. Some multi-turn encoders incorporate a gear-driven system that can be quite accurate but is complex and expensive to manufacture and prone to breakage and wear. Alternative non-contact multi-turns offer significantly increased encoder life by using a two-pole magnet and an array of reed switches to monitor revolutions and directional information.

Like single-turn absolute encoders, multi-turn encoders have a unique code for each position within 360° of rotation, dependent on the encoder's resolution, but providing unique codes for each revolution. As a result of these unique codes, an absolute multi-turn encoder will not lose its revolution count or angular position if power fails.

The newest absolute multi-turn encoders are available with 36-bit resolution – 18 bits over 360° and another 18 bits for counting revolutions. With this resolution, multi-turns can offer 262,144 positions over 262,144 revolutions, allowing the encoder to track 68,719,476,736 unique angular positions.

The simplest encoder technology available, the incremental encoder, is designed to monitor distance, speed, and direction. It features a pulse disk that consists of a double track of evenly spaced clear and opaque segments displaced by 90° electrical as shown in Figure 2.34. Defined as pulses per revolution (ppr), resolution is dictated by the number of clear and opaque lines on the disc. An incremental encoder with a resolution of 256 ppr would have 256 distinct lines on the disc.

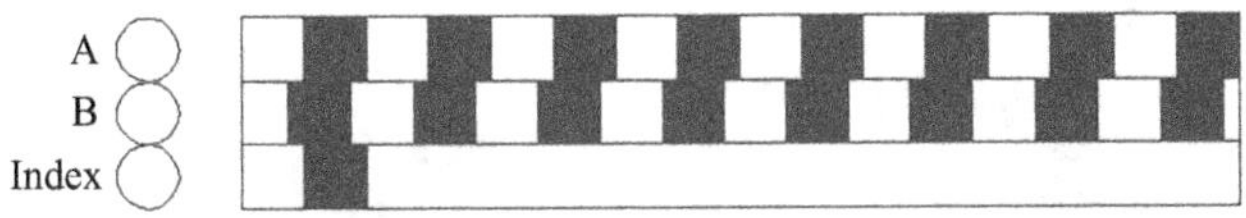

Figure 2.34 Encoder pattern for an incremental encoder

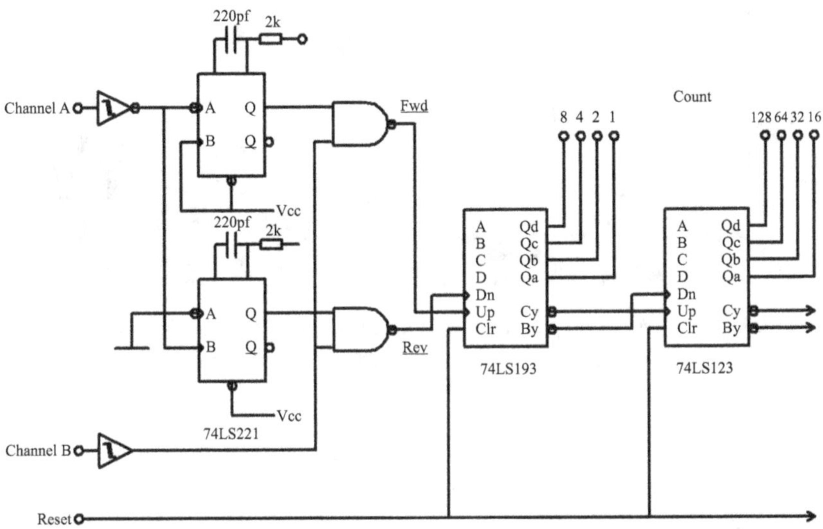

Figure 2.35 Counter circuitry for incremental encoders

With one pulse track on the encoder disc, incremental encoders deliver only on/off outputs that provide speed, distance and directional data to a monitoring or control device through the quadrature output. To measure direction, the controller's electronics monitor the A and B channels to determine which channel arrives first. The controller then verifies whether the encoder is turning clockwise or counter-clockwise. Distance is measured by assigning each pulse interval a distance value and counting the number of pulses with an up-down counter, as discussed earlier in this chapter.

In reality, incremental encoders do not have two separate rings for the A and B channels; instead, the two phototransistors are displaced by 90° electrical from each other, to give the same effect. If absolute position is required, there will always be a separate index pulse. Figure 2.35 shows a circuit that is used to output the absolute position derived from an incremental encoder. In general, similar circuitry is built into the more complex motion-control electronics modules used to drive motors.

In standard mode, an encoder counts the leading edge of one square wave signal. By counting both the leading and trailing edges of the signal, resolution is doubled, a process called 2 times interpolation.

Figure 2.36 Photograph of an incremental encoder with encoder disc exposed, reproduced courtesy of GPI with permission

By counting both edges on the A and B channels simultaneously, the output increases to four times the ppr, also known as 4 times interpolation. So, by interpolating a quadrature output by a factor of four, a 256-ppr encoder can offer 1024 ppr.

With this technology, resolutions of 40,000 ppr are possible.

For even higher resolution, some encoders replace square wave signals with sine wave outputs that allow for interpolation factors as high as 10 times. Through 10 times interpolation of a sine wave, a 256-ppr encoder is capable of 2,560 ppr.

To illustrate the benefits of particular resolutions, one could use an example of an incremental encoder in monitoring conveyor position. An encoder could be mounted to a conveyor drive wheel with a circumference of 150 mm at a 1:1 ratio such that each revolution of the encoder is equivalent to 150 mm of overall conveyor travel. By using an output of 40,000 ppr, the encoder is able to provide positional feedback accurate to 0.012 mm!

Unfortunately, in the event of a power failure, incremental encoders will usually lose positional information, and a reset or homing cycle must be performed to synchronise the encoder with the control device.

Figure 2.36 shows an example of the 7700 optical encoder and disc manufactured by Gurley Precision Instruments (GPI). It is 13 mm tall by 43 mm diameter with a resolution of up to 5000 ppr.

2.4.6 Ranging sensors

Non-contact distance measurement methods can be classified into three categories: interferometry, triangulation and time-of-flight. The method used by a particular sensor usually depends on the maximum range and the measurement accuracy required. For example, interferometric methods can be extremely accurate, but are prone to range ambiguity, while time-of-flight methods operate at longer ranges with poorer accuracy and triangulation sits somewhere in the middle.

2.4.6.1 Interferometry

Optical interferometry operates using the superposition of two monochromatic light beams so that extremely small displacements can be measured. The operational principle is best explained in conjunction with Figure 2.37.

Typically, an incoming beam of light will be split into two identical beams by a grating or a partial mirror. Each of these beams will travel a different path before they are recombined at a detector. The difference in the distance travelled by each beam creates a phase difference between them. It is this introduced phase difference that creates the interference pattern between the initially identical waves. If one of the paths is held constant as a reference, then any change in the distance to the mirror in the measurement path will result in a change in the relative phase of the two beams back at the detector.

Interferometers are ambiguous over distances of $\lambda/2$, but as with incremental encoders, a count can be made of the number of electrical cycles out of the detector to accommodate extremely accurate displacement measurement.

2.4.6.2 Triangulation sensors

As discussed by Brooker (Brooker 2021), triangulation sensors make good non-contact devices for measuring ranges from a few centimetres up to about a metre with reasonable accuracy. One of the most common uses is a collimated light-emitting diode (LED) source and a position-sensitive detector (PSD) to measure the

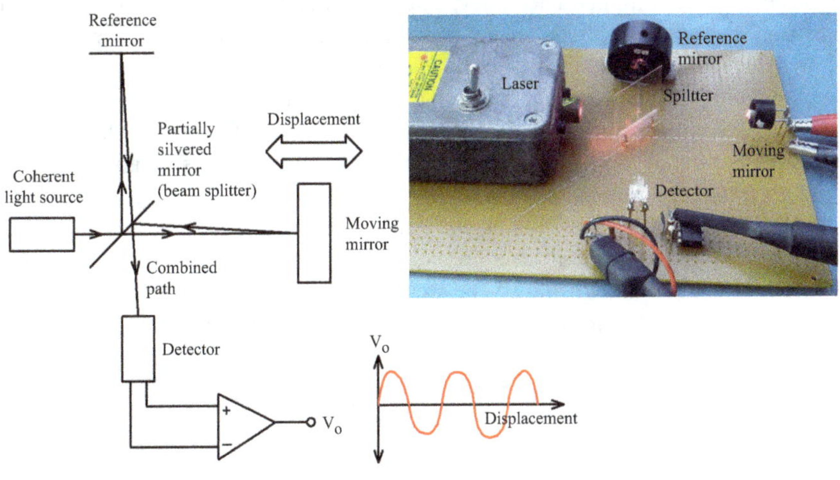

Figure 2.37 Schematic diagram of a Michelson interferometer and a photograph showing it used to measure the displacement of a piezo transducer (Brooker 2015)

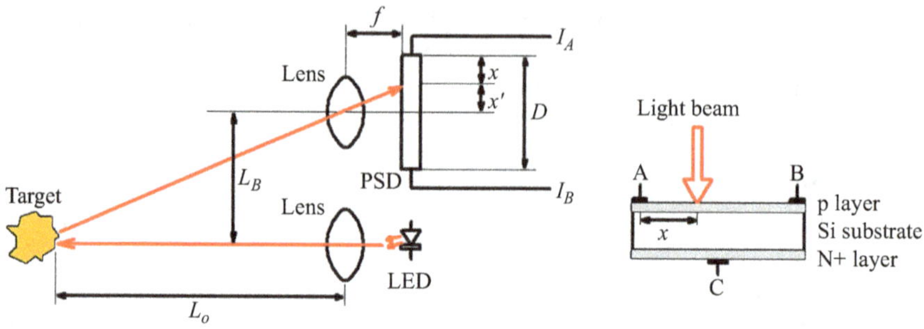

Figure 2.38 Measuring range using a PSD-based sensor (Brooker 2021)

direction of arrival of the reflected light beam. As shown in Figure 2.38, the relationship between the range to the target, L_o, and the distance between the transmit and receive apertures, L_B, is a function of the focal length, f, and the displacement, x', from the centre of the PSD.

$$L_o = f\frac{L_B}{x'} = f\frac{L_B}{D/2 - x} \tag{2.26}$$

An LED or laser source emits a narrow beam of infrared light in the direction of the target. The small amount of reflected light is focused onto the sensitive surface of the PSD which generates two output currents, I_A and I_B, that are each proportional to the displacement of the light spot from the centre of the device.

Though the current generated by the PSD depends on the intensity of the incident radiation, the ratio of the two currents does not, so the distance to the target can be determined with good accuracy. The PSD operates by exploiting the change in resistance of a doped silicon semiconductor that is proportional to the intensity of the incident light. The device is fabricated from a thin slab of high-resistance silicon with the top and bottom layers doped p and n+, respectively. Two electrodes, A and B, are placed at opposite ends of the upper layer and a common electrode, C, covers the bottom. The distance between the upper electrodes is D, and the corresponding resistance is R_D.

If the beam strikes the PSD at a distance x from electrode A, the resistance between that point and the electrode is R_x, and a photocurrent, I_o, proportional to the intensity of the light, will flow. The amount that flows to the electrodes A and B will be proportional to the relative distances from the incident beam; therefore

$$I_A = I_o\frac{R_D - R_x}{R_D} = I_o\frac{D - x}{D} \tag{2.27}$$

and

$$I_B = I_o\frac{R_x}{R_D} = I_o\frac{x}{D} \tag{2.28}$$

Taking the ratio of the two currents I_A and I_B

$$S = \frac{I_A}{I_B} = \frac{D}{x} - 1 \tag{2.29}$$

The distance, x, can then be written as

$$x = \frac{D}{S + 1} \tag{2.30}$$

Substituting into (2.26) gives

$$L_o = f\frac{2L_B}{D}\frac{S + 1}{S - 1} = k\frac{S + 1}{S - 1} \tag{2.31}$$

where k is called the module geometrical constant. Therefore, L_o, the distance from the sensor to the target, can be determined in terms of the ratio of the two currents out of the PSD (Fraden 2003).

A common IC that operates using this principle is the Sharp Electronics GP2Y0A21YF which has an operational range between 10 cm and 80 cm with the characteristics as illustrated in Figure 2.39.

In more critical applications, a laser source is used because of its brightness and good spatial coherence and a CCD array is often used as a receiver. On transmit, the width of the laser beam is diffraction-limited by the size of the exit aperture. On scattering from the target, the coherence of the laser beam is lost and so, to produce the smallest spot on the CCD array, it must be placed at the focal plane of a lens (Amman 2001).

Longer-range triangulation sensors can be used to obtain 2D or 3D images of the terrain by scanning the beam. Such scanning mechanisms must be constructed to ensure that both the transmitter and the receiver can see each point, so that as the

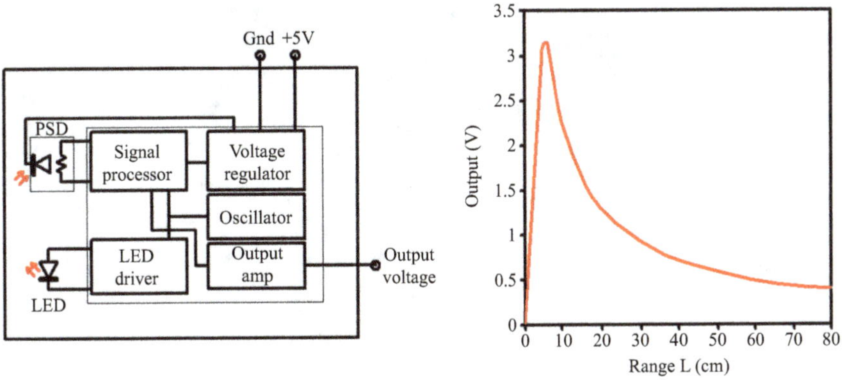

Figure 2.39 Sharp GP2Y0A21YF distance measurement IC schematic block diagram and voltage output as a function of range, courtesy Sharp Electronics

transmitted spot is swept across the target, a corresponding image is reflected onto the CCD array (Probert-Smith 2001).

Figure 2.40 shows a diagram of a sensor developed for 2D scanning (Livingstone and Rioux 1986).

2.4.7 Time-of-flight ranging

The basic principles of active non-contact range-finding are similar for electromagnetic (radar, laser etc.) and active acoustic sensing. A signal is radiated towards an object or target of interest, and the reflected or scattered signal is detected by a receiver and used to determine the range.

As shown in Figure 2.41, a source of radiation is modulated and fed to a transmit antenna, or aperture, which is usually matched to the impedance of the transmission medium to maximise power transfer. This can take the form of a horn for acoustic or radar sensors or an appropriately coated lens for a laser. The antenna also operates to concentrate the radiated power into a narrow beam so as to maximise the operational range and minimise the angular ambiguity of the measurement. When the transmitted beam strikes the target, a portion of the signal is reflected or scattered because the target has a different impedance, or refractive index, to the medium through which the signal is propagating. A small percentage of the reflected power travels back to the receiver (which is often co-located with the transmitter) where it is captured by the receiver antenna and converted to an electrical signal that can be filtered to remove extraneous noise, before being amplified and detected.

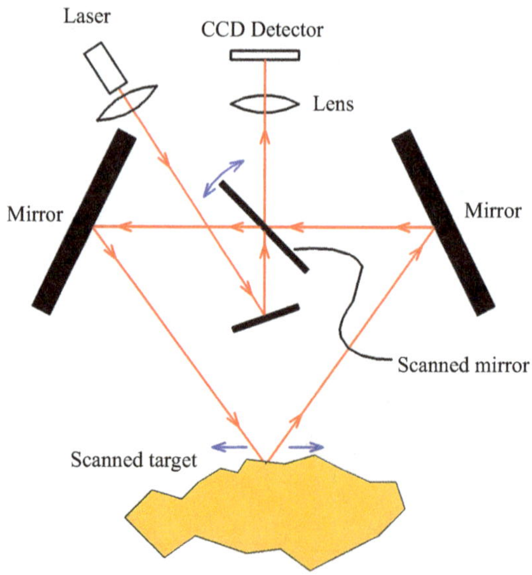

Figure 2.40 Triangulation-based line scanner [adapted from Livingstone and Rioux (1986)]

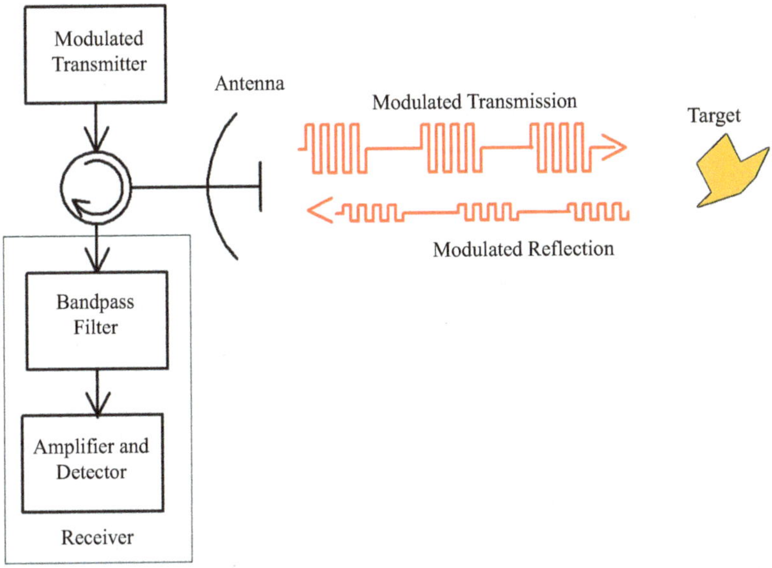

Figure 2.41 Operational principles of a generic time-of-flight sensor (Brooker 2021)

The time between the transmission of a pulse and the reception of an echo is used to provide range. Because the round-trip time is measured, there is a factor of two in the formula

$$R = \frac{v\Delta T}{2} \tag{2.32}$$

where R is the range (m), v is the propagation velocity (m/s) and T is the round-trip time (s).

Most ranging sensors transmit and receive from the same aperture, or from adjacent apertures, requiring special techniques that allow the transmission and receipt of signals simultaneously. In general, the transmit pulse is made sufficiently short to have been radiated before the echo is received. This limitation restricts the minimum distance that can be measured and requires that very short pulses are transmitted.

Because the velocity of sound is so much smaller than the speed of light, measuring short ranges with acoustic systems is not difficult, but for optical or microwave time-of-flight sensors it is only recently that advances in IC speed have allowed similarly short ranges and accuracies.

Single-chip time of flight sensors such as those made by ST microelectronics use FlightSense technology to measure the time taken for emitted pulses of infrared laser light to reach the target and be reflected back to a detector. The VL53L0 accurately determines the absolute distance to a target without the object's reflectance greatly influencing the measurement. It can report distances of up to 2 m with

1 mm resolution, but its effective range and accuracy (noise) depend heavily on ambient conditions and target characteristics like reflectance and size, as well as the sensor configuration. (The sensor's accuracy is specified to range from ±3% at best to over ±10% in less optimal conditions.).

2.4.8 Measuring rate and angular rate

2.4.8.1 Incremental encoders

Magnetic and optical incremental encoders and interferometers output a continuous stream of pulses at a frequency that is a function of the rate. Good estimates of the rate can be obtained by measuring the frequency over a reasonable interval. This is achieved by counting cycles. If the speed is low, better estimates are made by running a high-speed clock and measuring the interval between pulses as shown in Figure 2.42b.

2.4.8.2 Tacho-generators

Tacho-generators are small AC or DC generators that output a voltage in proportion to the rotational speed of a shaft. They are capable of measuring speed and direction of rotation but not position. They convert a rotational speed into an isolated analogue voltage signal that is suitable for remote indication and control applications. They are often used in servo systems to supply velocity or damping signals and may be mounted on or in the same housing as a servo motor.

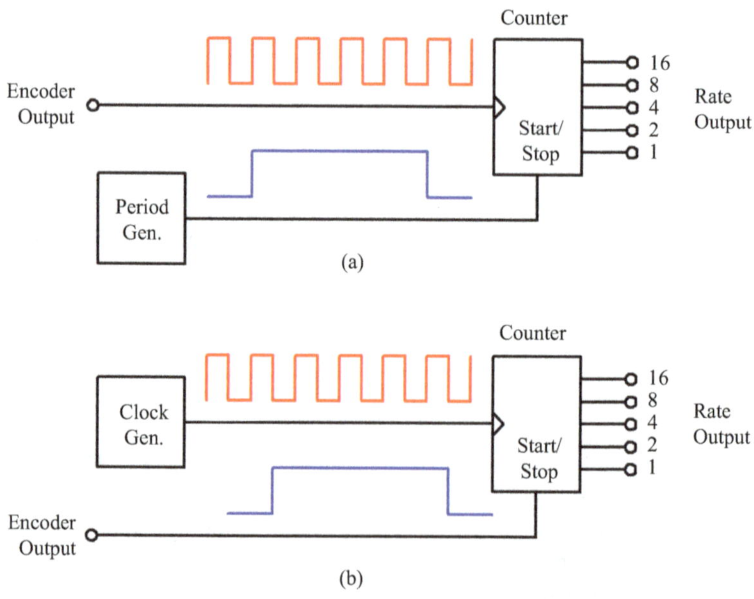

Figure 2.42 Estimating rate from the output of an incremental encoder for (a) high speeds and (b) low speeds

There are two basic types of tacho-generators: alternating current (AC) generators and direct current (DC) generators. Both types develop an output voltage that is proportional to rotational speed. In the case of DC generators, the polarity depends upon the direction of rotation while in the AC case, it is the relationship between phases that changes.

Performance specifications for these devices include generated voltage, accuracy, maximum speed and ripple. Generated voltage is measured in volts (V) per revolution per minute (rpm) over a range of speeds. Accuracy or linearity is the deviation of the voltage output signal expressed as a percentage, and ripple is determined as a percentage of the output voltage.

2.4.8.3 Rate gyros

In many applications, angular rate must be measured with reference to an inertial frame rather than relative to some physical object. Rate gyroscopes, commonly called rate gyros, use the conservation of angular momentum to keep one or more inertial axes pointed in one direction as the external frame translates and rotates. An output voltage is produced which is proportional to the rate of rotation of an axis perpendicular to the axis of the gyro.

Many high-accuracy rate gyros still consist of a spinning rotor mounted in a single gimbal, as shown in Figure 2.43(a). A gyro mounted in this manner has one degree of freedom; that is, it is free to tilt in only one direction. The rotor in a rate gyro is generally restrained from precessing by some means, usually a spring arrangement. This is done to limit precession and to return the rotor to a neutral position when there is no angular change taking place. The amount of precession of a gyro is proportional to the force that causes the precession.

If the gyro's plane of rotor spin is changed by rotating the case about the input axis, the gyro will precess as shown because turning the gyro case has the same

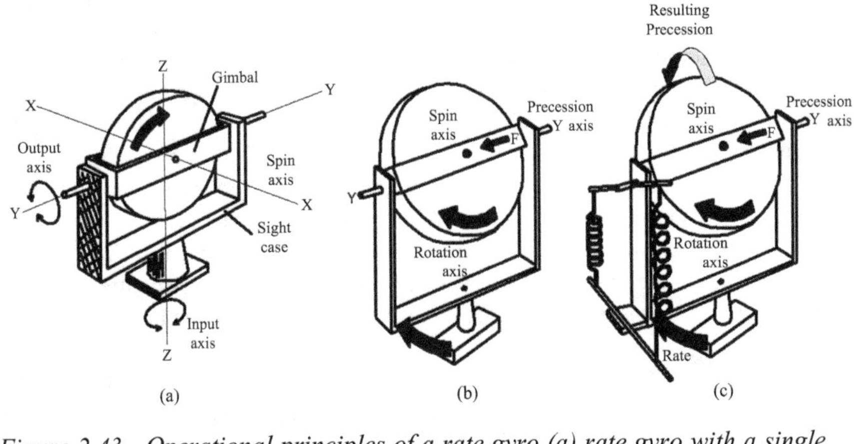

Figure 2.43 Operational principles of a rate gyro (a) rate gyro with a single degree of freedom, (b) rate gyro precession and (c) precession of a spring retained rate gyro [adapted from Neets (2000)]

effect as applying a torque on the spin axis. This is illustrated by arrow F in Figure 2.43(b). The direction of precession can be determined by using the right-hand rule.

The force applied at F will cause the gyro to precess at right angles to the force. Likewise, attempting to turn the gyro case will have the same result. The gyro will precess, as shown by the arrows, around the *Y–Y* axis (output axis). Since the rate of precession is proportional to the applied force, the precession can be increased by increasing the speed at which the gyro case is moved.

The amount of precession is proportional to the rate at which the gyro base is turned. This characteristic of a gyro, when properly used, fits the requirements needed to sense the rate of motion about any axis. Figure 2.43(c) shows a method of restraining the precession of a gyro to permit the calculation of an angle. Springs have been attached to the cross-arm of the output shaft. These springs restrain the free precession of the gyro.

As the gyro precesses, it exerts a precessional force against the springs that is proportional to the momentum of the spinning wheel and the applied force. This force will result in the precession increasing until it is just balanced by the force on the springs, and it will remain in the precessed position as long as the gyro base is rotated at the same constant speed. When rotation ceases, the force at F is removed and the gyro will stop trying to precess, so the spring returns the cross-arm to the neutral point. A low-friction device to measure the precession angle and output an electrical signal completes the rate gyro.

Spinning mass rate gyros are used in several areas, such as the guidance of intercontinental ballistic missiles (ICBMs) and submarine-launched ballistic missiles (SLBMs). They are also the technology of choice in low jitter pointing for the Hubble telescope and control and stabilisation of satellites. In addition, they can be found in legacy systems such as aircraft navigators and tactical missiles that were built before the advent of solid-state equivalents.

Most modern low-cost micro-electromechanical systems (MEMS) rate gyros operate using the Coriolis force. In the 1830s, Gaspard-Gustave de Coriolis discovered that an object moving in a rotating frame would cause the observer on the rotating frame to see an apparent acceleration of the object. In other words, if an object is moving in a straight line and is subject to a rotation, its path will deviate from that straight line.

In this type of rate gyro, a MEMS structure is caused to vibrate in one plane. If the structure is also rotating, then the Coriolis force will be generated at right angles to both the direction of vibration and the axis of rotation as shown in Figure 2.44.

If V (m/s) is the instantaneous velocity of the vibrating structure of mass m (kg), which is rotating at an angular rate ω (rad/s) then the Coriolis force is given by

$$F = 2m\omega V \tag{2.33}$$

As the velocity of the vibrating element is sinusoidal, the Coriolis force will introduce a lateral vibration at the same frequency. This can be converted to an electrical signal using a piezoelectric stress/bending mechanism or a variable capacitance.

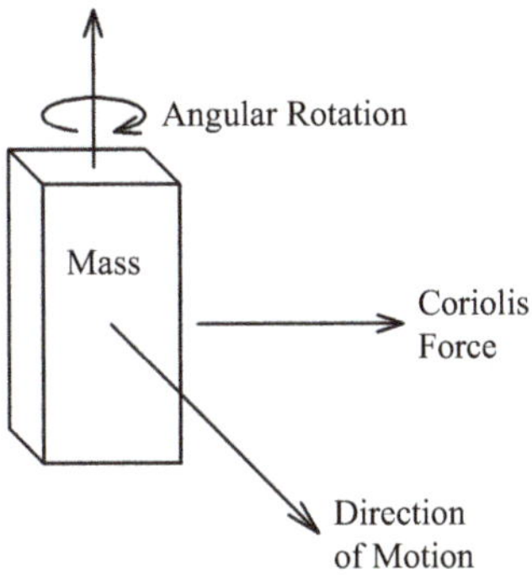

Figure 2.44 The Coriolis effect

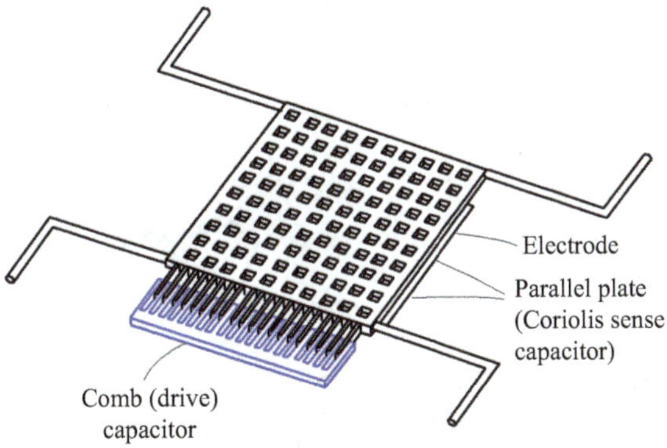

Figure 2.45 MEMS rate gyro with capacitive coupling

One implementation of a MEMS tuning fork gyro is shown in Figure 2.45. A proof mass attached to springs is forced to oscillate in the horizontal plane. A voltage is applied to a sensing electrode (sense plate) below the proof mass, creating an electrical field. The Coriolis force imparted by angular rotation causes the proof mass to oscillate vertically, which, in turn, changes the gap between the proof mass and the sense plate, as shown in Figure 2.46. This motion generates an AC current with amplitude proportional to the rotation rate.

Unfortunately, the bias voltage also generates an electrostatic force between the proof mass and the sense plate. This force acts to pull the proof mass towards the sense plate. In addition, since the gyro is a mechanical device, it is sensitive to external stimuli including vibration, acoustic excitation, and acceleration due to mechanical shock. In some applications, the acceleration levels at the gyro can be as large as 500 g. These stimuli will increase or decrease the sense gap depending on amplitude and frequency and must be considered when the rate gyro is used.

The main advantages of MEMS rate gyros are their small size and the fact that they are low power. In addition, they are extremely reliable and have a long mean

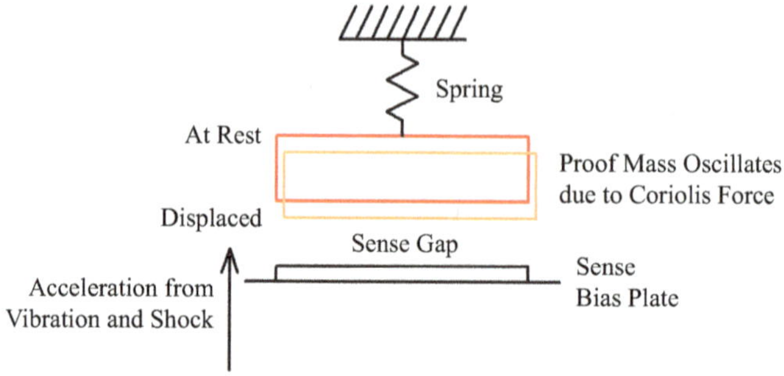

Figure 2.46 Capacitive mechanism to measure rate using the Coriolis force

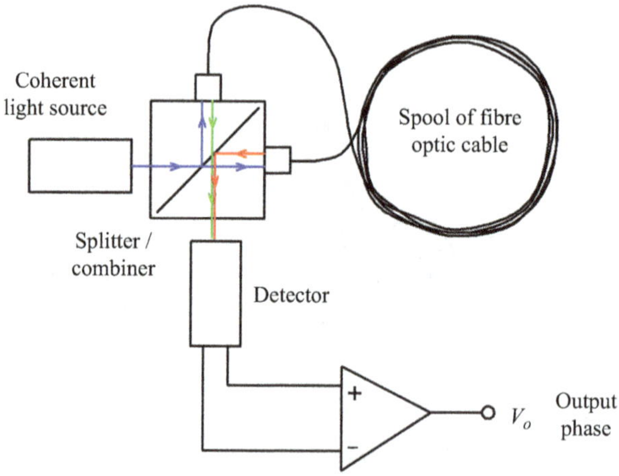

Figure 2.47 Fibre optic gyro (FOG) rate sensor

time between failures (MTBF), and because the mass of the vibrating element is so small, they are reasonably insensitive to external environmental factors like shock, vibration and acceleration.

MEMS rate gyros do suffer from drift and offset, and their sensitivity is lower than some of the other types. Because drift is often a function of temperature many packaged gyros include temperature compensation tables.

Another robust technology is the fibre optic gyro (FOG) rate gyro. These rate sensors use a fibre optic ring and a solid-state laser to measure rotation rates using the Sagnac effect. In 1913, Georges Sagnac showed that light sent around a closed loop in different directions would show a phase difference between the two beams when the loop was rotated (Fraden 1996).

The detection process is similar to that illustrated for the Michelson interferometer. As shown in Figure 2.47, a laser provides a coherent beam that is split and sent in opposite directions around a fibre optic ring (blue beam).

If the ring gyro is static, then the path lengths back to the detector are identical and the beams interfere constructively. However, if the ring is rotating, the path travelled in one direction (red beam) is slightly shorter than that travelled in the other (green beam) because the position of the detector has moved, with the result that the signals are no longer in phase. The phase difference is proportional to the difference in the two distances and the wavelength of the laser.

The Sagnac phase shift, ΔS, is

$$\Delta S = \frac{8\pi n A \omega}{c\lambda} \tag{2.34}$$

where A (m^2) is the cross-sectional area enclosed by the fibre optic coil, n is the number of turns of fibre optic cable around the ring, ω (rad/s) is the angular rate at which the ring is rotating in the sensitive plane, c (m/s) is the speed of light and λ (m) is the wavelength of the light.

Because all the other terms are known, FOG rate sensors can easily convert a measured phase shift ΔS, into an angular rate. This technique is very stable with temperature and the drift levels are very low. There are no moving parts and with the good reliability of modern laser diodes and detectors, the MTBF is very high.

There are many manufacturers of rate gyros as listed in Table 2.6. These range from precision military devices costing thousands of dollars to mass-produced MEMS devices for consumer electronics such as the Melexis MLX90609 single chip solution with a 20 × 20 mm footprint and a cost of $50

Table 2.6 Manufacturers of commercial rate gyros

Manufacturer	Manufacturer
Akustica	JAE
Analog Devices	Kenyon Laboratories
Applanix Corp	KVH Industries
Applied Technology Associates	Litton Industries
Atlantic Inertial Systems	MEMSCAP
Cloud Cap Technology	MEMSense
Colibrys	Micro Aerospace Solutions
Crossbow Technology	Murata Electronics
Flightline Electronics	Northrop Grumman
Flightline Systems	Sagebush Technology
Goodrich Sensor Systems	Silicon Sensing Systems
Honeywell Sensing and Control	Sagem Avionics
iMAR GmbH	Systron Donner
Instrumented Sensor Technology	STMicroelectronics
IMU Solutions	TSS International
InterSense	Ultra Electronics
InvenSense	Watson Industries

Figure 2.48 STMicroelectronics LPY503AL rate gyro breakout, courtesy SparkFun, creative commons

which is shown in Figure 2.48. The specifications of this gyro include a ± 5% drift in scale for a temperature range from −40 °C to +85 °C and a noise power spectral density of 0.03°/s/$\sqrt{}$Hz. The latter equates to a rms noise of 0.25°/s for the maximum gyro bandwidth of 75 Hz. Other MEMS rate gyro manufacturers quote biases over temperature of between 0.3 and 3°/s for their commercial grade products. In contrast, the long-term drift quoted for good FOGs is quoted to be around 3°/hr, more than one thousand times lower than that achievable from MEMS gyros.

The consumer grade rate gyro LPY503AL shown in Figure 2.48 is a low-power dual-axis (pitch and yaw) MEMS gyro with a ±30°/s maximum range. Options include integrated low-pass filters and embedded power-down and self-test features. Both x and z outputs are analogue.

In most applications, rate gyros are not used in isolation, but usually as part of an inertial measurement unit (IMU) which uses sensor modelling and filtering techniques to improve overall sensing accuracy.

2.4.9 Accelerometers

Accelerometers are sensors designed to measure continuous mechanical vibration such as bearing vibration, transitory vibration from blasts or impacts, or the lower frequency acceleration generated by bodies in motion. These devices are generally bonded firmly to a structure and aligned in the direction of interest as they are sensitive to motion along one axis only.

2.4.9.1 Theory

All accelerometers are based on the inertial effects associated with a mass connected to a moving object through a spring, damper and a displacement sensor as shown in Figure 2.49. When an object accelerates, there is relative motion between the housing and the seismic mass which is measured by the displacement sensor (Alciatore and Histand 2003).

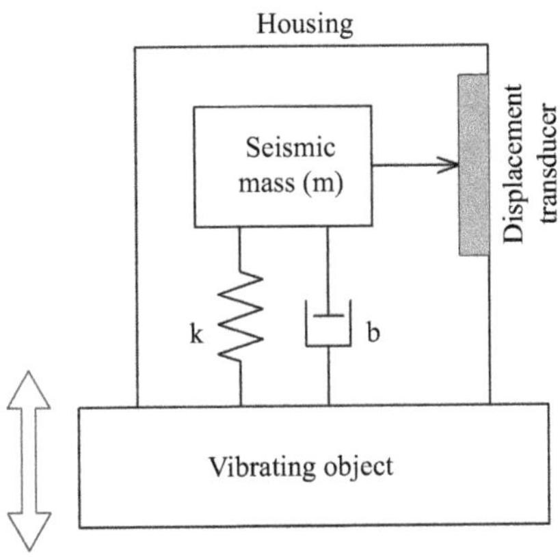

Figure 2.49 Conventional accelerometer construction

The important characteristics of an accelerometer are its sensitivity and dynamic range as well as its frequency response. The sensitivity determines the smallest acceleration that can be reliably measured. The dynamic range can then be used to find the maximum acceleration if the response must remain linear. The frequency response gives the range of frequencies that an accelerometer can measure.

If the relative displacement x_r (m) between the seismic mass and the object measured by a position transducer mounted between the seismic mass and the housing is given by

$$x_r = x_o - x_i \tag{2.35}$$

The spring force, F_k (N), can then be expressed in terms of this displacement and the spring constant k

$$F_k = k(x_o - x_i) = kx_r \tag{2.36}$$

and the damper force can be expressed in terms of the time derivatives, $\dot{x}_i$ and $\dot{x}_o$, and the damping constant b

$$F_b = b(\dot{x}_o - \dot{x}_i) = b\dot{x}_r \tag{2.37}$$

Applying Newton's second law

$$-F_k - F_b = m\ddot{x}_o \tag{2.38}$$

These forces have negative signs because they are in the opposite direction to the reference direction x_o.

Substituting equations (2.36) and (2.37) into (2.38)

$$-kx_r - b\dot{x}_r = m\ddot{x}_o \tag{2.39}$$

The output acceleration $\ddot{x}_o$ can be written as

$$\ddot{x}_o = \ddot{x}_r + \ddot{x}_i \tag{2.40}$$

and this can be substituted into equation (2.39) and rearranged to produce a second-order differential equation

$$m\ddot{x}_r + b\dot{x}_r + kx_r = -m\ddot{x}_i \tag{2.41}$$

This can be rewritten in terms of the natural frequency and the damping ratio as

$$\frac{1}{\omega_n^2}\ddot{x}_r + \frac{2\zeta}{\omega_n}\dot{x}_r + x_r = -\frac{1}{\omega_n^2}\ddot{x}_i \tag{2.42}$$

where the natural frequency is

$$\omega_n = \sqrt{\frac{k}{m}} \tag{2.43}$$

and the damping ratio is

$$\zeta = \frac{b}{2\sqrt{km}} \tag{2.44}$$

If the input displacement is limited to a sinusoidal term with amplitude X_i then because the system is linear, the output displacement will also be sinusoidal with the same frequency and with a phase shift, φ.

$$\begin{aligned} x_i(t) &= X_i \sin \omega t \\ x_o(t) &= X_r \sin(\omega t + \phi) \end{aligned} \tag{2.45}$$

From the known characteristics of second-order systems, the amplitude ratio can be written as

$$\frac{X_r}{X_i} = \frac{(\omega/\omega_n)^2}{\left(\left[1-\left(\frac{\omega}{\omega_n}\right)^2\right]^2 + 4\zeta^2\left(\frac{\omega}{\omega_n}\right)^2\right)^{1/2}} \tag{2.46}$$

and the phase shift will be

$$\phi = -\tan^{-1}\left(\frac{2\zeta(\omega/\omega_n)}{1-\left(\frac{\omega}{\omega_n}\right)^2}\right) \tag{2.47}$$

To determine the output displacement in terms of the input acceleration, equation 2.44a is differentiated twice

$$\ddot{x}_i(t) = -X_i\omega^2 \sin \omega t \tag{2.48}$$

Equation (2.46) can be rewritten as

$$H_a(\omega) = \frac{X_r\omega_n^2}{X_i\omega^2} = \frac{1}{\left(\left[1 - \left(\frac{\omega}{\omega_n}\right)^2\right]^2 + 4\zeta^2\left(\frac{\omega}{\omega_n}\right)^2\right)^{1/2}} \tag{2.49}$$

It is therefore possible to relate the measured output displacement amplitude in terms of the input acceleration amplitude as

$$X_r = \left(\frac{1}{\omega_n^2}\right) H_a(\omega)\left(X_i\omega^2\right) \tag{2.50}$$

where $(X_i\omega^2)$ is the input acceleration amplitude. It can be expressed in terms of the transfer function, $H_a(\omega)$, and the output amplitude, X_r

$$\left(X_i\omega^2\right) = \frac{X_r\omega_n^2}{H_a(\omega)} \tag{2.51}$$

If the accelerometer is designed so that $H_a(\omega) = 1$ over a large frequency range, then the input acceleration amplitude is given by the relative displacement amplitude scaled by a constant factor ω_n^2

$$\left(X_i\omega^2\right) = \omega_n^2 X_r \tag{2.52}$$

It can be seen from Figure 2.50 that the most linear response occurs with a damping ratio $\xi = 0.707$, and the natural frequency ω_n as large as possible. Looking back at equation (2.43) it can be seen that ω_n can be made large by choosing a small seismic mass and a large spring constant.

2.4.9.2 Piezoelectric accelerometers

Piezoelectric accelerometers use piezoelectric crystals to measure the force exerted by the seismic mass. Deformation of these crystals results in the generation of a charge across the opposite faces. As illustrated in Figure 2.51, a piezoelectric accelerometer includes a crystal in contact with the mass. In addition to the natural damping inherent in the crystal, additional damping is sometimes included by filling the housing with oil.

When the object is accelerated, relative displacement occurs between the housing and the seismic mass due to the inertia of the mass. The resulting strain within the piezoelectric crystal causes a charge to be developed on the two faces of the crystal that can be tapped by the conductive coating and measured using a charge amplifier (discussed in Volume 1, Chapter 5).

In general, piezoelectric accelerometers cannot measure constant or slowly changing accelerations because the crystal can only measure changes in strain. Low frequency response is typically a few hertz, while the high frequency response is determined by the mechanical characteristics of the accelerometer and the

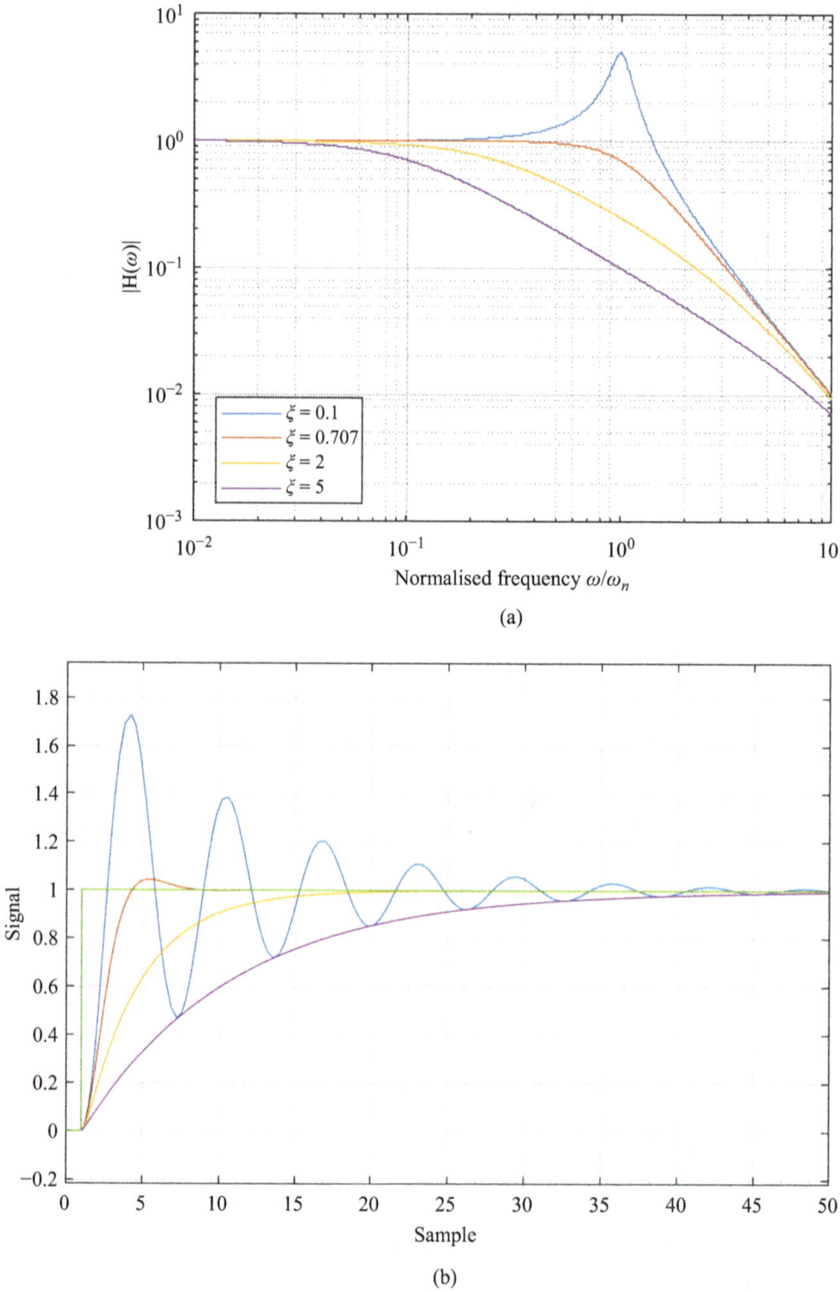

Figure 2.50 Ideal accelerometer (a) amplitude transfer function and (b) step response

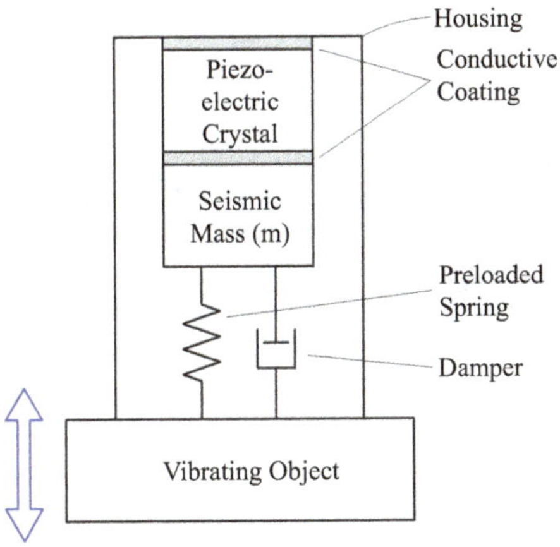

Figure 2.51 Piezoelectric accelerometer

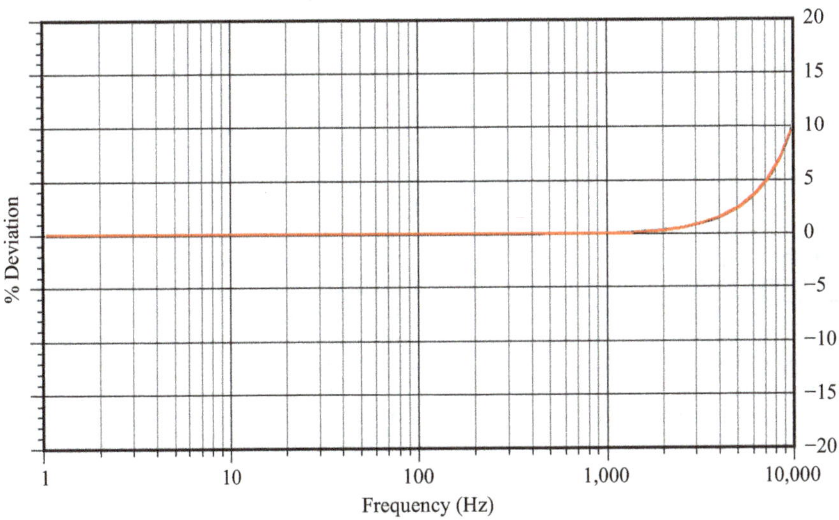

Figure 2.52 Frequency response of an Endevco 2273A piezoelectric accelerometer

mounting stiffness. This is normally close to the natural (resonant) frequency of the system, which is usually in the kHz range. Manufacturers generally specify this as a deviation from the nominal as shown in Figure 2.52.

Accelerometers come in all sizes and shapes for any conceivable application, as can be seen from Figure 2.53.

In addition to piezoelectric accelerometers, most manufacturers also make the piezoresistive types that can operate down to DC and are therefore useful for measuring static acceleration (gravity). These devices use a bridge configuration to output a voltage proportional to the acceleration.

A subset of the manufacturers of accelerometers are listed in Table 2.7.

Figure 2.53 An assortment of accelerometers, reproduced courtesy Endevco with permission

Table 2.7 Manufacturers of a selection of accelerometers

Manufacturer	Manufacturer
Advantech	Hansford Sensors
Amphenol Wilcoxon	Honeywell Sensing and Control
Analog Devices	IFM Electronic
ASC Sensors	Instrumented Sensor Technology
Althen Sensors	Invensense
Banner Engineering Corp	Jewell Instruments
Bosch Sensortech	Kionix
Columbia Research Labs	Kistler Instrument Corporation
Dataforth Corporation	Measurement Specialities
DJB Instruments	Pacific Instruments Inc
CEC Vibration Products	Pico Technology
Dytran Instruments	Reiker
Endevco Corporation	Rockwell Automation
Extech Instruments	Soltec Corporation
First Sensor AG	STMicroelectronics
	TE Connectivity

2.4.10 Tilt sensors

Pinball players all know at least one application of tilt sensors. At their most basic, tilt sensors consist of a pool of mercury and two contacts within a hermetically sealed glass container that complete a circuit if the device is tilted.

More sophisticated tilt sensors can be manufactured from accelerometers mounted on three orthogonal axes, and equating the relative gravity vector in each. However, some low-cost tilt sensors still exploit the bubble-level principle. Figure 2.54 shows one axis of a fluid-filled sensor tipped at $\sim 15°$. As the sensor tilts, the surface of the fluid remains level due to gravity. The fluid is electrically conductive, and the conductivity between the two electrodes is proportional to the length of the electrode immersed in the fluid. At the angle shown, for

(a)

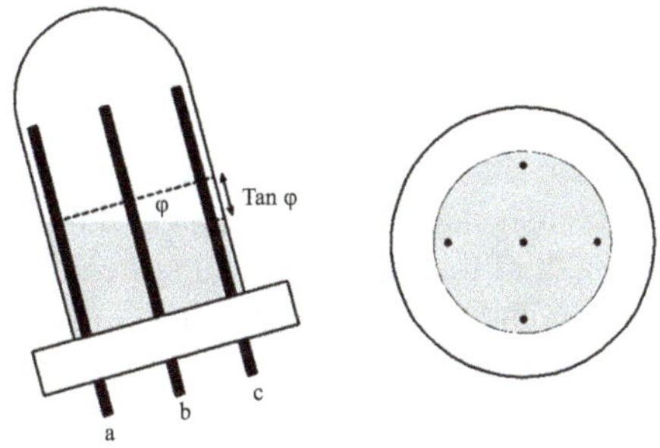

(b)

Figure 2.54 Electrolytic tilt sensor

example, the conductivity between pins *a* and *b* would be greater than that between *b* and *c*.

Electrically, the sensor is similar to a potentiometer, with resistance changing in proportion to tilt angle, and therefore the output can be determined using a Wheatstone bridge in its dynamic unbalanced configuration. For small angles, typically less than 20°, the output voltage is proportional to the tangent of the tilt angle. However, as the angle increases, nonlinearities become more pronounced, and more sophisticated microcontrollers are required to apply corrections to the measurements to maintain accuracy.

DC induces electrolysis in the fluid as positive ions in the fluid migrate to the cathode, where they combine with excess electrons and lose some of their charge. In a similar fashion, negative ions in the fluid propagate to the anode and combine with excess protons to lose their charge. If this is allowed to proceed, the reaction will eventually reduce the conductivity of the fluid to the extent that the sensor is no longer operational.

To prevent electrolysis, alternating current (AC) must be used to excite the sensor. The required frequency and symmetry of the AC waveform depend on the chemistry of the fluid and composition of the electrodes. The frequency must be high enough to make the process described above reversible, which for some electrolytes can be as low as 25 Hz while other solutions require a minimum of 1000 Hz to 4000 Hz.

Dual-axis sensors exhibit the same fluid characteristics as single-axis devices but have the added complexity of interaction between the axes. Both axes share the centre electrode with the four outer electrodes placed at the four corners of a square. Misalignment between electrodes gives rise to cross-axis coupling that can result in significant errors.

Two techniques can be used to derive independent measurements for each axis. The first is to excite only one axis at a time, alternating between pitch and roll at an appropriate rate, and the second technique requires two excitation frequencies, one twice the other. Here, all four pins are driven simultaneously, and the phase of the excitation determines which axis is being measured (Powell and Pheifer 2006).

2.4.11 Pressure measurement

Pressure is defined as the normal force per unit area exerted by a fluid (liquid or gas) on any surface. Only the force normal to the surface needs to be considered (Webster 1999). Blood pressure is one of the physiological variables that is easily measured, hence it has been used as an indication of general health, and particularly the health of the cardiovascular system, for many years. The measurement of blood pressure, and particularly the timely diagnosis of dangerously high blood pressure, hypertension, has saved many thousands of lives.

Three types of pressure measurements are commonly performed:

- Absolute pressure: Represents the pressure difference between the point of measurement and a perfect vacuum where the pressure is zero.
- Gauge pressure: This is the pressure difference between the point of measurement and the ambient (atmospheric) pressure.
- Differential pressure: The difference in pressure between two points, one of which has been chosen to be the reference.

Several common units are used for pressure measurement. They are kilopascal (kPa), pound per square inch (psi), inches or cm of water at 4 °C (cmH_2O), inches or mm of mercury at 0 °C (mmHg), millibar (mbar). Table 2.8 lists the conversion factors between these units.

A column of fluid is commonly used to measure pressure because there is a simple relationship between the pressure, P, the density of the fluid, ρ, the acceleration due to gravity, g, and the height of the column, h.

$$P = \rho g h \tag{2.53}$$

Biomedical pressure sensors are mostly used to measure blood or air pressure. They can also be used to measure the pressure in pneumatic and hydraulic actuators for control or monitoring purposes.

The conventional sphygmomanometer is an indirect method of measuring blood pressure. It consists of an inflatable cuff and a mercury manometer and is considered to be the "gold standard" for blood pressure measurement because it cannot be in error if operated correctly. However, its main disadvantage is that it is unable to provide a continuous reading of pressure variations, and its update rate is limited. In

Table 2.8 Conversion between different pressure measurement units

Unit	kPa	psi	in H_2O	cmH_2O	in Hg	mm Hg	mbar
kPa	1	0.145	4.015	10.2	0.2593	7.501	10
psi	6.895	1.0	27.68	70.31	2.036	51.72	68.95
in H_2O	0.2491	0.03613	1.0	2.54	0.07355	1.868	2.491
cmH_2O	0.09806	0.01422	0.3937	1.0	0.02896	0.7355	0.9806
in Hg	3.386	0.4912	13.6	34.53	1.0	25.4	33.86
mm Hg	0.1333	0.01934	0.5353	1.36	0.03937	1.0	1.333
mbar	0.10	0.01450	0.04015	1.020	0.02953	0.7501	1.0

addition, only the systolic and diastolic arterial pressures can be measured. The operational principles of the sphygmomanometer are quite simple. A cuff is placed around the upper arm and inflated, and arterial blood can only flow past it when the arterial pressure exceeds that of the cuff. At this stage, blood squeezing through the reduced aperture of the brachial artery generates turbulence which can be heard by a practitioner with a stethoscope. As the pressure in the cuff is reduced and monitored using a manometer, the first indication of this sound (called Korotkoff sounds) gives an indication of the systolic pressure. This reduction in cuff pressure continues until Korotkoff sounds disappear, and that is an indication of the diastolic pressure.

Direct pressure measurement devices consist of a chamber with a flexible diaphragm making up a portion of one wall, with the other side of the diaphragm at atmospheric pressure. A pressure differential across the diaphragm will cause it to deflect, and this deflection can be measured using a displacement sensor. Capacitance is the method of choice for measuring small displacements. If a second plate is placed within a few hundred microns of the diaphragm, changes in the relative separation result in changes in capacitance as discussed earlier in this chapter.

Today, such small displacements are generally measured using strain gauges in a bridge configuration. In the past, a mechanical assembly comprising the diaphragm, a force rod and four unbonded strain gauges was used. With increasing deflection, two of the strain gauges become progressively more relaxed and two are stretched (Gregory 1975). Modern pressure transducers are generally manufactured onto a single MEMS silicon chip where a portion of the chip is formed into a diaphragm and semiconductor strain gauges (piezoresistive bridge) are incorporated directly onto the diaphragm to form small inexpensive pressure sensors as shown in Figure 2.55. Such sensors are sufficiently low-cost that they can be used

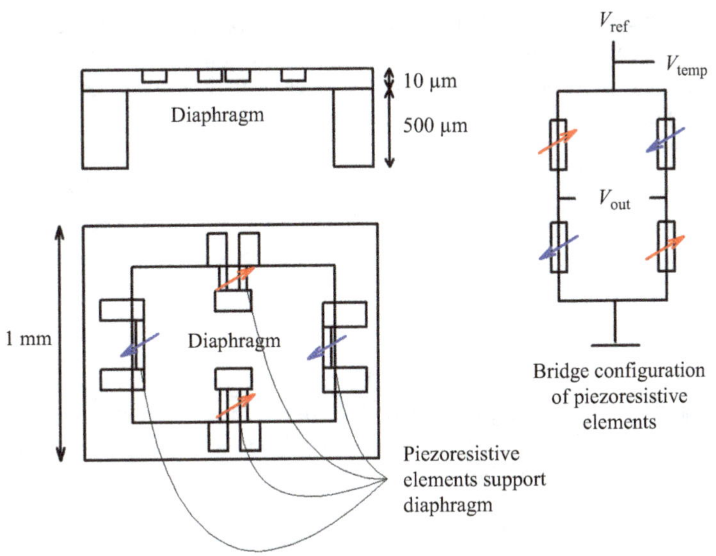

Figure 2.55 MEMS strain gauge pressure transducer

as disposable single-use devices for measuring blood pressure without the need for additional sterilisation. (Bronzino 2006).

Wire strain gauges usually have gauge factors of between two and four, while the semiconductor-based units have gauge factors of between 50 and 200. For silicon, the gauge factor is typically 120. However, in measuring pressure, sensitivities are generally quoted in microvolts per volt (applied to the strain gauge) per millimetre of mercury (Cromwell *et al.* 1973). Signal conditioning and display instruments for these sensors consist of a method of exciting the strain gauge bridge, a method of zeroing or balancing it followed by an amplifier and a display device or data logger.

In the pressure transducer shown in Figure 2.55, the total force, F (N), applied to the membrane is equal to the product of differential pressure, ΔP (Pa) and the area of the membrane, A (m^2)

$$F = \Delta P A \tag{2.54}$$

The stress introduced in the bridge elements is directly proportional to the applied force, and its effect on the semiconductor resistors is to alter their resistance by ΔR

$$\frac{\Delta R}{R} = (\alpha_l \sigma_l + \alpha_t \sigma_t) \tag{2.55}$$

where R (Ω) is the initial resistance, α_l and α_t are the piezoresistive coefficients in the longitudinal and transverse directions, respectively, and σ_l and σ_t are the stresses in these directions (Fraden 1996).

The magnitudes of these coefficients depend on the orientation of the silicon crystal, and the resistors are positioned on the diaphragm in such a manner (as shown in Figure 2.55) as to have the longitudinal and transverse coefficients of opposite polarities. This results in the resistances changing in opposite directions as the pressure alters

$$\alpha_l = -\alpha_t = \frac{1}{2}\alpha \tag{2.56}$$

Therefore

$$\frac{\Delta R_1}{R_1} = -\frac{\Delta R_2}{R_2} = \frac{1}{2}\alpha(\sigma_{1y} - \sigma_{1x}) \tag{2.57}$$

where σ_{lx} and σ_{ly} are the longitudinal stresses in the x and y directions, respectively.

For the full-bridge configuration shown in Figure 2.55, excited by a voltage V_{ref}, the output voltage will be

$$V_{\text{out}} = \frac{V_{\text{ref}}}{2}\alpha(\sigma_{1y} - \sigma_{1x}) \tag{2.58}$$

Because α is a function of temperature, the output voltage will change with temperature. The substrate temperature must therefore be measured, and the appropriate corrections applied.

For measuring blood pressure, the chamber containing the diaphragm is coupled via a thin plastic catheter to an artery. The catheter is filled with saline solution so that the arterial blood pressure is coupled directly to the diaphragm outside the patient. However, in this configuration it is important to realise that both ends of the catheter must be at the same height to avoid hydrostatic effects, and the tube must be sufficiently stiff to minimise compliance effects on the frequency response of the sensor. Air bubbles in the catheter and obstructions due to clotted blood or other materials can introduce distortions into the measured waveform due to resonance and damping effects.

It is now possible to obtain miniature pressure sensors that are located at the tip of the catheter and measure the pressure within the blood vessel (Cromwell *et al.* 1973, Allen 2002). As illustrated in Figure 2.56, these devices are sufficiently small to be inserted into an artery or a vein through the needle of a hypodermic syringe, and because of their small size have an insignificant effect on the pressure or flow at the measurement site.

Long-term stability of pressure monitors is a problem, and drift, particularly when measuring low pressures such as venous blood or cerebrospinal fluid, can result in significant errors. Pressure transducers need regular recalibration, which can pose significant problems, particularly if the sensor is implanted.

Table 2.9 shows a comparison of the properties of some pressure sensing technologies.

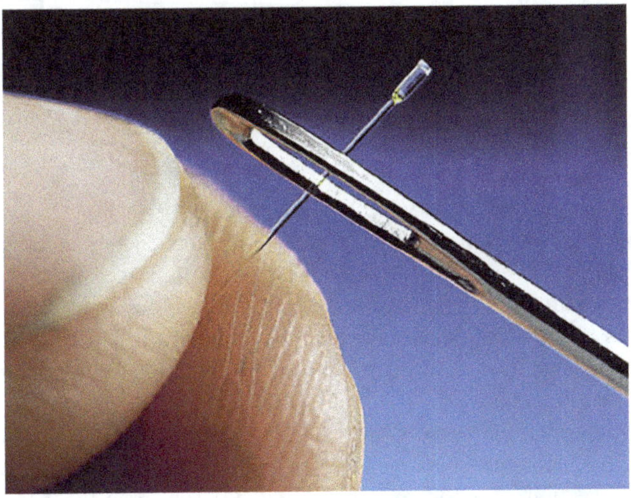

Figure 2.56 Fibre optic pressure sensor from Fiso Technologies employs a MEMS-based sensing catheter for in vivo physiological measurements (Allen 2002), reproduced courtesy of Fiso Technologies

Table 2.9 Comparison of properties of some pressure sensing technologies (Pons 2008)

Property	Capacitive	Potentiometric	Resistive	Piezoelectric
Maximum range	Good	Excellent	Excellent	Good
Minimum size	Good	Good	Excellent	Poor
Sensitivity	Excellent	Good	Poor	Good
Repeatability	Excellent	Poor	Good	Good
Temp. Stability	Excellent	Excellent	Excellent	Poor
Hysteresis	Low	High	High	Medium
Installation	Easy	Easy	Hard	Easy

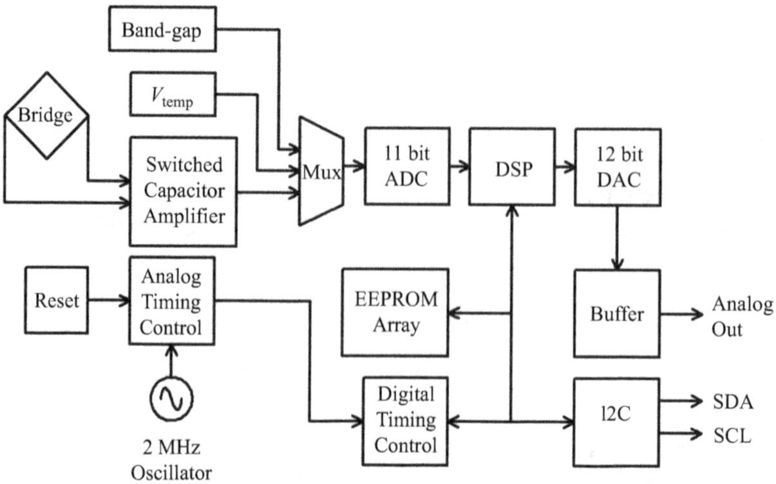

Figure 2.57 Block diagram of the Silicon Microstructures SM5822 pressure sensor processing structure

Measurement of respiratory processes is generally achieved by measuring expired volume using a spirometer or flow rate using a pneumotachograph as discussed in Volume 2, Chapter 2. However, air pressure is measured as part of the operation of negative and positive pressure respiratory devices. Sensors for measuring air pressure in these devices need not be miniaturised but must be capable of measuring pressures from less than 2 cm of H_2O up to 40 cmH_2O to assist or provide full ventilation.

Differential pressure sensors are often used to measure air flow using the obstructive-flow technique employed by some pneumotachographs, as discussed in the following section. A good example of such a sensor is the SM5822 manufactured by Silicon Microstructures which is designed to operate at pressure ranges from 0 to 0.6 PSI (0 to 42 cmH_2O) full scale.

As shown in Figure 2.57, the piezoresistive bridge is only the beginning of the pressure measurement process. The bridge output is filtered and then digitised along with the temperature and a band-gap reference voltage. These parameters are then processed by the onboard DSP to correct for pressure and temperature non-linearities before conversion back to an analogue voltage and output. An I2C digital interface is also provided.

Finally, as with all sensors in contact with body fluids or inserted within the body, pressure sensors must remain inert and mechanically stable within corrosive environments.

2.4.12 *Sound pressure*

Microphones play an essential role in making hearing aids and cochlear implants possible by converting acoustic signals propagating in the air as variations in sound pressure into electrical signals. At higher frequencies, "microphones," commonly known as ultrasound transducers, are used in sonar devices for sensory substitution systems discussed in Volume 1, Chapter 4 and in medical Doppler and imaging systems.

A large variety of physical mechanisms can be harnessed to perform the microphone function. These include electromagnetic induction discussed in Volume 1, Chapter 2, as well as optical interferometry and the piezoelectric effect discussed earlier in this chapter. However, since the advent of MEMS, millions of low-cost microphones have been produced that use changes in the capacitance between a fixed plate and a vibrating diaphragm to produce a varying electrical output.

2.4.13 *Flow*

The most common medical applications of flow measurement include blood flow through arteries and veins, and airflow into and out of the lungs. It also includes the measurement of drug dispensing through catheters and gas flow during anaesthetics among others. There are a large number of different sensors that can be used to measure flow by determining the rate of displacement of either mass or volume. However, whichever sensor is used, inherent difficulties in the measurement make the process complicated, and it is necessary to consider the natural characteristics of both the environment and the medium. These include pipe shapes and materials, and the temperature, pressure and viscosity of the fluid, among others. (Fraden 1996).

2.4.13.1 Differential pressure flowmeter

Flow is commonly measured using a differential pressure flowmeter across an orifice plate, a Venturi tube or a nozzle. The operation of these flowmeters is based on an observation by Bernoulli that if an annular restriction was placed in a pipe, then the velocity of the fluid through the restriction was increased. The increase in velocity at the restriction then causes the static pressure to decrease, and a pressure difference is built up. The difference between the pressure upstream and the pressure downstream of the restriction is related to the rate of flow through it, and therefore through the pipe. A differential pressure-based flow transducer consists of an obstruction to cause a pressure drop and a method of measuring the differential pressure across the obstruction (Webster 1999).

Bernoulli's equation which defines the relationship between fluid velocity (v), fluid pressure (p) and height (h) above some fixed point, for fluid flowing through a pipe of varying cross section is fundamental to quantifying the differential pressure measurement technique.

For the inclined, tapered pipe shown in Figure 2.58, Bernoulli's equation states that

$$\frac{p_1}{\rho g} + \frac{v_1^2}{2g} + h_1 = \frac{p_2}{\rho g} + \frac{v_2^2}{2g} + h_2 \quad (2.59)$$

The sum of the pressure head, the velocity head and the potential head is constant along a flow streamline. It is assumed that the flow is frictionless (no viscosity) and of constant density (incompressible).

If there is a restriction in a piece of pipe as shown in Figure 2.59, then since $h_1 = h_2$, Bernoulli's equation can be rewritten as

$$\frac{p_1 - p_2}{\rho} = \frac{v_1^2 - v_2^2}{2} \quad (2.60)$$

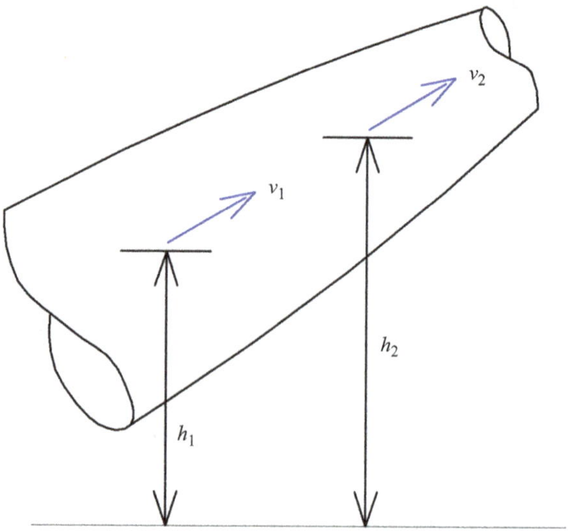

Figure 2.58 Flow through an inclined tapered pipe

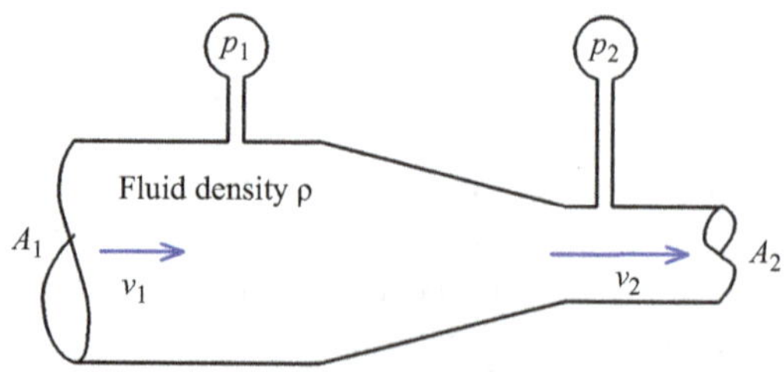

Figure 2.59 Using a restriction in a piece of pipe to measure flow

And the conservation of mass requires that

$$v_1 A_1 \rho = v_2 A_2 \rho \tag{2.61}$$

Rearranging and substituting for v_2 in equation (2.54) gives the volumetric flow rate, Q, in terms of the drop in pressure across the restriction in the pipeline

$$Q = v_1 A_1 = \frac{A_2}{\sqrt{1 - \left(\frac{A_2}{A_1}\right)^2}} \sqrt{\frac{2(p_1 - p_2)}{\rho}} \tag{2.62}$$

The assumption of incompressibility is reasonable for a liquid, but not a gas, but the assumption that the fluid has no viscosity, resulting in a flat velocity profile, is generally not justified. This requires the application of correction factors to compensate.

The orifice plate, shown in Figure 2.60(b), is the simplest and cheapest type of differential pressure flowmeter. It is simply a plate with a hole of the specified diameter that is clamped into the pipe. Unfortunately, after passing through the plate, the jet continues to contract until it reaches a minimum diameter, before

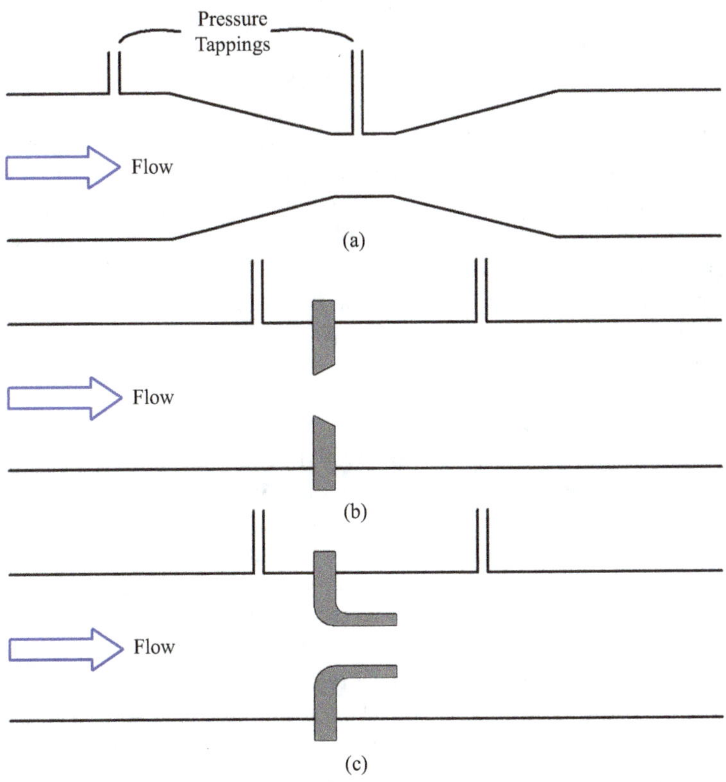

Figure 2.60 Common differential pressure flowmeters, (a) Venturi, (b) orifice plate and (c) nozzle

expanding again. Because A_2 should be this diameter, which is unknown, correction factors must be applied to equation (2.57). This modified equation is

$$Q = \frac{C}{\sqrt{1-\beta^4}}\varepsilon\frac{\pi}{4}d^2\sqrt{\frac{2(p_1-p_2)}{\rho}} \tag{2.63}$$

where ρ is the density of the fluid upstream from the plate, d is the diameter of the hole in the orifice plate and $\beta = d/D$ where D is the upstream diameter of the interior of the pipe. The empirically determined correction factors are C, which is affected by diameter ratio, Reynolds number, pipe roughness etc, and ε which is the expansibility factor.

The Venturi tube, shown in Figure 2.60(a), is the oldest type of differential pressure flowmeter. Because the change in diameter is more gradual, this better approximates the theoretical result, C, in this case is close to 0.95. Its major disadvantages are the lower differential pressure for a given diameter ratio, and the expense of manufacture.

The nozzle method shown in Figure 2.60(c) combines the best features of the other two techniques making it both reasonably cheap to manufacture and accurate. Pneumotachographs, discussed in Volume 2, Chapter 2, use a variation of the nozzle method in which a restrictive element made from a perforated plate, or a large number of narrow-bore pipes, is used. These offer very little resistance to the airflow and hence the pressure drop is small.

It follows from equation (2.58) that the pressure gradient technique requires the use of either a single differential pressure sensor or a pair of absolute sensors. Additionally, because the flow rate is proportional to the square root of the pressure difference, if a linear representation of the flow is required, a square root extraction circuit or algorithm must be applied.

The primary advantage of this technique is the absence of any moving components and the ready availability of standard pressure sensors. The main disadvantage is the requirement that the flow may be affected by the use of a restrictive device.

2.4.13.2 Temperature flowmeters

An improved method of measuring flow would be to mark the fluid in some way and then to detect the movement of the mark. Marking techniques could be as simple as mechanical floats or as complicated as radioactive elements, or a dye that changes the optical properties of the medium. In medicine, the dye dilution method is used for studies in haemodynamics. Under most circumstances, however, placing any foreign material into the fluid is impractical or forbidden.

A non-invasive marker is temperature, and sensors called hot-wire or thermo anemometers use this principle. One sensor type that is often used to measure blood flow consists of a small, isolated heating element immersed in the fluid between two similarly immersed temperature probes. If the medium is not flowing, then diffusion will result in the two probes reading the same temperature, but if there is flow, the downstream probe will be warmer than the upstream one. In general, the heating element is placed closer to the downstream element to improve sensitivity, and it must always be heated to above the highest temperature reached by the fluid under normal conditions. In another method, a thermistor placed in the bloodstream is kept at a constant temperature using current feedback. The amount of energy required to maintain the temperature is proportional to the flow rate because the temperature of the blood and its thermal conductivity remain constant (Cromwell *et al.* 1973)

2.4.13.3 Ultrasound flowmeters

Another nonintrusive method to measure flow is to use ultrasound. Sensors can measure changes in the effective velocity in the medium, or they can use the Doppler shift. In Figure 2.61, two transducers are placed on either side of a pipe directed through the fluid towards each other. If a pulse is transmitted from the one, the time, T (s), taken to reach the other will be

$$T = D/c \tag{2.64}$$

where D (m) is the distance between the transducers and c (m/s) is the speed of sound in the medium if the fluid is still.

If the fluid is flowing then the propagation speed is altered by the flow and equation (2.59) becomes

$$T = \frac{D}{c \pm v_c \cos \theta} \tag{2.65}$$

where v_c (m/s) is the average fluid velocity and the $\pm$ refers to the direction of flow relative to the direction of the ultrasound signal.

It has been shown that for laminar flow $v_c \approx 4v_a/3$ and for turbulent flow $v_c \approx 1.07v_a$, where v_a (m/s) is the flow averaged over the cross-sectional area (Fraden 1996).

By taking the time difference, ΔT between the downstream and upstream measurements

$$\Delta T = \frac{2Dv_c \cos\theta}{c^2 + v_c \cos^2\theta} \approx \frac{2Dv_c \cos\theta}{c^2} \tag{2.66}$$

which is true for most situations because $c >> v_c \cos\theta$.

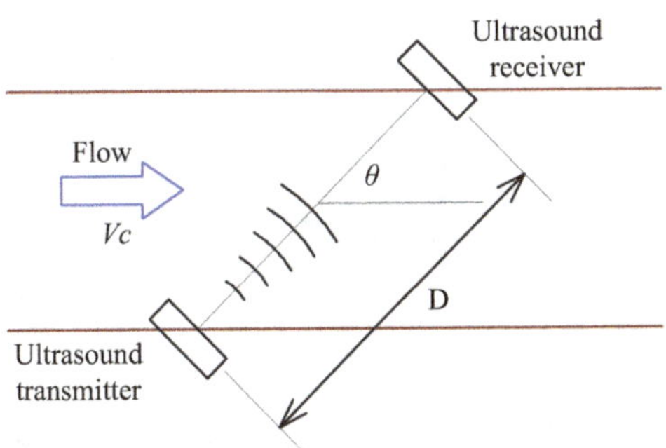

Figure 2.61 Ultrasonic flowmeter

Typically, sensors of this type transmit short pulses at high frequency (f = 3 MHz), and alternate between the transmitter being the upstream and the downstream transducer. An alternative is to measure the phase difference between the transmitted and the received signals.

2.4.13.4 Magnetic flowmeters

Magnetic flowmeters are based on the principle of magnetic induction. When an electrical conductor cuts through a magnetic field, a voltage is induced in the conductor, proportional to its velocity. This principle can be applied when the conductor is a column of conductive fluid flowing through a tube in a magnetic field. For measuring blood flow, a magnet that generates a magnetic field orthogonal to the direction of flow is placed around the blood vessel. Electrodes measure the voltage on either side of the blood vessel orthogonal to both the magnetic field and the direction of flow. The measured voltage is directly proportional to the flow rate so long as the conductivity of the blood remains constant. Figure 2.62 shows an example of such a flow measurement sensor.

2.4.13.5 Target flowmeters

Target meters measure the force caused by liquid impacting on a target or drag-disk suspended in the liquid stream. A direct indication of the liquid flow rate is obtained by measuring the force exerted on the target. In its simplest form, the meter consists only of a hinged, swinging plate that moves outward, along with the liquid stream. A more sophisticated version uses a precision force-sensing element. For example, the small displacement of the target caused by the liquid flow can be sensed by a strain gauge. The output signal from the gauge is indicative of the flow rate. Target meters are useful for measuring flows of dirty or corrosive liquids.

Variable-area meters, often called rotameters, consist of a tapered tube and a float as shown in Figure 2.63. Although classified as differential pressure units, they are constant differential pressure devices. Flanged-end fittings provide an easy means for installing them in pipes. When there is no liquid flow, the float rests freely at the bottom of the tube. As liquid enters the bottom of the tube, the float begins to rise. The position of the float varies directly with the flow rate. Its exact

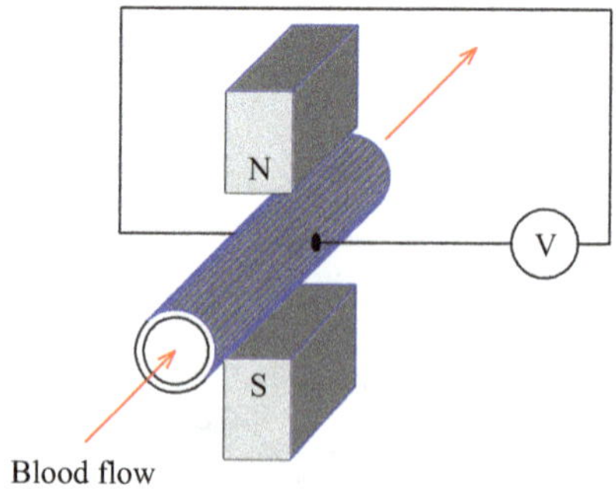

Figure 2.62 Schematic diagram of a magnetic blood-flow sensor

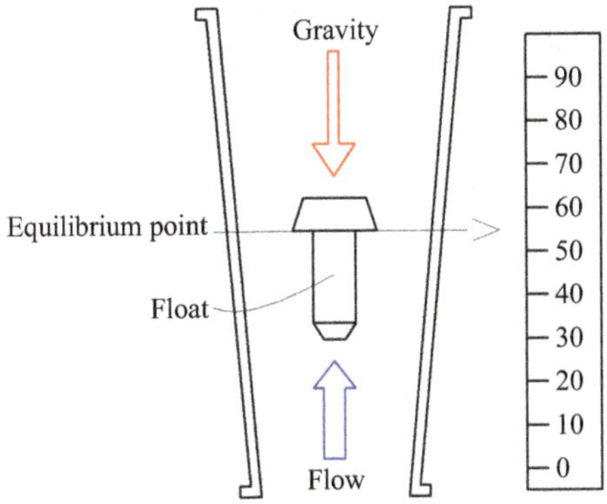

Figure 2.63 Flow measurement using a variable-area flowmeter (rotameter)

position is at the point where the differential pressure between the upper and lower surfaces balances the weight of the float.

As stated above, at equilibrium, the weight of the float, F_G (N) is balanced by the sum of its buoyancy, F_B (N) and the drag force F_D (N)

$$F_G = F_B + F_D \tag{2.67}$$

The weight of the float is given by

$$F_G = V_f \rho_f g \tag{2.68}$$

where V_f is the volume of the float (m^3), ρ_f is its density (kg/m^3) and g is the acceleration due to gravity (9.8 m/s^2).

The buoyancy of the float is

$$F_B = V_f \rho_{fl} g \tag{2.69}$$

where ρ_{fl} is the density of the fluid displaced by the float (kg/m^3).

And finally, the drag is

$$F_D = \frac{C_D A_f \rho_{fl} v^2}{2} \tag{2.70}$$

where C_D is the drag coefficient, A_f is the cross-sectional area of the float (m^2) and v is the velocity of the fluid (m/s)

Substituting into (2.67) and solving for the liquid velocity (m/s)

$$v = \sqrt{\frac{2V_f g\left(\rho_f - \rho_{fl}\right)}{C_D A_f \rho_{fl}}} \tag{2.71}$$

The volumetric flow rate, Q (m^3/s) is just the product of the velocity of the liquid and the cross-sectional area of the tube, A_m (m^2)

$$Q = A_m v = A_m \sqrt{\frac{2V_f g\left(\rho_f - \rho_{fl}\right)}{C_D A_f \rho_{fl}}} \tag{2.72}$$

Because the flow rate can be read directly on a scale mounted next to the tube, no secondary flow-reading devices are necessary. However, if desired, a sensor can be used to measure the float's level and transmit a flow signal.

Unfortunately, at low Reynolds numbers, the drag coefficient for a sphere (a common form of float in a rotameter) is inversely proportional to the flow rate, but for $1000 > R_e > 4000$ it becomes linear.

2.4.13.6 Turbine flowmeters

Many medical appliances still use the basic turbine or vane to measure flow because, if properly installed and calibrated, they provide the highest accuracies attainable for any currently available flowmeter for both liquids and gases. The unit consists of a multiple-bladed rotor mounted with a pipe perpendicular to the liquid flow. The rotor spins as the liquid passes through the blades as shown in Figure 2.64. The rotational speed is a direct function of flow rate and can be sensed by magnetic pick-up, photoelectric cell, or gears. As an alternative, a tachometer can be used. The number of revolutions or electrical pulses counted for a given period is directly proportional to flow volume. Turbine meters, when properly specified and installed, have good accuracy, particularly with low-viscosity liquids.

Assuming that the blades of the turbine are straight and thin with the radius of the shaft at the roots of the blade, a (m) and the radius of the turbine at the blade tips, R (m). As shown in the figure, the width of the blades in the direction of flow is c (m) and the average spacing s (m). For an incoming flow with velocity v (m/s), the turbine will rotate with angular velocity ω (rad/s).

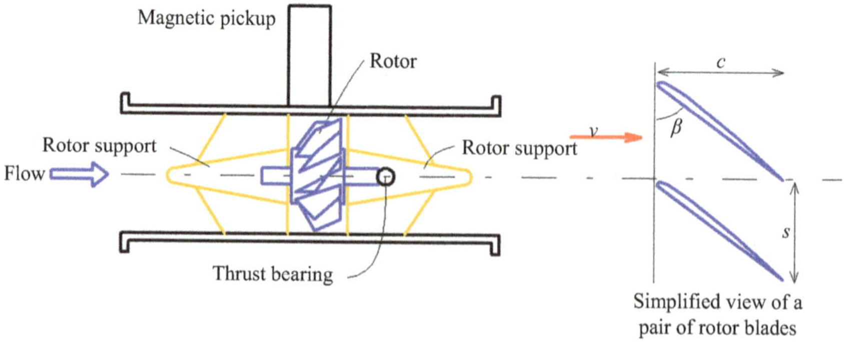

Figure 2.64 Turbine flowmeter

If there is no drop in velocity across the turbine blades, the ideal angular velocity is ω_i.

$$\bar{r}\omega_i = v \tan\beta \tag{2.73}$$

where $\bar{r}$ (m) is the rms value of the inner and outer radii of the blades and β (rad) is the angle between the shaft axis (direction of flow) and the turbine blade.

$$\bar{r} = \sqrt{\frac{R^2 + a^2}{2}} \tag{2.74}$$

The flow rate v (m/s) will be directly proportional to the ideal angular velocity

$$v = \frac{\bar{r}}{\tan\beta}\omega_i \tag{2.75}$$

and for the cross-sectional area of the pipe, A (m^2), the volume flow rate, Q (m^3/s) will be

$$Q = Av = \frac{A\bar{r}}{\tan\beta}\omega_i \tag{2.76}$$

In reality, there is a decrease in the flow velocity across the turbine, as some torque is required to spin it, then the relationship between the volume flow rate, Q (m^3/s) and the angular velocity of the turbine ω (rad/s) becomes much more complicated

$$Q = \frac{\omega\bar{r}^2 A^2}{\bar{r}} A \tan\beta - 0.037\mathrm{Re}^{-0.2} n(R + a)S \sin\beta \tag{2.77}$$

where Re is the Reynolds number and n is the number of blades.

A major concern with turbine meters is bearing wear. Bearingless options have been developed using the liquid itself for lubrication. These are particularly important in medical applications, particularly implanted ventricular-assist pumps discussed in Volume 2, Chapter 4.

2.4.13.7 Impeller flowmeters

Impeller flowmeters, also called paddle-wheel meters, are one of the more commonly used types. They are a direct offshoot of the old undershot water wheels. In-line meters are constructed as a unit that includes inlet and outlet orifices, while the insertion type can be installed into existing pipes through a round hole as shown in Figure 2.65.

In-line meters are more sensitive than axial turbine flowmeters at low flow rates because the blade incidence angle is much larger. They are also relatively insensitive to the flow regime (laminar or turbulent).

As with the turbine meters, paddle rotation is usually sensed using a magnetic pickup, and the flow rate is determined by measuring the pulse rate.

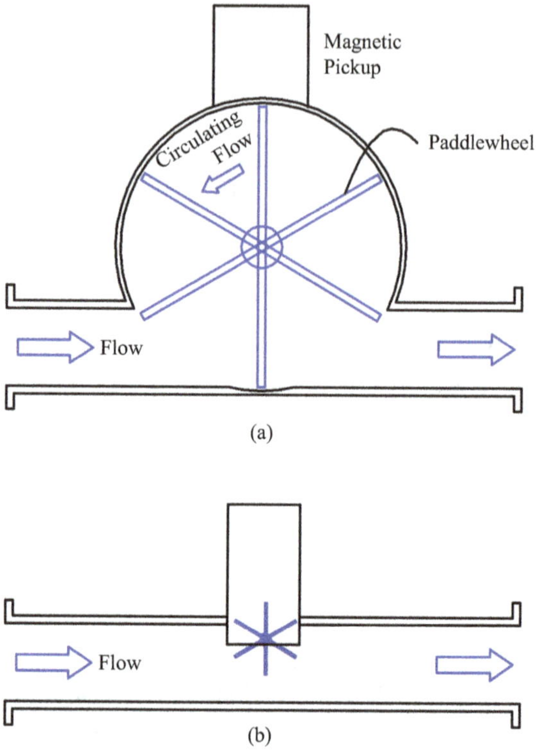

Figure 2.65 Diagram of (a) an in-line and (b) an insertion impeller flowmeter

2.4.14 Temperature sensors

The measurement of temperature is probably one of the most common sensing requirements in the medical field. The human organism can only function effectively over a small range of temperatures, so almost all processes associated with the organism are temperature controlled. These extend from incubators for premature babies through heart–lung machines and even respirators.

Mechanical measurement of temperature relies on the expansion of the working material. This can be a liquid in the case of a normal thermometer or a solid in the case of a bimetallic strip.

2.4.14.1 Bimetallic thermostats

It can be shown (Webster 1999) that the radius of curvature of a uniformly heated bimetallic strip will change from a radius R_o to R (m) with a temperature change from T_o to T (°C) is

$$\frac{1}{R} - \frac{1}{R_o} = \frac{6(1+m)^2(\alpha_2 - \alpha_1)(T - T_o)}{t\left[3(1+m)^2 + (1+mn)(m^2 + 1/mn)\right]} \tag{2.78}$$

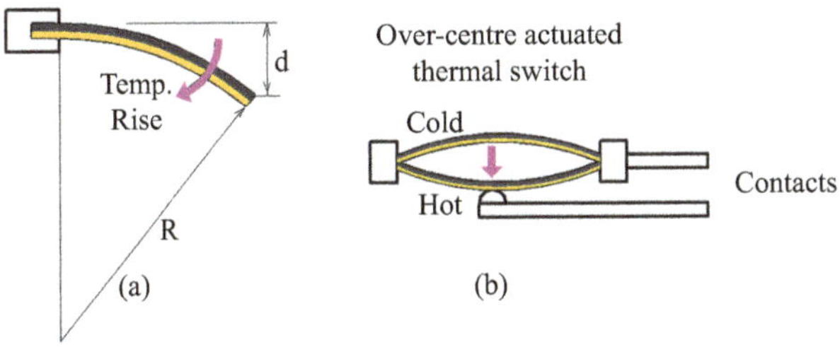

Figure 2.66 Bimetallic thermostat (a) mechanism and (b) over-centre switch

where $1/R_o$ initial curvature of the strip at temperature T_o; α_1 and α_2 are the coefficients of expansion of the two materials $\alpha_1 > \alpha_2$; n is the E_1/E_2 respective Young's moduli of the two materials; m is the t_1/t_2 respective thicknesses of the two materials and t is the $t_1 + t_2$ total thickness of the strip.

It can be seen that the ratio of the Young's moduli hardly makes any difference, therefore $n = 1$. In addition, most industrial bimetallic strips are made from strips in which the materials are the same thickness, therefore $m = 1$. Finally, if the strip starts out flat, then

$$\frac{1}{R} = \frac{3}{2}\frac{(\alpha_2 - \alpha_1)(T - T_0)}{t} \tag{2.79}$$

It is interesting to compare this equation to that defined for bimorph piezoelectric actuators discussed in Volume 2, Chapter 1.

Bimetallic strips are mostly used in thermostats which are in turn used for temperature control by switching heaters or refrigeration plants on or off, as illustrated in Figure 2.66.

2.4.14.2 Resistance temperature detector (RTD)

Most other thermometer types are based on materials whose electrical properties change with temperature. Possibly the most accurate of these are RTDs. Standard platinum RTDs can be manufactured with accuracies of ±0.0001 °C, while simple industrial RTDs are generally accurate to about ±0.1 °C.

The sensitive portion of an RTD is a coil of high-purity wire or a thin film deposited onto a ceramic substrate. The metal used is usually platinum, copper or nickel depending on the temperature range and accuracy requirements, as can be seen in Figure 2.67. Platinum is the material of choice because it does not oxidise even when subjected to very high temperatures.

In operation, a constant current (0.8 mA to 1 mA typically) is passed through the coil. As the temperature increases, the resistance will increase (the temperature coefficient is positive), and the voltage that is developed across the coil will reflect this. An accurate voltmeter calibrated to measure temperature can be used to display this value. Most modern temperature sensors use an accurate analogue-to-digital converter to digitise the voltage so that it can be further processed (curve fitting), then incorporated into some control or display application.

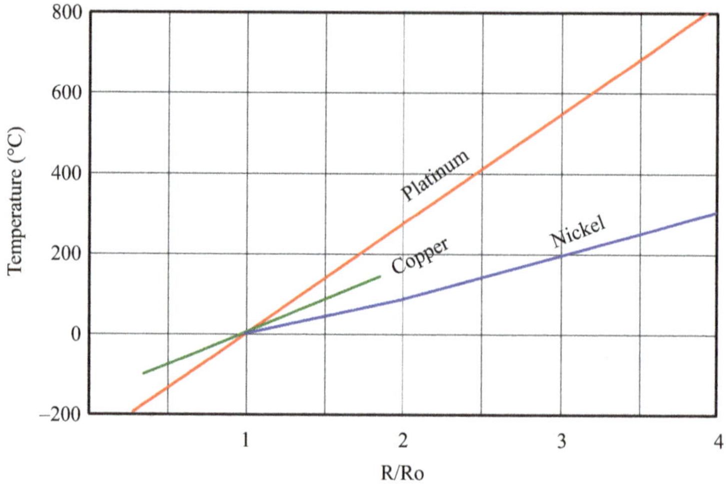

Figure 2.67 Resistance characteristics of a number of different RTDs

Platinum RTDs can be operated from about –184° C to +649 °C. They have good temperature to resistance linearity, good repeatability and long life. Nickel can be used over a smaller range because the relationship between temperature and resistance becomes very nonlinear at temperatures above 300 °C. It is, however, the most sensitive of the commonly used materials. Copper has a very linear resistance to temperature relationship, but it oxidises at moderate temperatures so it cannot be used at above 150 °C.

The temperature coefficient of resistance, α, of copper is 0.00427, that of nickel is 0.00672 and that of platinum is 0.003902, where α is defined as the change in resistance per degree C per ohm over the range from 0 °C to 100 °C

$$\alpha = \frac{R_{100} - R_0}{100R_0} \tag{2.80}$$

2.4.14.3 Thermistors

Thermistors are generally semiconductor materials that exhibit a higher temperature coefficient of resistance than pure metals, but their linearity is generally worse. Silicon positive temperature coefficient (PTC) thermistors rely on the bulk properties of doped silicon and have temperature coefficients of between 0.07 and 0.08. These are often built into electronics for temperature compensation.

Negative temperature coefficient (NTC) thermistors are made from metal oxides, and they exhibit a monotonic decrease in resistance with increasing temperature. They are very nonlinear as can be seen from Figure 2.68 (Maxim 2001).

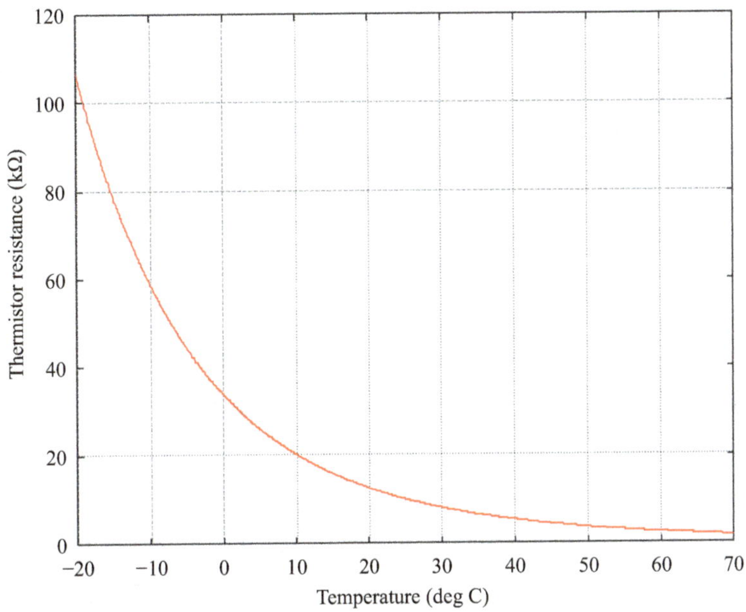

Figure 2.68 Resistance characteristics of a typical NTC thermistor, R_{25} = 10 kΩ, β = 3965 K

The standard formula that describes the resistance of an NTC thermistor as a function of temperature is

$$R_T = R_{25}\exp\left\{\beta\left[\left(\frac{1}{T+273}\right)-\frac{1}{298}\right]\right\} \tag{2.81}$$

where R_{25} is the resistance at 25 °C, β is the thermistor's material constant (K) and T is the actual temperature of the thermistor (°C). R_{25} and β are generally published in the manufacturer's data sheet; R_{25} can range from 22 Ω to 500 kΩ and β typically from 2500 to 5000 K.

Linearisation of this function can be achieved using resistance or voltage mode techniques. Resistance mode linearisation involves placing a normal resistor in parallel to the thermistor. If the resistance is chosen to equal R_{25} then the region of relatively linear resistance operation will be symmetrical around room temperature, as shown in Figure 2.69.

In voltage mode linearisation, the thermistor is connected in series with a normal resistor to form a voltage divider. For a fixed bias voltage and a resistor equal to R_{25}, the region of linear voltage will be symmetrical around room temperature. The result of this linearisation process is shown in Figure 2.70.

2.4.14.4 Semiconductor sensors

A semiconductor PN junction of a diode or transistor exhibits quite a strong temperature dependence. If the junction is connected to a constant current source, the voltage

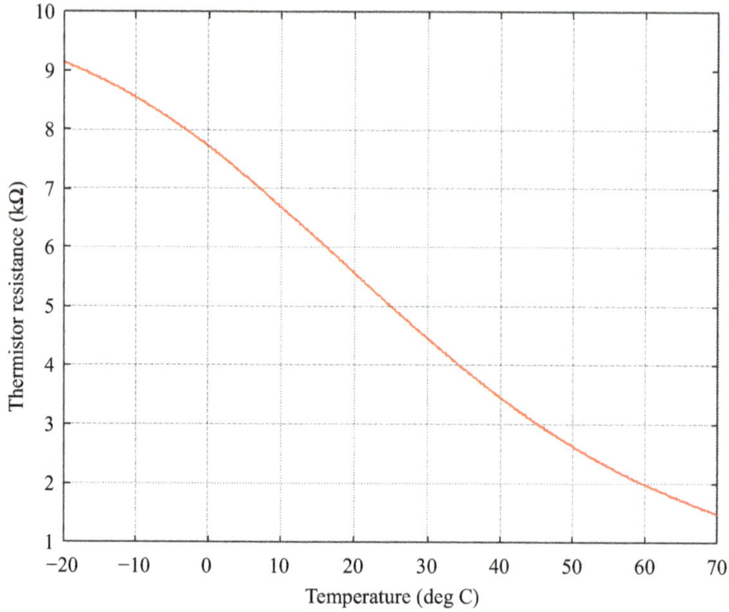

Figure 2.69 Resistance mode linearisation of a thermistor

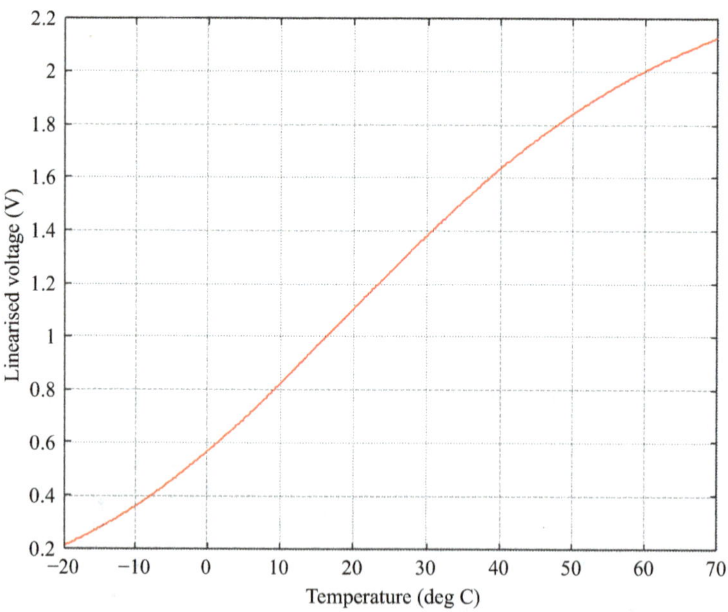

Figure 2.70 Voltage mode linearisation of a thermistor

becomes a measure of the junction temperature. This relationship is very linear and is therefore used for accurate three-terminal temperature-sensing integrated circuits.

The LM35Z temperature sensor has a linear output internally trimmed for the Celsius scale with a sensitivity of 10 mV per °C and a nonlinearity error confined within ±0.1 °C. The output can therefore be modelled as

$$V_{out} = V_0 + \alpha T \tag{2.82}$$

where T (°C) is the temperature, V_0 (volts) should equal zero but can be as large a 10mV resulting in an offset error of up to 1 °C. The slope α (mV per °C) can vary between 9.9 and 10.1.

2.4.14.5 Thermocouples

A thermocouple consists of a combination of two different materials bonded together that will generate a potential difference, ΔV, proportional to the temperature difference, ΔT, between the hot and cold terminals. This characteristic is known as the Seebeck effect (after Thomas Seebeck) and can be mathematically expressed as

$$\Delta V = \alpha_s \Delta T \tag{2.83}$$

where α_s is the Seebeck or thermoelectric coefficient, and its value depends on the difference between the thermoelectric coefficients of the two materials used.

A typical thermocouple consists of wires made from the two materials and a method of measuring the potential difference as shown in Figure 2.71. The designated reference temperature T_{ref} is usually supplied by an ice-water bath in a laboratory, but in the field that is not practical so the ambient temperature is used as the reference. Where the reference junction cannot be held at 0 °C, the observed value must be adjusted by the temperature difference.

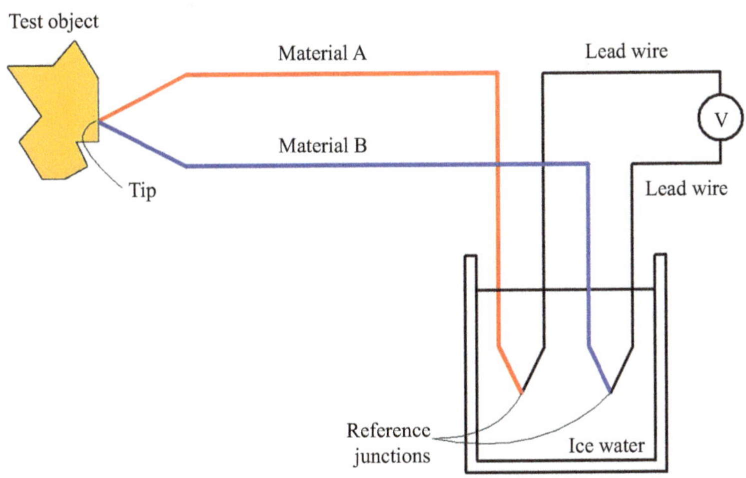

Figure 2.71 Single temperature reference thermocouple

If the thermoelectric coefficients, α_{SA} and α_{SB}, are nearly constant across the targeted temperature range, then the temperature at the probe tip T_{tip} can be determined from the reference temperature T_{ref} and the difference between the coefficients of the two materials.

$$V_{out} = (\alpha_{SA} - \alpha_{SB})(T_{tip} - T_{ref}) \tag{2.84}$$

Therefore, the temperature of the probe tip can be determined as

$$T_{tip} = T_{ref} + \frac{V_{out}}{\alpha_{SA} - \alpha_{SB}} \tag{2.85}$$

In practice, manufacturers generally provide calibration functions for their products. These functions are usually high-order polynomials and are calibrated with respect to a certain reference temperature. Suppose that the coefficients of the calibration polynomials are $a_0, a_1, a_2, \ldots, a_n$. The temperature at the probe tip can then be related to the voltage output as

$$T_{tip} = a_0 + a_1 V_{out} + a_2 V_{out}^2 + \cdots + a_n V_{out}^n \tag{2.86}$$

Note that the above formula is effective only if the reference temperature, T_{Ref}, in the measurement remains equal to the reference temperature specified on the data sheet.

Materials should have a high thermoelectric coefficient, low thermal conductivity and low resistivity. Unfortunately, as shown in Table 2.10, materials like gold, silver and copper that have low resistivity also have poor thermoelectric coefficients, while those with high thermoelectric coefficients, like bismuth (Bi) and antimony (Sb), have high resistivities (Fraden 2003).

Table 2.10 Thermoelectric coefficients of some common materials at 0 °C

Material	Thermoelectric coefficient, α_S (μV K^{-1})
Aluminium	3.5
Antimony	47
Bismuth	−72
Cadmium	7.5
Carbon	3.0
Constantan	−35
Copper	6.5
Germanium	300
Gold	6.5
Iron	19
Lead	4
Mercury	0.6
Nichrome	25
Nickel	−15
Platinum	0
Potassium	−9
Rhodium	6
Selenium	900

Table 2.10 (Continued)

Material	Thermoelectric coefficient, α_S (μV K^{-1})
Silicon	440
Silver	6.5
Sodium	−2
Tantalum	4.5
Tellurium	500
Tungsten	7.5

2.4.15 Tactile sensing

Tactile feedback is one of the critically important aspects of any successful hand prosthesis. In addition, it also plays a role in improving haptic feedback in remotely monitored medical examinations and surgery (Pawluk *et al.* 1998).

Tactile feedback relies on contact-based effects including contact stresses, slippage, heat transfer and hardness. These properties, in a grasped object, can be classified into geometric and dynamometric types (Webster 1999). Among the geometric properties are presence, location in relation to the end-effector, shape, dimensions and surface conditions. Among the dynamometric parameters associated with grasping are force distribution, slippage, elasticity and hardness as well as friction.

Though tactile sensing requires sophisticated sensors, it is also reliant on the processes through which the device interacts with the explored object. These include controlling contact force and end-effector position and orientation. This leads to active tactile sensing which requires a high degree of complexity in the acquisition and processing of tactile information.

Tactile sensing generally involves the interaction of a rigid object with the compliant cover layer of the tactile sensor, as shown in Figure 2.72. The indentation of the tactile layer can be analysed from two distinct perspectives. The first is the measurement of the contact stresses (force distribution) in the layer, which is relevant to controlling manipulation tasks. The second is the deflection profile of the layer which is important in recognising geometrical features of the object.

In 1980, a survey was conducted to determine the general specifications for tactile sensors (Harmon 1982), and these have been used as guidelines by many tactile sensor designers:

- Spatial resolution 1–2 mm,
- Array sizes of 5×10 to 10×20 elements,
- Sensitivity between 0.5×10^{-2} N and 1×10^{-2} N for each sensing element,
- Dynamic range 1000:1,
- Stable behaviour with no hysteresis,
- Monotonic response, but not necessarily linear,
- Compliant interface, rugged and inexpensive.

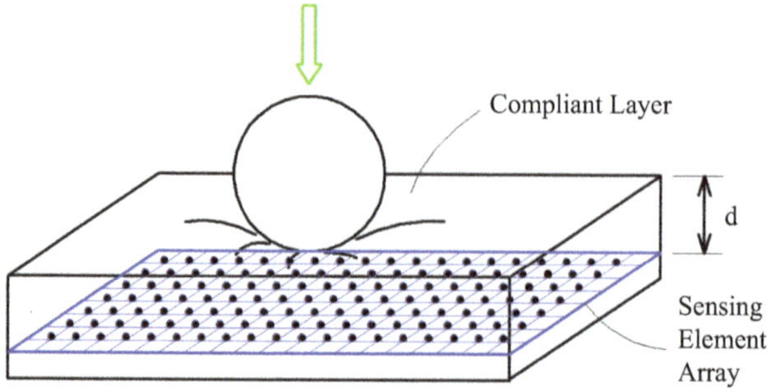

Figure 2.72 Interaction between a rigid object and the compliant covering of a tactile sensor

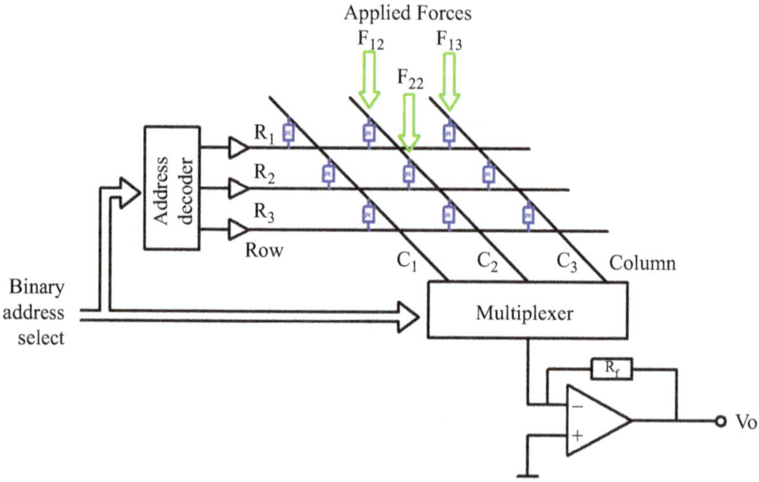

Figure 2.73 Configuration of a resistive tactile sensor

2.4.15.1 Resistive tactile sensors

These sensors rely on materials whose resistance changes with increases in applied force. Conductive elastomers manufactured by embedding conductive particles in natural or silicone rubber were amongst the first. These can operate using changes in the bulk resistance of the material or changes in the contact resistance. Piezoresistive materials have also been used, as have embedded pressure sensors.

A typical array will include conductive strips connecting the rows and columns of the grid so that the individual element resistances may be sensed, as shown in Figure 2.73.

A binary address selection connects a voltage, V_i, to a single row of resistive elements, while simultaneously the address controls a multiplexer which connects a single column to an op-amp inverter. The voltage output, V_o, is proportional to the resistance of a single element

$$V_o = -\frac{R_f}{R_{RC}} V_i \tag{2.87}$$

where R_{RC} is the resistance of the element at row R and column C and R_f is the resistance of the feedback resistor.

2.4.15.2 Capacitive tactile sensors

If the row and column conductive strips are placed on either side of a compliant dielectric material, then the intersection of each forms a tiny capacitor, as illustrated in Figure 2.74. The capacitance of a parallel plate capacitor is directly proportional to the product of the plate area, A, and the dielectric constant, ε, and inversely proportional to the distance between them, h.

$$C = \frac{\varepsilon A}{h} \tag{2.88}$$

An applied force that reduces the distance between the plates will result in an increase in capacitance that can be measured by an external circuit.

If stray capacitances between elements are ignored and the multiplexers are ideal, then the capacitors function as an AC voltage divider and V_o will be a function of the element capacitance

$$V_o = \frac{C_{RC}}{C_d + C_{RC}} V_i \tag{2.89}$$

Where C_d is the detector capacitance and C_{RC} (e.g. C_{3E}) is the capacitance of the element addressed by row $R = 3$ and column $C = E$.

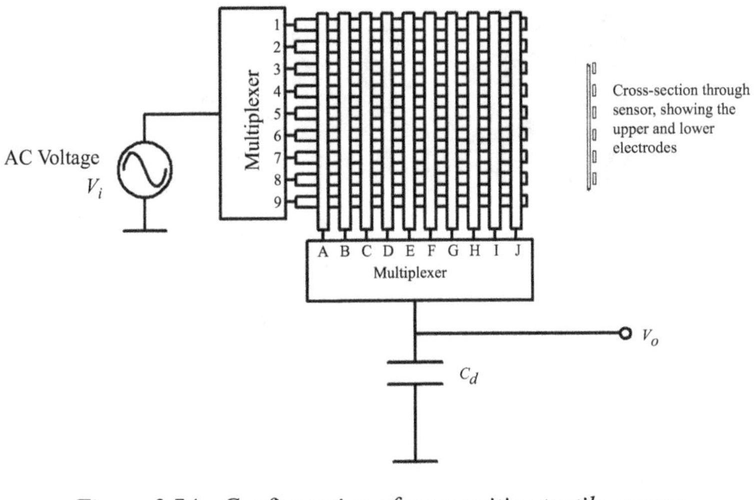

Figure 2.74 Configuration of a capacitive tactile sensor

2.4.15.3 Piezoelectric tactile sensors

As discussed earlier in this chapter, a piezoelectric material is one that will develop a charge, Q, across opposite faces when subjected to a force or deformation. Because the material is an insulator, and a dielectric, it also forms a capacitor, and so a voltage can be measured across each element

$$V = \frac{Q}{C} = \frac{Qh}{\varepsilon A} \tag{2.90}$$

where h is the thickness of the piezoelectric material, A is the area of each element and ε is the dielectric constant.

The piezoelectric material most commonly used is PVF because it has a strong piezoelectric effect; therefore, it can be made very sensitive. It is also flexible and can be made into small sensor elements. Unfortunately, it is sensitive to temperature, and because the output only occurs when the elements are flexed, the response does not extend right down to DC.

Sliding force sensors can be constructed using thin slivers of piezoelectric materials orientated in different directions. These are situated within a silicone rubber skin, and will respond differently as the sensor is moved in different directions across a rough surface. According to Fraden (Fraden 1996), they are capable of detecting surface discontinuities or bumps as small as 50 μm high.

2.4.15.4 Other tactile sensors

Exotic materials such as carbon nanotubes (CNTs) and carbon microcoils (CMC) can also be embedded into a rubber matrix to make tactile sensors as shown in Figure 2.75.

Unlike normal resistive sensors, because of the coiled nature of the conductive material, a significant inductive component also exists. These sensors are therefore

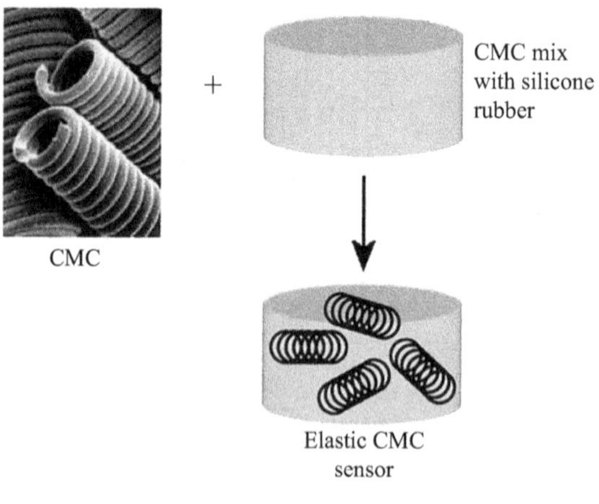

Figure 2.75 Tactile sensor made from carbon microcoils [adapted from (Nishikawa et al. 2005)]

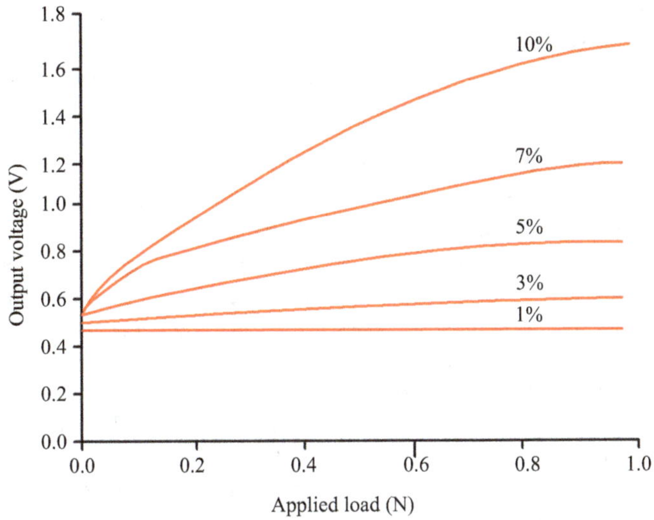

Figure 2.76 Measured sensitivity of a CMC-based tactile sensor with different concentrations of CMC [adapted from Nishikawa et al. (2005))]

excited by an AC signal at a frequency of between 100 and 400 kHz to give the best response. In addition, the sensitivity can be controlled by adjusting the concentration of the CMC material compared to the silicone rubber as shown in Figure 2.76 (Nishikawa *et al.* 2005).

High-density (8 × 8) CNT arrays were developed by Weichandt (2024) to measure EMG signals on the upper arm. A high-speed mixer containing the CNT material and part A of the silicone was run for 20 min prior to the catalyst (part B) being added and mixed for a further five minutes before casting. Contact resistances of the electrodes, Surface resistances varied between 3.6 and 1.7 MΩ depending on the proportion of CNT.

In addition to these resistive methods, various other mechanisms are used including optical and pneumatic methods.

2.4.16 Chemical sensors

Chemical sensors are sensitive to stimuli produced by various chemical compounds or elements. Their most important property is selectivity (the ability to discriminate between types), while a secondary one is their relationship to concentration. Chemical sensors are very important in medical applications; for instance, the ability to monitor oxygen concentration in the air or in solution is crucial during surgery.

Biosensors are considered to be a special class of chemical sensors because they have a much greater selectivity and sensitivity than other chemical sensors.

Man-made biosensors often use enzymes that have evolved over millions of years to be extremely sensitive to specific molecules. For these sensors, it is critically important that the active material on the sensing element be immobilised on the physical transducer, and that it remains active for the lifetime of the sensor.

2.4.16.1 Enzyme and catalytic sensors

Enzymes have two remarkable properties: they are extremely selective to a given substance and very effective in increasing the rate of chemical reactions. The sensing element can be a heated probe, an electrochemical sensor or an optical one. The enzyme is immobilised into a hydrogel into which the chemical diffuses. The enzyme-assisted reaction alters the characteristics of the hydrogel in a manner that can be detected by the associated sensor.

Catalytic sensors are a subset of enzyme sensors in which the catalytic reaction releases heat, and the temperature change can be measured by the sensor. These sensors have been developed specifically to measure low concentrations of inflammable gases (particularly in mines). As shown in Figure 2.77, a typical catalytic sensor comprises a temperature sensor such as an RTD or a thermistor surrounded by the appropriate catalyst. As the concentration of the target molecule increases, the rate of conversion increases as does the temperature.

In general, a pair of sensors are used together in a Wheatstone bridge configuration as discussed earlier in this chapter. The one sensor is immersed in the fluid containing traces of the molecule to be detected, while the other is placed in an inert solution. The two solutions are in physical contact so the initial temperatures of the two thermistors start out equal.

2.4.16.2 Electrochemical sensors

These are the most versatile and best developed of all the chemical sensor types. They measure voltage, current or resistivity and generally consist of a pair of electrodes as part of a closed circuit.

In voltage-based sensors, a redox reaction at the electrode-electrolyte interface of the electrodes results in a half-cell potential developing at each. One of the reactions involves the molecule of interest, while on the other hand, a known

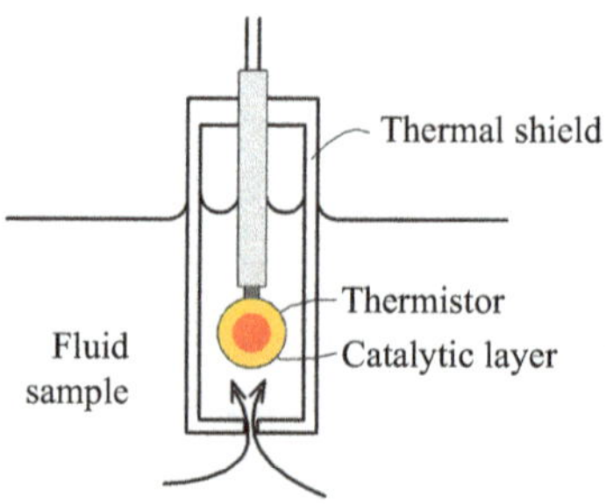

Figure 2.77 Catalytic thermal sensor [adapted from Fraden (1996)]

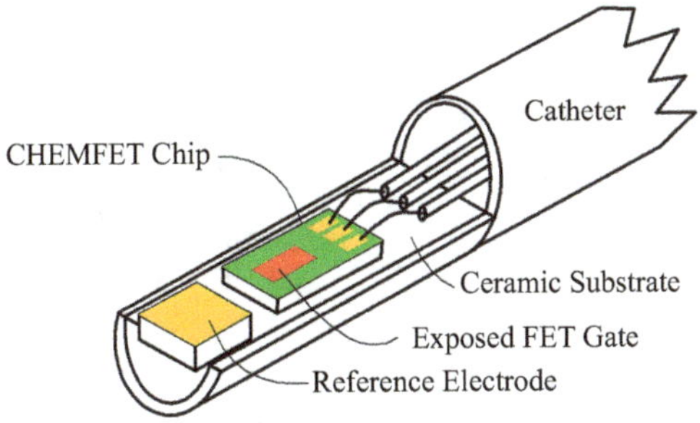

Figure 2.78 Catheter tip with a CHEMFET pH sensor for medical applications [adapted from Fraden (1996)]

reaction occurs. To maintain equilibrium, the current flow should be minimised, so a very high-impedance amplifier is used to measure the potential difference across the two electrodes. This potential difference is a function of the concentration of the molecule of interest. These sensors (shown in Figure 2.78) can be made using field effect transistors (FETs) and are known as CHEMFETs. They are sensors in which the gate of the transistor is coated with an appropriately sensitive gate insulator material which is, in turn, exposed to the electrolyte. As the charge around the sensing area changes, the conductivity of the FET is altered, and this can be measured. Ion-selective membranes can be deposited on top of the gate insulator to provide a large selection of different chemical sensors.

In resistive sensors (also known as conductive sensors), the conductivity of the electrolyte is dependent on the chemical concentration of the molecule of interest. As with the voltage types, these sensors are incorporated into one arm of a Wheatstone bridge.

2.4.16.3 Resistive chemical sensors

To detect the presence of a liquid, the sensor must be specific to a particular molecule at a specific range of concentrations. A typical resistive sensor discussed by Fraden (Fraden 1996) consists of a silicone rubber-carbon mixture in a thin film to maximise its surface-to-volume ratio. In the presence of a polar solvent (hydrocarbon fuel), the polymer matrix swells, pushing the carbon particles further apart and increasing the resistance significantly. This change can be measured in the normal way.

2.4.16.4 Oscillating chemical sensors

Measurements of minute masses cannot use conventional scales because they are not sufficiently sensitive. Oscillating sensors have the required sensitivity because they measure the shift in resonant frequency of a small piezoelectric crystal. A piezoelectric crystal cut to oscillate at ultrasonic frequencies will have a resonant mode that is a function of its mass. The surface of the crystal is covered by a thin layer of material that has an affinity for the molecule of interest, so if any is present, it will attach to the surface and alter the mass – and consequently the resonant frequency.

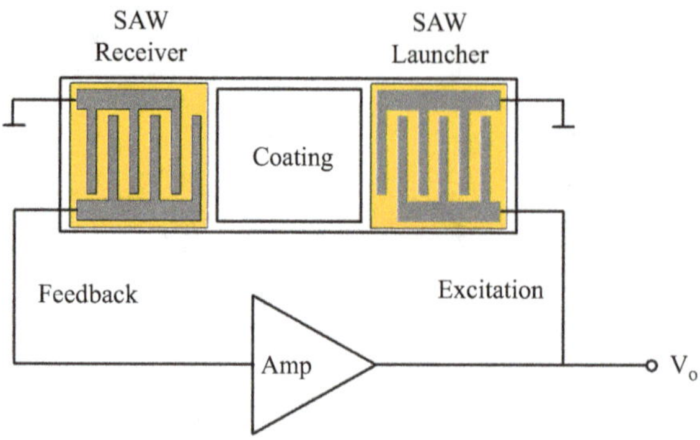

Figure 2.79 Flexing plate gas sensor [adapted from Fraden 1996)]

Oscillating sensors can be extremely sensitive, typically 5 MHz cm^2/kg. This means that a 1 Hz frequency shift corresponds to about 17 ng/cm^2 added weight. The response is very linear with mass and the dynamic range is quite broad – up to 20 μg/cm^2.

A second type of oscillating sensor uses surface acoustic waves (SAW). These sensors comprise a SAW transmission line (membrane) covered by a chemically sensitive coating situated between an acoustic launcher and a receiver as shown in Figure 2.79.

An oscillator drives the launcher that stimulates the production of a mechanical wave that travels along the transmission line to the receiver where it is converted back to an electrical signal for analysis. In the gas sensor shown, oscillation is maintained by feedback from the receiver that will be a function of the transmission time. This is in turn sensitive to the mass of the membrane that depends on the concentration of the molecule of interest.

The theoretical sensitivity of such sensors operating at 2.6 MHz is of the order of 900 cm^2/g, so if a sensor with a sensitive area of 0.2 cm^2 captures 10 ng the oscillator frequency is shifted by

$$\begin{aligned}\Delta f &= -900 \times 2.6 \times 10^6 \times 10^{-8}/0.2 \\ &= -117\ \text{Hz}\end{aligned}$$

SAW sensors are very versatile and can be adapted to measuring a wide variety of chemical compounds just by changing the coating on the membrane. Some of these are listed in Table 2.11.

2.4.16.5 Microbalance odour sensors

Odour sensors are gas sensors that have sensitivities approaching that of the human nose. Human odour sensors are covered with a phospholipid bilayer membrane, and it is believed that odorant adsorption into the membrane induces nerve impulses that travel to the brain for interpretation. The manufactured odour sensors shown in Figure 2.80 use a composite sensor made from polyvinyl chloride (PVC), a

Table 2.11 SAW chemical sensors (Fraden 1996)

Compound	**Chemical coating**	**SAW substrate**
Organic vapour	Polymer film	Quartz
SO_2	Triethanolamine (TEA)	Lithium niobate, silicon
H_2	Pd	Quartz
NH_3	Pt	Lithium niobate
H_2S	WO_3	Lithium niobate
Water vapour	Hygroscopic material	Lithium niobate, quartz
NO_2	Phthalocyanine	Lithium niobate
NO_2, NH_3, SO_2, CH_4	Phthalocyanine	Lithium niobate
Vapour explosives, drugs	Polymer	Quartz
SO_2, CH_4	None – thermal conductivity	Lithium niobate

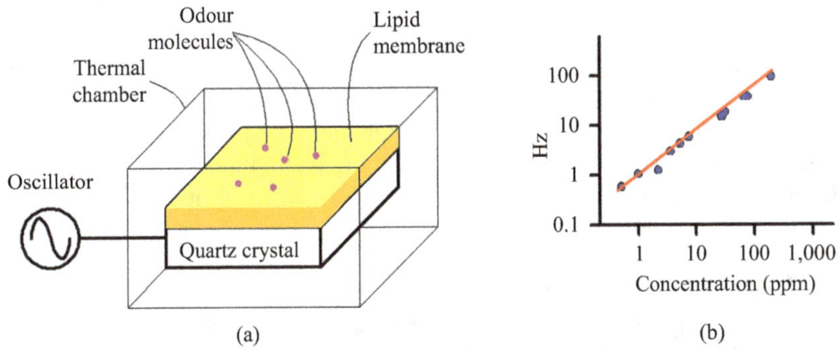

Figure 2.80 Microbalance odour sensor (a) and the transfer function (b) for amylacetate gas [adapted from Fraden 1996)]

plasticiser and a synthetic lipid. The lipid molecules are randomly oriented in the polymer matrix.

A 200 μm thick membrane of this polymer composite is spread on one side of the quartz crystal oscillator that is designed, along with the membrane blend, to have a resonant quality factor, $Q > 5 \times 10^4$.

As with the other oscillating sensors, odour molecules embed into the membrane and alter the resonant frequency because of the increase in mass. Experimental results indicate a sensitivity starting at 1 ppm (approximately the human threshold) and extending in a linear fashion up to 3000 ppm.

2.4.17 *Optical chemical sensors*

Optical methods are amongst the oldest techniques for sensing biochemical concentration. In principle, these sensors consist of a light source tuned to a frequency that will interact with the molecule of choice, a method of directing the light into the medium and a photodetector for processing the optical signal.

Optical sensors are usually based on optical fibres or planar waveguides, and there are three main methods of sensing at the surfaces of these devices:

- The molecule directly affects the optical properties of the waveguide such as evanescent waves[1] or surface plasmons[2].
- An optical fibre is used to convey light to and from the sample. Changes in the optical properties of the medium containing the molecule of interest are sensed by an external spectrophotometer.
- An indicator or chemical reagent placed near the tip of the optical fibre reacts with the molecule of interest to produce an optical signature that can be detected. These include absorption spectroscopy and fluorimetry.

Light sources range from a wide variety of coherent laser sources through LEDS extending from the IR through the visible as well as broad spectral incandescent lamps.

Optical elements include optical fibre, lenses, prisms, beam splitters, polarisers and diffraction gratings amongst others. Optical fibres and lenses convey light to and from the test area while prisms and diffraction gratings provide wavelength selection for frequency-selective processing. Polarisers are useful in identifying or selecting the light polarisation for transmission and reception.

Photodetection is most often performed using photodiodes that are sensitive to the frequency band of interest. Most systems use a pair of identical diodes – one in the active channel and the other to function as a reference.

2.4.17.1 Evanescent wave spectroscopy

When light propagates along an optical fibre, it is not completely confined to the core region but penetrates a short distance (typically one wavelength) into the cladding. Therefore, any compounds close to the surface will have an effect on this wave. Typically, a section of fibre is left unclad and in contact with a solution of the molecule of interest, or with a reagent that reacts with the molecule. Multiple internal reflections along this section of the fibre contribute to the overall absorption, or fluorescence that can be measured at the end of the fibre.

2.4.17.2 Surface plasmon resonance

In this technique, the fibre is clad with a metallised layer, typically gold or silver. When monochromatic light from the laser reaches this layer, it is absorbed by the plasma generated by the conduction electrons of the metal. This results in a phenomenon called surface plasmon resonance (SPR) which results in absorption of the light. The resonance depends on the angle, wavelength and polarisation state of the incident light and the refractive index of the metal film and material adjacent to it. Therefore, SPR is an extremely sensitive method of measuring changes in the refractive index of the medium.

[1]Electromagnetic waves generated in the medium outside the optical waveguide when light is reflected from within.

[2]Resonances induced by an evanescent wave in a thin film deposited on a waveguide surface.

2.4.17.3 Optical fibre sensors

Optical fibres are small and low cost. In addition, because electrical potentials are not involved, there is no electrical risk to the patient, and interference from electric and magnetic fields does not occur.

Most measurements are based on spectral or absorption changes in the medium determined by the active molecule, either directly, or through an indicator-mediated reaction. Most chemicals of medical interest such as hydrogen, oxygen, carbon dioxide and glucose require the use of reagents.

In optical fibre-based sensors, light travels efficiently to the end of the fibre where it exits into the medium and interacts with the molecule or reagent before returning via the same or a different optical fibre to a detector for analysis and interpretation. Typical reagent-mediated optical fibre-based sensors are shown in Figure 2.81.

2.4.17.4 Measurement of blood oxygen concentration

This is commonly known as oximetry and generally relies on a colour change in the blood as the relative amounts of deoxyhaemoglobin (Hb) and oxyhaemoglobin (HbO_2) vary.

Measurements are performed at two specific wavelengths, the first at 660 nm, where there is a large difference between the relative absorption of the two molecules, and the second at ≈ 805 nm where the absorption is independent of blood oxygenation. The oxygen saturation, O_{2sat}, can then be determined from the ratio of the two absorption levels, $\gamma_{\lambda 1}$, and $\gamma_{\lambda 1}$

$$O_{2sat} = a - b\frac{\gamma_{\lambda 1}}{\gamma_{\lambda 2}} \tag{2.91}$$

where a and b are sensor-dependent constants.

In vivo fibre-optic oximeters can be manufactured within catheters that are inserted into a vein or artery, or as non-invasive sensors clamped to the finger or ear where blood flow is close to the surface.

The latter are referred to as pulse oximeters and only monitor differences in the absorption as a function of time. This relationship is dependent on the blood volume within the sensitive area and will therefore be a function of arterial pressure

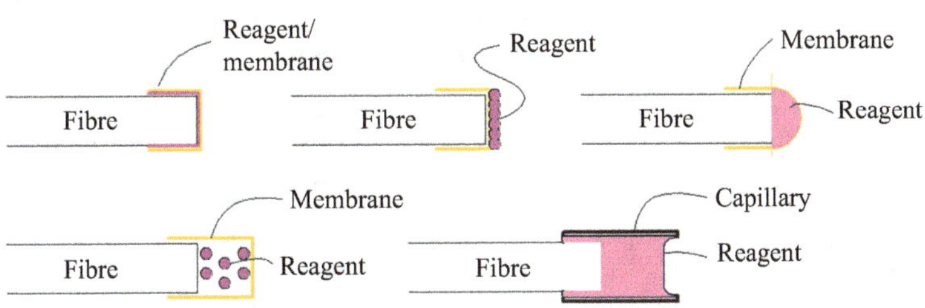

Figure 2.81 Configurations of different reagent-mediated fibre optic sensors [adapted from Bronzino (2006)]

and heartbeat. A normalisation process determines the ratio of the changing component, the alternating current (AC), and the fixed current (DC) component of the measured signal to produce a stable and reliable signal.

Pulse oximeters consist of a pair of small and inexpensive LEDs, one operating in the red, and the other in the infrared band. A single, highly sensitive phototransistor monitors the signal level transmitted through a finger or ear lobe, or the backscatter from any other well-vascularised portion of the body.

2.4.17.5 Measuring other molecules

Fibre-optic sensors have been developed to measure other blood constituents as well as oxygen partial pressure (PO_2). These include CO_2 partial pressure (PCO_2) and blood pH, both of which are essential to clinical diagnosis and the management of respiratory and metabolic problems. In addition, glucose sensors have been developed, and these are playing an increasingly important role as the incidence of diabetes increases.

2.5 Electrodes

Biological functions often have some electrical activity associated with them. The activity can be a constant DC potential, or a time varying one. Even though almost all the organs in the body generate some voltage, most of the signals are small and hard to measure. However, there are a number that have proved to be useful for biological assessment. These are listed in Table 2.12.

The mechanism of electrical conductivity in the body involves ions as charge carriers. Therefore, picking up bioelectric signals involves converting the ionic currents into electric currents that will flow through wires. This conversion process is carried out by electrodes that consist of electrical conductors in contact with the aqueous ionic (electrolyte) solutions in the body. At the interface between the electrode and the electrolyte solution, an electrochemical reaction needs to take place in order for a charge to be transferred.

When no current is flowing between the electrode and the organism, a potential, known as the half-cell potential exists across the boundary. If current flows,

Table 2.12 Useful bioelectric signals (Bronzino 2006)

Bioelectric signal	Abbreviations	Source
Electrocardiogram	ECG	Heart – seen from the body surface
Cardiac electrogram		Heart – seen from within
Electromyogram	EMG	Muscle
Electroencephalogram	EEG	Brain
Electrooptigram	EOG	Eye dipole field
Electroretinogram	ERG	Eye retina
Action potential		Nerve or muscle
Electrogastrogram	EGG	Stomach
Galvanic skin reflex	GSR	Skin

then the potential may drop, an effect known as polarisation, and the sensitivity of the electrode may be reduced.

Polarisable electrodes pass a current between the electrode and the electrolytic solution by changing the charge distribution in the solution near the electrode. No actual current crosses the electrode-electrolyte interface. Non-polarised electrodes allow current to pass freely across the interface without changing the charge distribution in the electrolytic solution.

Electrodes made from noble metals such as platinum are highly polarisable, and the charge distribution in the electrolyte adjacent to the electrode is different from that of the bulk. This limits the ability of such electrodes to measure DC or low-frequency signals. In addition, if the electrode moves with respect to the electrolyte solution, the charge distribution in the solution adjacent to the electrode surface will change and this will generate a voltage change that will appear as a motion artefact in the measurement. For these reasons, non-polarisable electrodes are preferred for most biomedical measurements.

The silver–silver chloride electrode is a good example of a non-polarisable electrode that is suitable for many biomechanical applications. Such electrodes have a silver base with a silver/silver chloride matrix on the surface. They do not polarise and so are more capable of conducting low frequency and DC signals and are also less prone to generating motion artefacts.

A simplified equivalent circuit of an electrode is a parallel resistor and capacitor with the resistor describing the DC and low frequency impedance of the electrode, and the capacitor describing the higher frequency AC component. Typical surface electrodes have impedances of between 2 and 10 kΩ, with larger electrodes having lower impedances, and needle or micro electrodes having much higher values (Cromwell *et al.* 1973).

2.5.1 Body-surface biopotential electrodes

This category of electrodes includes those that can be placed on the body surface for recording bioelectric signals. The integrity of the skin is not compromised, and they can be used for short- or long-duration applications. The earliest bioelectric potential measurements relied on immersion electrodes, which were simply buckets of saline solution into which the patient placed a hand and a foot as shown in Figure 2.82. As expected, this type of electrode presented many difficulties in regard to restrictions on the patient position and movement as well as the possibility of spillage.

Plate electrodes, first introduced in 1917, were a great improvement on immersion electrodes. They were originally separated from the skin by cotton pads soaked in saline to emulate the immersion electrode mechanism. Later, an electrolytic paste was used in place of the pad with the metal in contact with the skin.

2.5.1.1 Metal plate electrodes

The basic metal plate electrode consists of a flat metal plate with a thin film of conductive electrolyte gel between the plate and the skin to establish contact. These

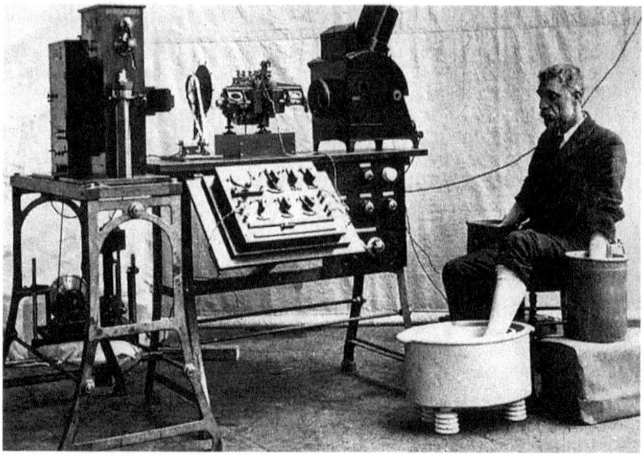

Figure 2.82 Early ECG system using immersion electrodes, reprinted from BMJ 1950;1:720, British Medical Association, London with permission

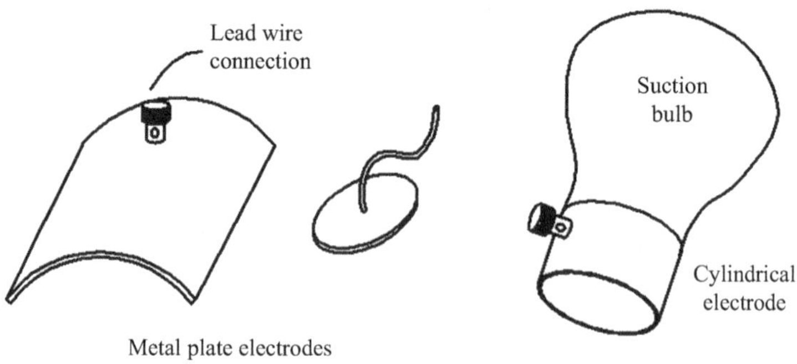

Figure 2.83 Examples of different skin electrodes [adapted from Bronzino (2006)]

are usually made from German silver (a nickel-silver alloy), platinum, gold or silver. They can be made from foil so that they are flexible, or they can even be formed with a suction cup to facilitate attachment as shown in Figure 2.83. These electrode types are primarily used for diagnostic recordings of ECG and EEG signals.

Conical gold-plated disc electrodes that include a hole at the apex of the cone for the insertion of electrolyte gel are frequently used for EEG monitoring, as the electrolyte can be replaced while the electrode is in situ.

These days, most electrodes are made from silver–silver chloride. They are easy to attach, and because they are disposable do not need to be cleaned after use.

A common application for skin electrodes is the electroencephalograph (EEG). Typically 64 or more electrodes are supported by a cap and in contact

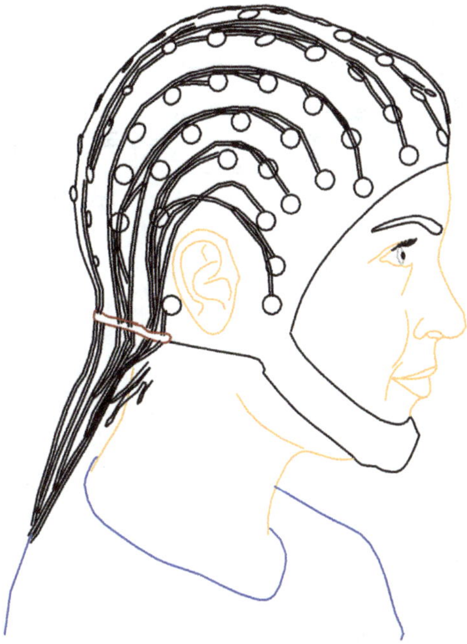

Figure 2.84 Cap-mounted 64-electrode electroencephalograph

with the scalp, as shown in Figure 2.84. The instrument records the voltages between pairs of electrodes that are the result of collective neural activity in the brain. Because the signals are so complex, machine learning is the only effective method for analysis.

2.5.1.2 Electrodes for long-term use

Long-term monitoring of biopotentials such as the ECG performed by cardiac monitors places a number of constraints on the electrodes shown in Figure 2.85. These must provide a stable interface, making the nonpolarisable variety preferable. In addition, mechanical stability, with the resultant reduction of motion artefacts, is important. One method of achieving this is to reduce motion between the electrode and the coupling electrolyte, and this can be done by recessing the electrode into the bottom of a cup full of electrolyte gel. Movement of the skin with respect to the electrode is then buffered by several millimetres of electrolyte, with the result that motion artefacts are reduced.

The advantages of this recessed electrolyte design can be realised in a much simpler design that lends itself to mass production. The electrodes consist of a conductive metal electrode bonded to a layer of open-cell foam that is impregnated with a high-viscosity electrolyte gel. Frequently, these electrodes are attached to a

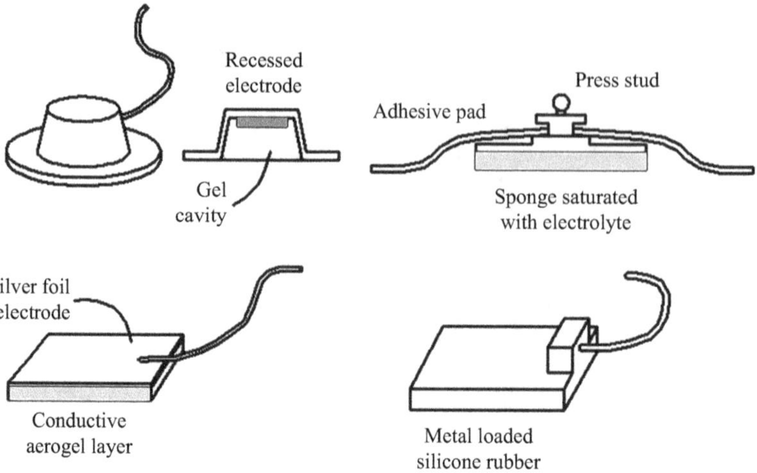

Figure 2.85 Electrodes for chronic patient monitoring [adapted from Bronzino (2006)]

press-stud through an insulated adhesive disc that holds the electrode in place against the skin.

A modification of this design replaces the electrode with a thin film of foil that can deform for a better fit on a curved body. The sponge can then be replaced with a thinner hydrogel film saturated with an electrolyte gel. This electrolyte/hydrogel mix is very sticky, so no further adhesive is required.

Very thin electrode films can be used in place of the foil electrodes if they are backed by a strong polymer film. The advantage of these is that if the metal film is thin enough (typically 1 μm or less), they are X-ray transparent. This is particularly useful for monitoring premature babies with sensitive skins as the repeated application and removal of electrodes can be a severe skin irritant (Bronzino 2006).

Special problems encountered in monitoring the ECGs of astronauts during long durations in space and under conditions of perspiration and considerable movement led to the development of spray-on electrodes. In these, a small spot of conductive material is sprayed or painted directly onto the skin which has previously been treated with an electrolyte coating (Cromwell *et al.* 1973).

Electrodes that do not use externally applied electrolyte gels are known as "dry." They can easily be applied and held in place using a rubber band or tape. They are made from graphite or metal-powder impregnated silicone rubber to produce a flexible conductive material similar to those discussed in the section on tactile sensors. These are applied directly to the skin, and the electrolyte layer is formed by a film of sweat that forms under the contact area.

Dry electrodes are used in home medical monitoring devices and on consumer goods like exercise bikes and treadmills to pick up an ECG signal. The signals picked up by these electrodes are generally much noisier than those obtained from the wet types. This requires that more consideration is taken with the signal processing to compensate for these shortcomings.

2.5.1.3 Electrodes for internal use

The internal electrodes shown in Figure 2.86 are generally smaller than their surface counterparts and don't require a gel layer because they are in intimate contact with the organism's own electrolyte solution. These are either needle electrodes introduced through the skin or cavity electrodes that can be implanted surgically or through a catheter. Good examples of the latter are the electrodes for a cardiac pacemaker that are introduced into the heart through a vein and barbed to ensure that good contact is maintained even after millions of contractions.

A needle electrode consists of a solid stainless-steel needle with a sharp point. An insulating layer covers most of the needle, exposing only a millimetre or so of the point. When this device is inserted into skeletal muscle, electrical signals are picked up by the exposed tip. Needle electrodes can also be made that are inserted through the hollow shaft of a standard hypodermic needle. These fine electrodes can then remain in place for long periods of time. Fine coiled-wire electrodes have also remained in skeletal muscles for a number of years without adverse effects.

For myoelectric applications, advances in electrode design allow complete rice-sized sensors to be implanted into muscles where they communicate wirelessly with an external receiver. A magnetic coil surrounding the implant region (usually the stump of an arm) powers the devices (Weir *et al.* 2009).

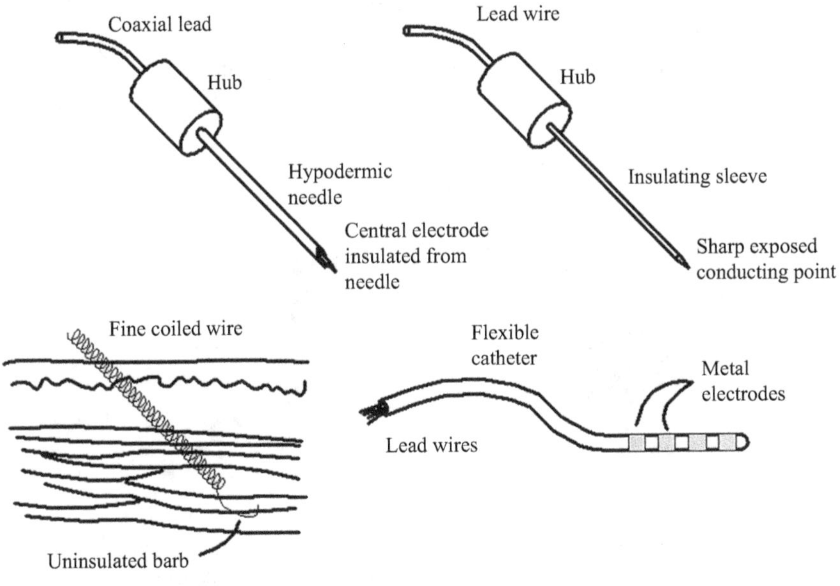

Figure 2.86 Internal electrodes [adapted from Bronzino (2006)]

2.5.1.4 Penetrating electrode arrays

Starting in 1983, work by Brummer, and subsequent work by Robblee, Rose, Cogan, and others at EIC Laboratories eventually resulted in microelectrodes made from activated iridium that could be formed into low-density arrays. The implantation of 38 microelectrodes in a human volunteer, at NIH, in 1994, provided the motivation for the development of higher-density arrays.

Researchers at the University of Utah, Salt Lake City, and the University of Michigan, Ann Arbor, have devised methods by which complex three-dimensional microelectrode arrays can be built. The Utah Electrode Array (UEA) is an example of such a neural interface and is shown in Figure 2.87a. A three-dimensional electrode array developed by researchers at the University of Michigan was built using more conventional integrated circuit technologies shown in Figure 2.87b. It has multiple electrode sites that are distributed along a number of electrode shanks, with the planes of the electrode shanks integrated into a single electrode array (McAlpine 2016).

The UEA consists of one hundred 1.5 mm-long microneedles that were designed to be inserted into the cerebral cortex to a depth of 1.5 mm, the level of normal neural input to the cerebral cortex. The electrodes of the UEA are built on a square grid with 400 μm spacing. One hundred gold bond pads are deposited on the back surface of these arrays, and one hundred thin insulated gold wires are bonded to these pads and to a percutaneous connector for connection to external electronics. The tip of each microneedle is metallised with iridium oxide to facilitate electronic to ionic transduction, and the entire array, with the exception of the tip of each microneedle, is insulated with a biocompatible polymer.

Similar to the UEA, the Utah Slanted Electrode Array (USEA) also has up to 100 microneedles, but their lengths are graded from 0.5 mm to 1.5 mm along the length of the array. The graded lengths of the USEA ensure that when it is inserted into a peripheral nerve, the electrode tips uniformly populate the nerve, with most nerve fibres being no more than 200 μm away from an active electrode tip (Medscape 2008).

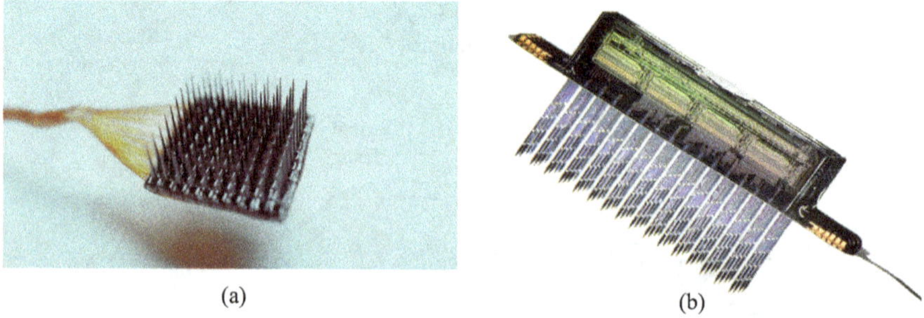

(a) (b)

Figure 2.87 Photographs showing (a) the Utah electrode array and (b) the Michigan array – © 2024 Blackrock Neurotech, LLC, reproduced with permission

In the first animal trials, it was found that these arrays containing multiple electrodes could not be pushed into the cortex as they just indent it (like a fakir on a bed of nails). A high-velocity pneumatic insertion tool was developed that could insert the array in 200 μs, corresponding to a velocity of 7 m/s.

The current state-of-the-art electrode arrays for recording neural activity are the Neuropixels 2.0. It consists of four pointed silicon shanks about 10 mm long and 70 μm wide, each containing 1280 electrodes in four rows along its length (Jun *et al.* 2017).

More details regarding electrode types and their performance can be found in the chapter on Neurovisual and Brain Interfaces, Volume 2, Chapter 3.

References

Alciatore, D. and M. Histand (2003). *Introduction to Mechatronics and Measurement Systems*, 2nd Ed. Boston: McGraw Hill.

Allen, R. (2002). "MEMS: Laying the Foundation for Exciting Applications." *Electronic Design.* Retrieved June 2008 from http://electronicdesign.com/Articles/Index.cfm?AD=1&ArticleID=2138.

Amman, M. (2001). "Laser ranging: a critical review of usual techniques for distance measurement." *Optical Engineering* 40: 10–19.

Bronzino, J., Ed. (2006). *Medical Devices and Systems.* Boca Raton, FL: CRC Press.

Brooker, G. (2021). *Sensors for Ranging and Imaging*, 2nd Ed. Stevenage: The Institution of Engineering and Technology (IET).

Brooker, G. M. (2015). "An adjustable radar cross section Doppler calibration target." *IEEE Sensors Journal* 15(1): 476–482.

Buchmann, I. (2005). "Battery University." Retrieved July 2008 from http://www.batteryuniversity.com/index.htm.

Campbell, M. (2010). *Heartbeat Generator Could Power Implanted Sensors. New Scientist.* London: Sunita Harrington.

Chen, X., X. Shiuou, Y. Nan and S. Yong (2010). "Nanogenerator for mechanical energy harvesting using PZT." *NanoLetters* 10(6): 2133–2137.

Cheng, T., J. Shao and Z. L. Wang (2023). "Triboelectric nanogenerators." *Nature Reviews Methods Primers* 3(1): 39.

Cromwell, L., F. Weibell, E. Pfeiffer and L. Usselman (1973). *Biomedical Instrumentation and Measurements.* Englewood Cliffs, NJ: Prentice-Hall, Inc.

Crownhart, C. (2023). How sodium could change the game for batteries. *MIT Technology Review.*

Dubček, T., D. Moreno-Garcia, P. Haag, *et al.* (2024). "In-sensor passive speech classification with phononic metamaterials." *Advanced Functional Materials* 34(17): 2311877.

Fett, T., D. Munz and G. Thun (1999). "Tensile and bending strength of piezoelectric ceramics." *Journal of Material Science Letters* 18(1999): 1899–1902.

Fraden, J. (1996). *Handbook of Modern Sensors*. New York, AIP Press: Springer-Verlag.

Fraden, J. (2003). *Handbook of Modern Sensors: Physics, Designs and Applications*. Switzerland: Springer Verlag.

Gregory, B. (1975). *An Introduction to Electrical Instrumentation*. London: Macmillan.

Harmon, L. (1982). "Automated tactile sensing." *International Journal on Robotics Research* 1(2): 196–212.

Horowitz, P. and W. Hill (1989). *The Art of Electronics*, 2nd Ed. Cambridge: Cambridge University Press.

Johnston, H. (2008). "Knee Brace Harvests 'Negative Work'." Retrieved June 2010 from http://physicsworld.com/cws/article/news/32812.

Jun, J. J., N. A. Steinmetz, J. H. Siegle, *et al.* (2017). "Fully integrated silicon probes for high-density recording of neural activity." *Nature* 551(7679): 232–236.

Livingstone, R. and M. Rioux (1986). "Development of a large field of view 3D vision system." *Optical techniques for Industrial Inspection*.

Martin-Garcia, G., G. Vazquez-Guzman, J. Sosa, A. Lopez, P. Martinez-Rodruigez and D. Langarica (2020). "Battery types and electrical models: A review." *IEEE International Autumn Meeting on Power Electronics and Computing (ROPEC 2020)*. Ixtapa, Mexico.

Maxim. (2001). "App note 817: Using Thermistors in Temperature Tracking Power Supplies." Retrieved June 2008, from http://www.maxim-ic.com/appnotes.cfm/appnote_number/817/.

McAlpine, K. (2016). *The Michigan Probe: Changing the Course of Brain Research*. ECE Magazine, University of Michigan.

Medscape. (2008). "The Neural Interface: The Utah Electrode Arrays." Retrieved September 2008 from http://www.medscape.com/viewarticle/560817_2.

Neets. (2000). "Module 15-principles of synchros, servos, and gyros." *Electrical Engineering Training Series*. Retrieved June 2009 from http://www.tpub.com/content/neets/14187/.

Nishikawa, K., P. Young-Kwang, M. Aizuddin, *et al.* (2005). "Development of carbon microcoils (CMC) sensor system with high sensitivity for effective acquisition of tactile information." *Intelligent Robots and Systems, 2005. (IROS 2005). 2005 IEEE/RSJ International Conference on*.

Pawluk, D., J. Son, P. Wellman, W. Peine and R. Howe (1998). "A distributed pressure sensor for biomechanical measurements." *Journal of Biomechanical Engineering* 120(2): 302–305.

Pepitone, J. (2023). "Be your own power plant – Energy harvesting tech for wearables is hitting its stride." *IEEE Spectrum*. 5–7.

Pons, J., Ed. (2008). *Wearable Robots - Biomechatronic Exoskeletons*. Chichester: John Wiley & Sons.

Powell, W. and D. Pheifer. (2006). "The Electronic Tilt Sensor." Retrieved September 2008 from http://archives.sensorsmag.com/articles/0500/120/index.htm.

Probert-Smith, P., Ed. (2001). *Active Sensors for Local Planning in Mobile Robotics*. Singapore: World Scientific.

Starner, T. and J. Paradiso (2004). Human Generated Power for Mobile Electronics. *Low-Power Electronics*. C. Piguet:, Ed. Boca Raton, FL: CRC Press.

Webster, J., Ed. (1999). *The Measurement, Instrumentation and Sensors Handbook*. Boca Raton, FL: CRC Press.

Weichandt, L. (2024). Conductive CNT Silicone in High-Density Electrode Arrays for Control of a Prosthetic Arm. BE, University of Sydney.

Weir, R. F., P. R. Troyk, G. A. DeMichele, D. A. Kerns, J. F. Schorsch and H. Maas (2009). "Implantable myoelectric sensors (IMESs) for intramuscular electromyogram recording." *IEEE Transactions on Biomedical Engineering* 56(1): 159.

Wu, Y., Y. Zhou, W. Asghar, *et al.* (2021). "Liquid metal-based strain sensor with ultralow detection limit for human–machine interface applications." *Advanced Intelligent Systems* 3(10): 2000235.

Chapter 3
Actuators

3.1 Introduction

Most biomechanical systems involve some sort of motion or action. This can range from the articulation of a large exoskeleton through the mechanical stimulation of the tiny bones in the middle ear. These actions are created by a force or torque that results in acceleration and displacement. In most cases, these actuators operate by the conversion of electrical power. However, in biomechatronics, pneumatic and hydraulic devices offer some advantages that are also discussed.

Electrical devices that produce or trigger a physical or physiological response by stimulating the body's musculature or nerves are also considered to be actuators. These include pacemakers, defibrillators, transcutaneous electrical nerve stimulation (TENS) devices and other electrode arrays such as retinal, neural or cochlear implants. They are discussed in Volume 2.

3.2 Electromechanical actuators

When a current-carrying conductor is placed in a magnetic field, a force is produced in a direction perpendicular to both the direction of the current and the magnetic field. This is the Lorentz force law, and can be stated, in vector form as (Alciatore and Histand 2003)

$$\vec{F} = \vec{I} \times \vec{B} \tag{3.1}$$

where $\vec{F}$ (N/m) is the force vector per unit length of conductor, $\vec{I}$ (A) is the current vector and $\vec{B}$ (Weber/m^2) is the magnetic vector or flux density. The relationship between these vectors can be determined using the right-hand rule which states that if the index finger is pointing in the direction of the current flow, and the middle finger is aligned with the magnetic field, then the thumb, extended perpendicular to the two fingers, points in the direction of the force.

The magnitude of the force, F (N), depends on the total length, l (m), of the conductor in the field, the magnetic flux, B (Weber/m^2), and the current, I (A), if all the components are orthogonal

$$F = BIl \tag{3.2}$$

This equation is extended later in this chapter to determine the force generated by a solenoid and the torque generated by an electric motor.

A second consideration of importance is the relationship between the magnetic flux, B, produced by a conductor wound into a coil (known as a solenoid), the current, I, flowing through the solenoid and the number of turns, N. If the solenoid is long compared to its diameter, then the field in its centre can be approximated by

$$B = \mu_o \frac{NI}{L} \tag{3.3}$$

where $\mu_o = 4\pi \times 10^{-7}$ (Weber/amp-m) is the permeability of free space and L (m) is the length of the solenoid.

The flux density drops off towards the ends of the coil, and within one radius, it reduces to 15% as the flux leaks out. This effect is illustrated in Figure 3.1.

Consider a 0.5 m long solenoid with a diameter of 30 mm consisting of five layers of windings of 600 turns each and carrying a current of 3 A. The magnetic flux at the centre of the solenoid is

$$\begin{aligned} B &= \mu_o \frac{NI}{L} \\ &= 4\pi \times 10^{-7} \frac{600 \times 5 \times 3}{0.5} \\ &= 0.0226 \text{ weber/m}^2 \end{aligned}$$

Note that the solenoid diameter and the number of layers of turns do not enter the equation so long as the length-to-diameter ratio remains large.

If the air core within the solenoid is replaced with a metal (usually iron), the magnetic field is increased in proportion to the relative permeability, μ_r, of the material

$$\begin{aligned} B &= \mu_r \mu_o \frac{NI}{L} \\ &= \mu \frac{NI}{L} \end{aligned} \tag{3.4}$$

where μ is the permeability of the material.

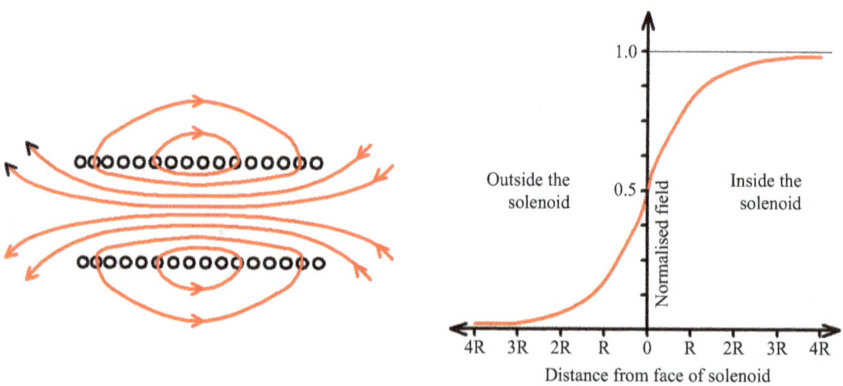

Figure 3.1 Flux density along the length of a coil

The relative permeability is dependent on the magnetic field and is also a function of frequency and saturation. Typical values of μ_r are 200 for magnetic iron and 100 for nickel at a magnetic flux density of 0.002 Weber/m^2. Special alloys can have much higher permeability with permalloy (78.5% Ni and 21.5% Fe) reaching 8000 and Mu-metal (75% Ni, 2% Cr, 5% Cu and 18% Fe) reaching 20,000.

If the air core of the solenoid in the previous example is replaced with an iron core with $\mu_r = 200$, the magnetic flux at its centre will increase to $0.0226 \times 200 = 4.52$ Weber/m^2.

When a conductor of length, l (m), cuts through a magnetic field with strength B (Weber/m^2), at a velocity v (m/s), a voltage ε (V) is developed across the conductor if the respective directions of the various components are orthogonal

$$\varepsilon = Blv \tag{3.5}$$

Once again, the direction of the induced current is determined by the right-hand rule, where the direction of the magnetic field is defined by the thumb, the direction of motion by the middle finger, and the direction of current flow (if the circuit is closed) by the index finger.

One final relationship that is important is Faraday's law. It states that the voltage, ε (V) induced in a coil is equal to the product of the number of turns, N and the rate of change of the magnetic flux $d\varphi/dt$, where φ is measured in Webers

$$\varepsilon = N\frac{d\varphi}{dt} \tag{3.6}$$

It is not important whether the change in flux is induced by moving a permanent magnet within the coil or changing the current through the primary winding of a transformer to induce a changing flux in the secondary.

3.2.1 *Solenoids and voice coils*

3.2.1.1 Solenoid actuator

Solenoid actuators consist of a fixed core and a movable plunger that completes a magnetic circuit broken by a small gap as shown in Figure 3.2. When the coil is energised by passing a current through it, the core/plunger combination forms an electromagnet, and the plunger is drawn downward to close the gap.

As a first-order approximation of how this works, a number of assumptions must be made. It is assumed that all the magnetic flux is constrained within the core and within the gap and that all of the resistance to magnetic flux flow (magnetic reluctance) occurs within the gap. The field is assumed to be uniform in the gap, and zero everywhere else and that the core material does not saturate.

The magneto motive force (*mmf*) set up by the coil is defined by the number of turns on the coil, N, and the current I (A), flowing through it

$$mmf = NI = \int H.dl \tag{3.7}$$

where H is the coercive force on the flux path, which is nonzero only in the gap.

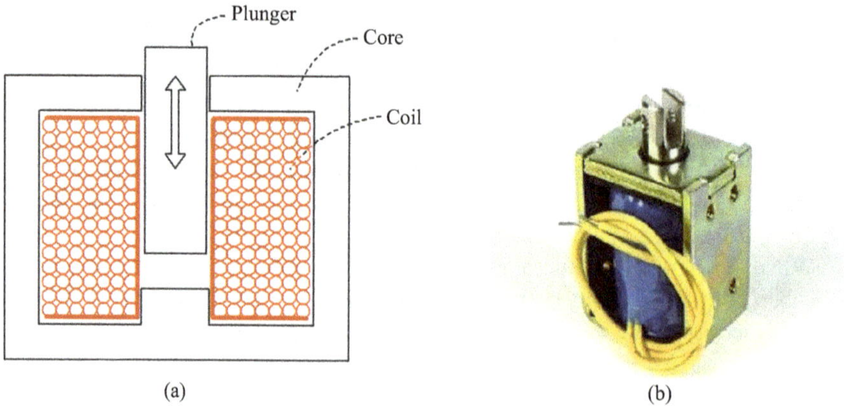

Figure 3.2 Solenoid actuator (a) schematic and (b) photograph

Because the field is considered to be uniform and the gap length l_g (m) is constant, then equation (3.7) can be simplified to

$$mmf = NI = Hl_g \tag{3.8}$$

therefore, the coercive force will be

$$H = \frac{NI}{l_g} \tag{3.9}$$

The energy W_{mag} (J) stored in the volume of space occupied by a magnetic field is

$$W_{\text{mag}} = \frac{1}{2}\int BHdV \tag{3.10}$$

where B (Weber/m^2) is the flux density.

Substituting for equation (3.9) and solving for a specific cross-sectional area A_g (m^2)

$$W_{\text{mag}} = \frac{\mu_o H^2 A_g l_g}{2} = \frac{\mu_o N^2 I^2 A_g}{l_g} \tag{3.11}$$

The mechanical energy is determined by the product of the force and the distance travelled

$$W_{\text{mech}} = \int F dl \tag{3.12}$$

Therefore, the force can be described by the change in energy per unit change in the length of the gap

$$F = \frac{dW_{\text{mech}}}{dl_g} \tag{3.13}$$

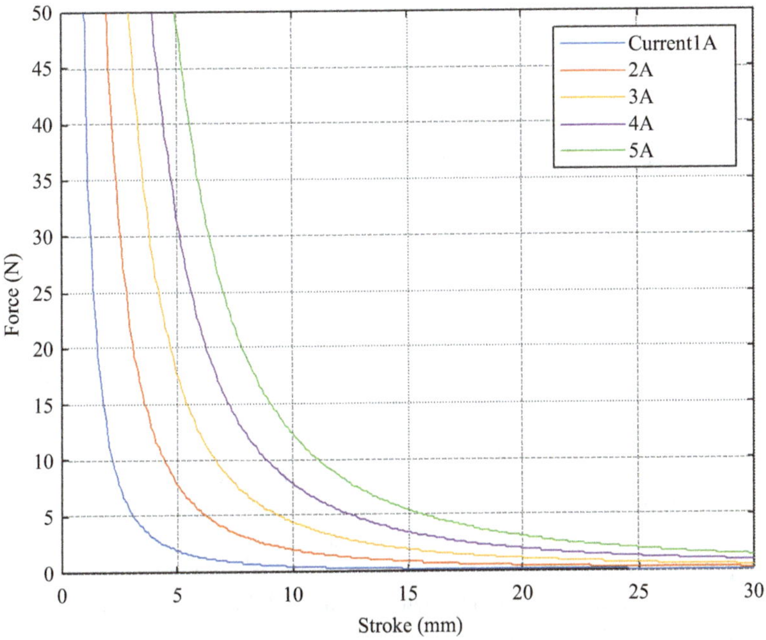

Figure 3.3 Relationship between the displacement and force of a solenoid with N = 1000 and the current as a parameter

and this must be equal to the magnetic energy given up. Hence

$$F = \frac{dW_{\text{mag}}}{dl_g} = \frac{\mu_o N^2 I^2 A_g}{2l_g^2} \tag{3.14}$$

From this result, it can be seen that as the gap is reduced, the force increases proportionally to the square of the gap length, with the result that the relationship between the force and the solenoid displacement is very non-linear. The first order approximation generated using the MATLAB® code below is shown in Figure 3.3 for a solenoid with $N = 1000$, $I = 1$ to 5 A and $A_g = 7.85 \times 10^{-5}$ m^2 (plunger diameter of 10 mm).

```
% solenoid force with current and gap
% solenoid1.m
degrad = pi/180;
N = 1000;                     % turns
d = 1.0e-02;                  % diameter of plunger

ag = pi*d*d/4;
mu0 = 4*pi*1.0e-07;
```

```
lg = (0.1:0.1:30)/1000;        % gap (m)
lg1 = lg*1000;                 % gap (mm)
f = [];

for i = 1:5                    % current (A)
 f(i,:) = mu0*N*N.*i.^2.*ag./(2.*lg.*lg);
end
 plot(lg1,f(1,:),lg1,f(2,:),lg1,f(3,:),lg1,f(4,:),lg1,f(5,:));
 grid
 axis([0,30,0,50]);
 xlabel('Stroke (mm)')
 ylabel('Force (N)')
 legend('Current 1 A','2 A','3 A','4 A','5 A')
```

This inverse-squared relationship of the solenoid force with distance is not the ideal relationship for an actuator. At the beginning of the stroke, where the load is to be accelerated, there is very little force, and at the end of the stroke, where the plunger should be decelerating, the force reaches a maximum.

The force-distance characteristics of the solenoid can be altered slightly by shaping the gap. For example, if the plunger tip is made in the form of a truncated cone, and the pin is made in the form of a cup as shown in Figure 3.4. When the gap is large, there is very little difference between the flux paths of this design and the square-ended design, so the attractive force will be similar. However, as the gap gets smaller and the cone is inserted within the cup, the cross-sectional area increases, but because the total flux remains unchanged, the flux density is reduced, and the force reduces in proportion.

Most solenoids operate against return springs to ensure that once the current has been disconnected, the plunger returns to the extended position. To ensure that

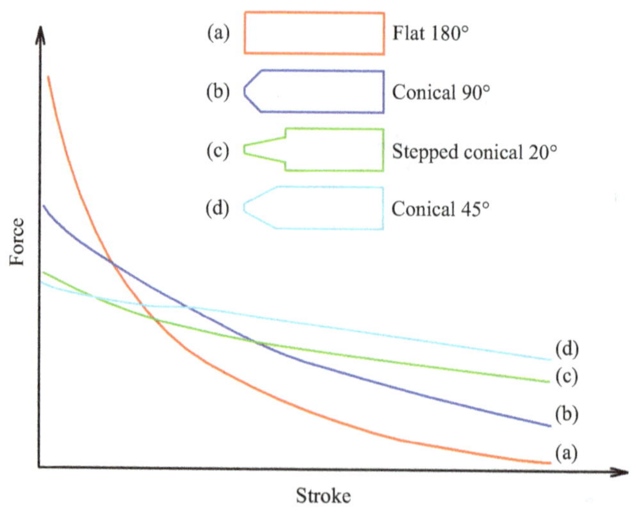

Figure 3.4 Effects on the Force Stroke curve of a number of differently shaped plungers

this occurs, a residual gap must remain when the solenoid is energised; otherwise, residual magnetism within the core can hold the plunger against the spring.

A frequently attempted experiment is to adjust the current supply to a solenoid to balance the load at some adjustable position. Unfortunately, this does not work – either the solenoid does not move, or it moves all the way to the inside stop. It is sometimes possible to achieve an unstable equilibrium, but there is no correcting force so any motion results in positive feedback and the plunger moves quickly to one of the extremes.

3.2.1.2 Voice coil actuators

Voice coil actuators can move an inertial load at extremely high accelerations (> 20 g) and relocate it to an accuracy of better than 10^{-5} mm over a limited range of travel. Motion can be in a straight line (linear actuators) or in an arc (swing-arm actuators). Most voice coil actuators are used to position the heads of disc drives, but they are also found in shaker tables, medical equipment, lens focusing applications and servo systems.

Linear voice coil actuators come in two forms, moving coil and moving magnet types as illustrated in Figure 3.5. Unlike solenoids, in voice coils, a reasonably uniform magnetic field is generated across a fixed gap using a powerful permanent magnet. The coil lies within this space and is held in neutral position by a spring mechanism.

When current is applied to the coil, a force is generated that is proportional to the product of the flux density, the length of the conductor in the magnetic field and the current as described in equation (3.2). Unlike solenoid actuators, there is a linear relationship between the current and the force, and hence between the current and the displacement against a spring. An additional characteristic of voice coil actuators that differs from solenoid actuators is that the direction of motion is determined by the direction of the current flow.

Linear voice coil actuators range from the minute, with peak forces from 0.7 N and strokes of 1 mm, to huge with forces of 2000 N and strokes of up to 50 mm. A typical voice coil actuator in the middle of the range is the LA10-12-027A manufactured by BEI Kimco. Its specifications are shown in Table 3.1.

Because of end effects, the force sensitivity, K_f (N per A), is not completely constant over the stroke, and will therefore result in some nonlinearity in the displacement when restrained by a spring, as shown in Figure 3.6.

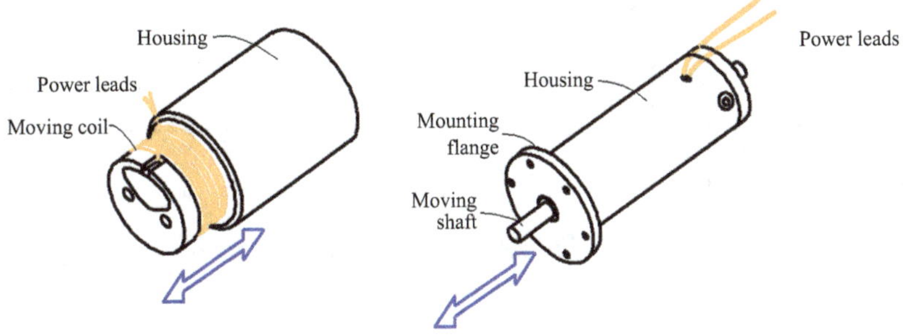

Figure 3.5 Voice coil actuators (a) moving coil and (b) moving magnet

Table 3.1 Specifications of the BEI Kimco LA10-12-027A linear voice coil actuator

Winding Constants	**Units**	**Tolerance**	**Symbol**	**Value**
DC resistance	Ohms	+/−12.5%	R	11
Voltage	Volts	Nominal	V_p	25.1
Current	Amperes	Nominal	I_p	2.28
Force sensitivity	Newton/Amp	±10%	K_f	5.87
Back EMF constant	V per m/s	±10%	K_b	5.87
Inductance	Milli Henry	±15@	L	3.05
Actuator parameters	**Units**		**Symbol**	**Value**
Peak Force	N		F_p	13.3
Actuator constant	N $W^{-0.5}$		K_a	1.77
Electrical time constant	Microsecond		τ_e	277
Mechanical time constant	Millisecond		τ_m	4.43
Power from I^2R	Watt		P_p	57.2
Stroke	mm		-	4.57

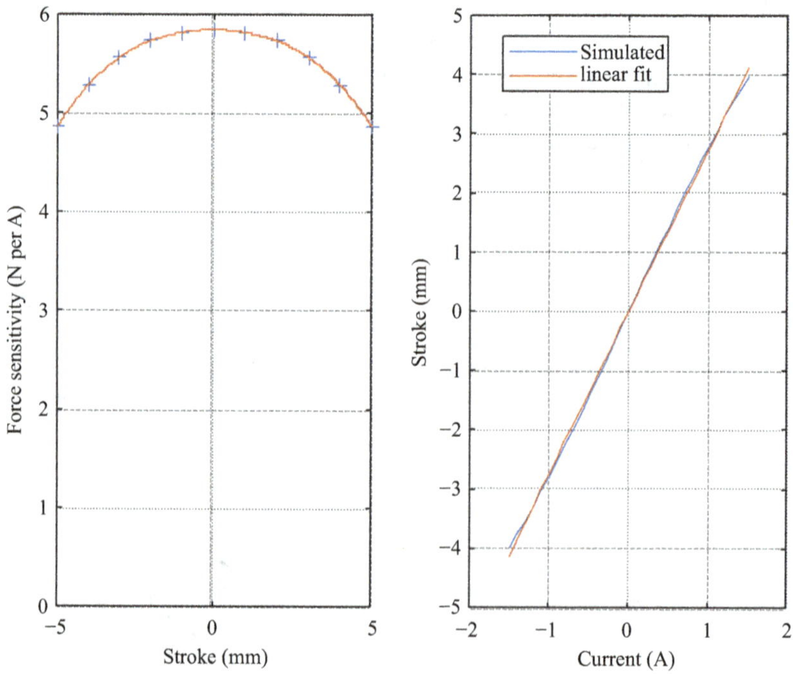

Figure 3.6 BEI Kimco LA10-12-27A linear voice coil actuator sensitivity for a spring constant k_s = 2 N/mm

```
% voice coil simulation
% voice_coil.m
% lookup table
% displacement stroke (mm)
% force sensitivity kf (N/A)
% spring constant ks (N/mm)
%
stroke = ([-5,-4,-3,-2,-1,0,1,2,3,4,5]);
kf = ([4.865,5.282,5.56,5.73,5.81,5.84,5.81,5.73,5.56,5.282,4.865]);
ks = 2;

% fit a spline through the points on the force sensitivity curve
strokei=(-5:0.01:5);
kfi = spline(stroke,kf,strokei);

% plot the force sensitivity vs stroke displacement
subplot 121
plot(stroke,kf,'+',strokei,kfi);
axis([-5,5,0,6]);
grid
xlabel('Stroke (mm)')
ylabel('Force Sensitivity (N per A)')

% becuase the force sensitivity is a function of the stroke, the nonlinear
% equation is solved by iteration
xout = [];
iout = [];
k = 0;

for i = -1.5:0.1:1.5                      % determine the stroke for a number
of currents
    k=k+1;
    xj = 0;                                % start with xj=0
    err = 10;

    while err > 0.0001                     % loop while the error is >
0.0001mm
        kfj = interp1(strokei,kfi,xj);  % lookup the force sensitivity
        fj = kfj*i;                        % the magnetic force is then
        xl = fj/ks                         % the new disp determined from
spring const
        err = abs(xl-xj);                  % check the error
        xj = xl;                           % update the estimate for the
stroke
    end
    xout(k)=xj;                            % store the results
    iout(k)=i;
end

% fit a linear function through the curve
p = polyfit(iout,xout,1);
% Evaluate the fitted polynomial p
xfit = polyval(p,iout);

% plot the stroke as a function of current
subplot 122
plot(iout,xout,iout,xfit,'-' );
grid
xlabel('Current (A)')
ylabel('Stroke (mm)')
legend('simulated','linear fit')
```

Because the relationship between the current is slightly nonlinear, really accurate positioning requires the addition of a linear displacement measurement sensor such as an LVDT or a linear potentiometer as discussed in Chapter 2. However, for most applications, the open-loop linearity is sufficiently good.

This description explains how the actuator can be positioned with good accuracy, but it does not describe the dynamics of reaching that equilibrium point.

When a conductor moves through a magnetic field, a voltage is induced across the conductor in a direction to oppose the motion (if current can flow); this is the back EMF as defined by equation (3.5). A voice coil will continue to accelerate, if the stroke is sufficiently long, until the back EMF is equal to the applied voltage, at which time the voice coil will have reached terminal velocity. In general, however, the velocity is limited to a value far below this.

In the case of the voice coil, the length of the conductor is determined by the radius of the coil, a (m), and the number of turns, N, making the back EMF, ε_b (V)

$$\varepsilon_b = 2\pi NaBv \tag{3.15}$$

When voltage is suddenly applied to a voice coil, the inductance of the windings limits the rate of increase in the current and the response of the device can be sluggish. To overcome this, a copper sleeve fixed to the magnet structure surrounds the coil. Eddy currents are induced in it by changes in the current through the coil. These currents cause a magnetic field that opposes the original field and cancels much of the apparent coil inductance, with the result that the response is much faster and more linear.

Should the high frequency response of the voice coil be required when designing a control system, the electrical and mechanical time constants as well as the inductance must be considered.

Figure 3.7 shows a range of voice coil actuators including both linear and rotary types, manufactured by USAS Motion.

Figure 3.7 Range of voice coil actuators manufactured, reproduced courtesy of USAS Motion

3.2.1.3 Biomechatronic applications

As shown earlier in this chapter, the relationship between force and stroke for solenoid actuators is very nonlinear. This limits their use in typical biomechatronic applications to operating latches engaging gears or driving pump diaphragms – applications that require only two states (in or out). In contrast, voice coil actuators provide a reasonably efficient method of generating linear motion, with open-loop positioning accuracies good to fractions of a millimetre. They can therefore be used to actuate prosthetic limbs, drive middle ear implantable hearing devices (MEIHDs), power ventricular assist devices (VADs) and even to control the focus in miniature optical prostheses.

3.2.2 DC motors

For a current-carrying conductor in a magnetic field, the magnitude of the force, F (N), depends on the total length, l (m), of the conductor in the field, the magnetic flux, B (Weber/m^2), and the current, I (A)

$$F = BIl \sin \alpha \tag{3.16}$$

where α (rad) is the included angle between B and I.

If this single conductor is replaced by a single turn coil mounted on a rotational axis O – O', the interaction of the current in coil segment AB with the magnetic field B will create a force F, shown downward in Figure 3.8. In the same way, the coil segment CD will result in a similar force but in the opposite direction because the direction of the current flow is reversed in this case.

Note that the current in the coil segment BC is parallel to the magnetic field making the net force on this section zero. It is also obvious that there is no net force on the coil because the two forces are the same size, but there is a resultant moment (torque) around the rotational axis.

If current continues to flow into the coil, it will rotate about the axis O–O' by 90° until the coil is orientated vertically and then stop because the forces in the

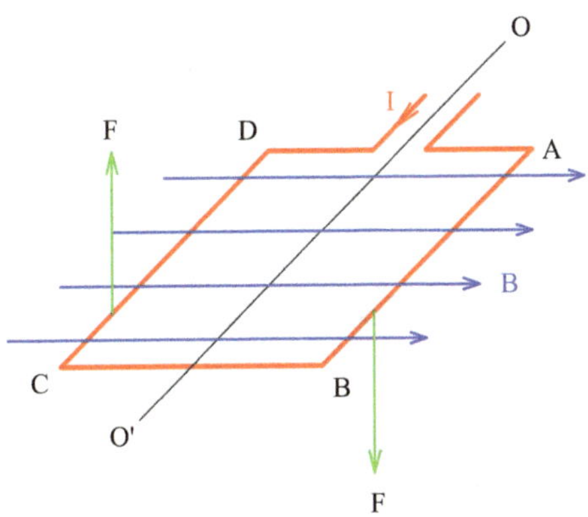

Figure 3.8 Application of Faraday's law for a single loop coil

conductors will pass through the rotation axis and the turning moment will have reduced to zero.

If the current was reversed, and the coil was just past the equilibrium point, then it would rotate by a further 180° and then stop again. To allow for continuous motion, some form of switching or commutation is required to automatically reverse the direction of the current at the correct angle. In brushed motors, these are performed using a mechanical method, while in brushless motors the position of the coil is sensed electronically using a hall switch (see Chapter 2), and commutation also occurs electronically.

3.2.2.1 Single coil DC motor

Mechanical commutation using brushes is shown in Figure 3.9. It can be seen that a pair of brushes remains in contact with the coil over angles from 0° to 90°. As with the previous figure, the force on the segment AB is downward, and the force on segment CD is upward. This produces a torque in the clockwise direction. There is a short period in the region around 90° where the brushes are disconnected, and the motor must rely on its momentum to carry it past this region. The sense of the current into the coil is now reversed with the force on CD being downwards and that on AB being upwards, which continues to produce a small torque in the clockwise direction. As the angle increases, the torque reaches a

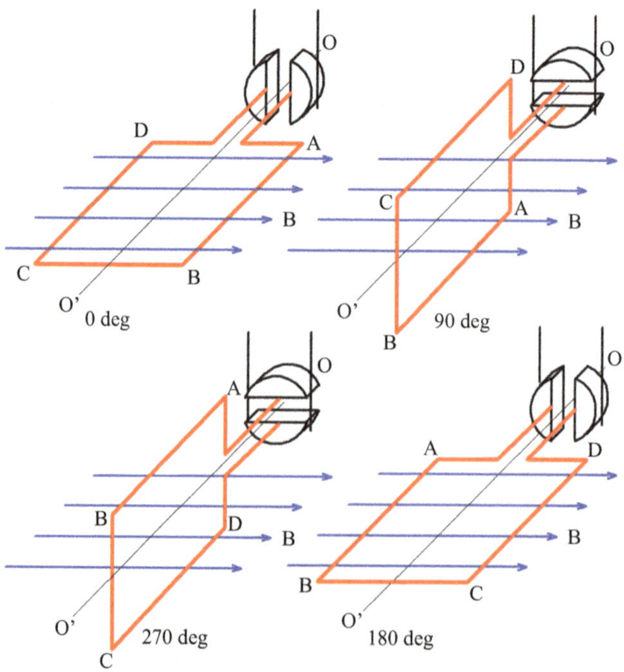

Figure 3.9 Commutation of a single-coil DC motor

maximum at 180° before decreasing to zero at 270° where the commutation switches again.

This crude motor is impractical for a number of reasons, the most important of which is the large variation in the torque experienced, as shown graphically in Figure 3.10.

3.2.2.2 Multiple coil DC motor

If a second coil with its commutation mechanics is added at 90° to the single coil, the amount of ripple will be reduced significantly, as can be seen in Figure 3.11. In real motors, the number of coils is generally increased well beyond two coil segments with the result that there is very little torque ripple.

A second problem with these motors is the fact that they rely on a single turn, with the result that the amount of torque generated is limited. Real motors have multi-turn coils to maximise the torque output.

For a coil with N turns, the maximum force generated by N parallel but insulated conductors running along the segment AB is

$$F = NBIl \tag{3.17}$$

The peak torque generated in the coil as a result of this force, F (N), is the product of the force and the distance from AB to the rotational axis O–O'. If this

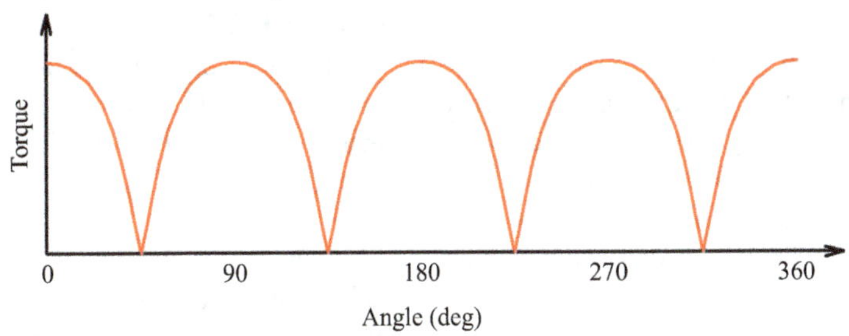

Figure 3.10 Torque variation in a single-coil DC motor

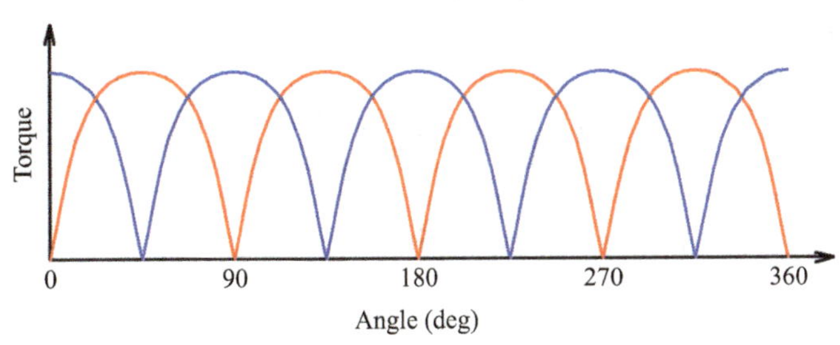

Figure 3.11 Torque variation in a two-coil DC motor

distance is R (m), then the peak torque, τ_p (Nm), is

$$\tau_p = RNBIl \tag{3.18}$$

Because the same amount of torque is developed by conductors running along segment CD, the peak torque will be doubled

$$\tau_p = 2RNBIl \tag{3.19}$$

For any manufactured motor, R,N,B and l are fixed, leaving the user control of the current I (A), only. Therefore, it is convenient to specify a motor in terms of the relationship between the current and the torque. This is known as the torque constant, K_m (Nm/A)

$$K_m = \frac{\tau}{I} = 2RNBl \tag{3.20}$$

One further characteristic of DC motors is important. It was shown earlier that when a conductor of length, l (m), cuts through a magnetic field with strength B (Weber/m^2), at a velocity, v (m/s), an EMF, ε (V), is developed across the conductor.

$$\varepsilon = Blv \tag{3.21}$$

If the same assumptions are adopted as were for the development of the motor, it can be shown that an EMF will develop across the brushes if an external torque is applied to rotate the coils physically. The magnitude of this voltage is dependent on the motor configuration and the angular velocity. This allows a constant of proportionality that describes the relationship between the angular velocity and the voltage to be determined. It is generally known as the back EMF constant, K_e (V per rad/s).

When a DC motor that is not driving a load is supplied with a voltage V_i (V), its speed will increase until the back EMF just equals the applied voltage. This is known as the no-load speed, ω_n (rad/s). As the load increases and the motor is required to provide an increasing amount of torque, the motor speed will decrease in a reasonably linear manner until the motor stalls and the torque is a maximum. This is known as the stall torque, τ_s (Nm). In other words, there is a trade-off between how much torque a DC motor can generate and how fast it spins, as shown in Figure 3.12.

It is important to realise that the graphs shown in Figure 3.12 are for a specific applied voltage. If the applied voltage is altered, then the slope of the line remains the same, it is just displaced vertically as shown in Figure 3.13.

Given the two points at the extremes of the graph, an equation can be written describing the torque in terms of the rotation speed, or the motor speed in terms of the torque

$$\tau_{\text{mot}} = \tau_s - \frac{\omega \tau_s}{\omega_n} \tag{3.22}$$

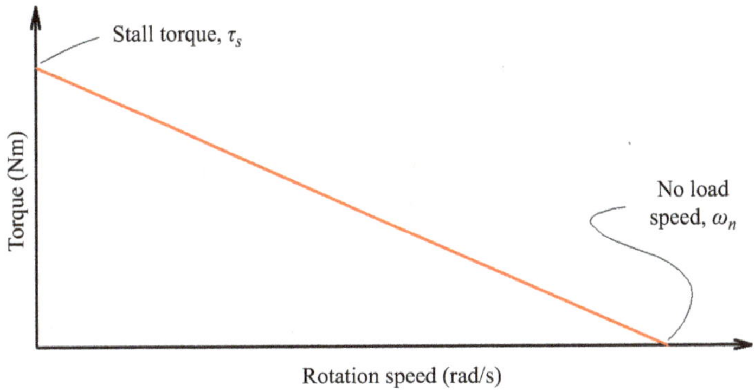

Figure 3.12 Relationship between motor speed and torque for a DC motor

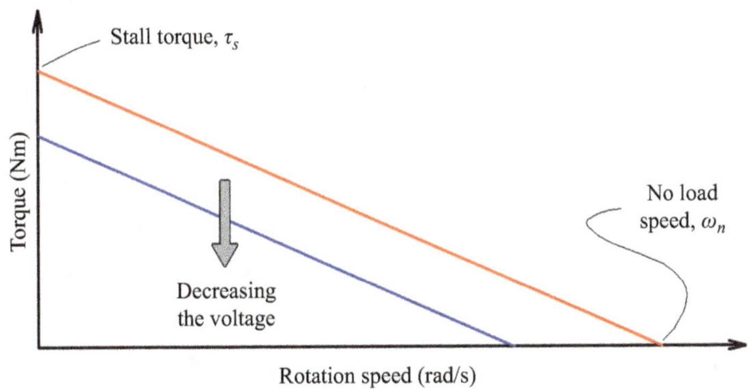

Figure 3.13 Effect of changing the applied voltage

and

$$\omega_{\text{mot}} = \frac{(\tau_s - \tau)\omega_n}{\tau_s} \tag{3.23}$$

The output power P_{mot} (W) supplied by the motor is equal to the product of the torque and the speed.

$$P_{\text{mot}} = \omega\tau \tag{3.24}$$

Therefore, at the two extremes, no power is output, and a maximum occurs at the centre of the graph where $\omega = \omega_n/2$ and $\tau = \tau_s/2$. This relationship which equates to the area of a rectangle below the curve is shown graphically in Figure 3.14.

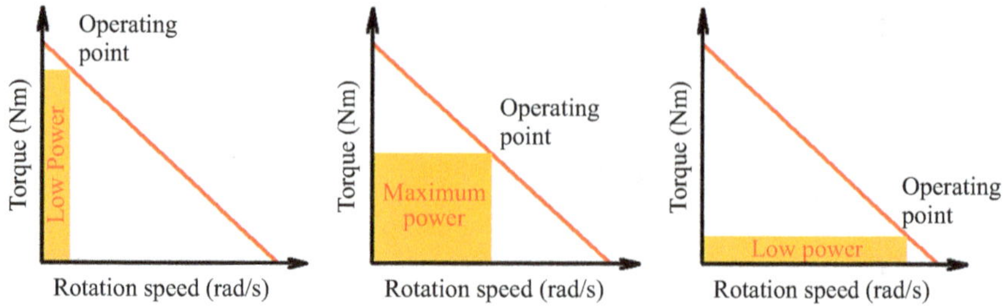

Figure 3.14 Relationship between speed, torque and output power

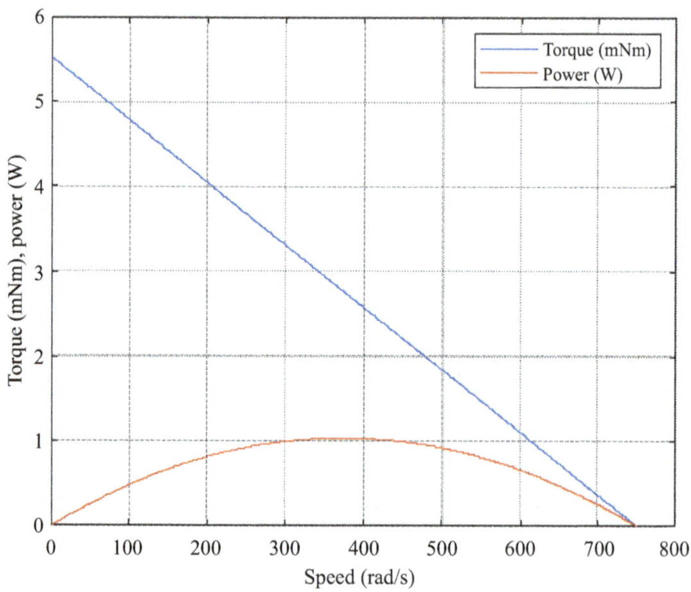

Figure 3.15 Maxon 2 W motor characteristics

3.2.2.3 Real motor characteristics

A 2 W Maxon motor designed to operate from 15 volts has a no-load speed of 7160 rpm (750 rad/s) and a stall torque of 5.53 mNm. The relationship between the torque, output power and rotation speed is shown in Figure 3.15. Note that the actual output power for this motor reaches a maximum of just over one watt when $\omega = \omega_n/2$ and $\tau = \tau_s/2$ as predicted.

```
% DC motor torque speed and power
% motor.m
rpm2rads = 2*pi/60;
taus = 5.53e-03;              % stall torque Nm
wn = 7160 * rpm2rads;         % no load speed (rad/s)
R = 53.3;                     % winding resistance (ohm)
Kt = 19.6/1000;               % torque constant (Nm/A)
V = 15;                       % Voltage (V)
Io = 4.91e-03;                % No load current (A)

% calculate the torque and power as a function of the speed
w = (0:1000)*wn/1000;         % points along the speed axis
tau = taus - w.*taus./wn;     % torque (Nm)
pmot = tau.*w;                % mechanical power (W)

% Plot the torque speed and power speed curves
plot(w,tau*1000,w,pmot)
grid
xlabel('Speed (rad/s)')
ylabel('Torque (mNm), Power (W)')
legend('Torque (mNm)','Power (W)')

% calculate the speed and power as a function of the torque
figure
tau = (0:1000)*taus/1000;     % points along the torque axis (Nm)
w = (taus-tau).*wn./taus;     % speed (rad/s)
pmot = tau.*w;                % mechanical power (W)

% Plot the speed torque and power torque curves
plot(tau*1000,w,tau*1000,pmot*1000)
grid
ylabel('Speed (rad/s), Power (mW)')
xlabel('Torque (mNm)')
legend('Speed (rad/s','Power (mW)')
```

The governing equation for the power curve can be obtained by substituting equation (3.11) or (3.12) into equation (3.13).

$$P_{\mathrm{mot}}(\omega) = -\frac{\tau_s}{\omega_n}\omega^2 + \tau_s\omega \tag{3.25}$$

$$P_{\mathrm{mot}}(\tau) = -\frac{\omega_n}{\tau_s}\tau^2 + \omega_n\tau \tag{3.26}$$

The current drawn by the motor is proportional to the output torque. The motor specifications include the no-load current I_o = 4.91 mA and the starting current, I_a = 281 mA.

The no-load current is equivalent to the friction torque τ_f

$$\tau_f = K_t I_o \tag{3.27}$$

where K_m (Nm/A) is the torque constant described earlier.

Motors develop the highest torque when starting, and this is much larger than the torque available when running. Therefore, the current required on start-up is also a maximum

$$\tau_s = K_m I_a \tag{3.28}$$

From the motor specifications, the torque constant K_m =19.6 mNm/A. This is in fact just the ratio of the stall torque and the starting current, $K_m = \tau_s / I_a$.

The friction torque can be calculated from the no-load current and the torque constant.

$$\begin{aligned}\tau_f &= K_m I_o \\ &= 19.6 \times 4.91 \times 10^{-3} \\ &= 0.096 \text{ mNm}\end{aligned}$$

The motor efficiency, η, describes the relationship between the mechanical power output by the motor and the electrical power supplied.

$$\eta = \frac{\omega \tau}{V_i I} = \frac{\omega K_m}{V_i} \tag{3.29}$$

It can be seen that the efficiency should increase with increasing speed (decreasing torque) for a fixed supply voltage. However, at low torque (high speed), friction losses become increasingly significant, and efficiency decreases rapidly. Maximum efficiency, $\eta_{\max}$, is calculated using the starting current and the no-load current and is dependent on the supplied voltage

$$\eta_{\max} = \left(1 - \sqrt{\frac{I_o}{I_a}}\right)^2 \tag{3.30}$$

For the specified motor

$$\begin{aligned}\eta_{\max} &= \left(1 - \sqrt{\frac{I_o}{I_a}}\right)^2 \\ &= \left(1 - \sqrt{\frac{4.91}{281}}\right)^2 \\ &= 0.75\end{aligned}$$

which is close to the specified efficiency of 76%. As a rule of thumb, the maximum efficiency occurs at roughly one seventh of the stall torque. This means that maximum efficiency and maximum output power do not occur at the same torque.

In addition to frictional losses (mechanical losses), there are copper losses due to the resistance of the windings, and in iron-core motors, losses due to magnetisation effects (electrical losses). Iron losses do not occur in modern coreless DC motors; therefore, the power balance can be written as

$$P_{\mathrm{el}} = P_{\mathrm{mot}} + P_{\mathrm{dis}} \tag{3.31}$$

where P_{dis} (W) are all the electrical losses that do not contribute to the generation of a torque.

This power balance can be expanded

$$V_i I = \omega\tau + I^2 R \tag{3.32}$$

For the 2 W Maxon motor, the resistance of the coil windings is R = 53.3 Ω, so the power dissipated can be determined. This is a maximum at start-up when the motor is still stalled

$$\begin{aligned} P_{dis} &= I_a^2 R \\ &= 281 \times 10^{-3} \times 53.3 \\ &= 4.2\ \mathrm{W} \end{aligned}$$

It exceeds the power rating of the motor by a factor of two; therefore, the motor cannot be operated in this regime for long periods without overheating.

If the motor operates at the peak power point where $\tau = \tau_s/2 = 5.53/2 = 2.765$ mNm, the current will be the sum of the load current and the no-load current. The load current is determined from the output torque and the torque constant

$$I = \frac{\tau}{K_m} + I_o \tag{3.33}$$

$$\begin{aligned} I &= \frac{2.765}{19.6} + 4.91 \times 10^{-3} \\ &= 0.1411 + 4.91 \times 10^{-3} \\ &= 0.1460\ \mathrm{A} \end{aligned}$$

This makes the electrical power dissipated in the coil windings

$$\begin{aligned} P_{dis} &= I^2 R \\ &= 0.146^2 \times 53.3 \\ &= 1.13\ \mathrm{W} \end{aligned}$$

The power dissipated in the windings should, in theory, be equal to the power output as mechanical energy at this point. The efficiency can be determined as the ratio of the output power to the total power

$$\begin{aligned} \eta &= \frac{P_{out}}{P_{out} + P_{dis}} \\ &= \frac{1.04}{1.04 + 1.13} \\ &= 0.48 \end{aligned}$$

The temperature rise is determined from the thermal resistance between the windings and the air. For the 2 W motor, this comprises two components: the thermal resistance between the housing and the air, R_{th1} = 33 °C/W, and that between the windings and the housing, R_{th2} = 7 °C/W. Under steady-state conditions, the temperature difference between the motor windings and the ambient is

$$\Delta T = P_{dis}(R_{th1} + R_{th2}) \tag{3.34}$$

The temperature of the windings will then be the sum of the ambient air temperature and the temperature rise

$$T_{mot} = T_{amb} + \Delta T \tag{3.35}$$

In the case of the stalled motor

$$\begin{aligned}\Delta T &= P_{dis}(R_{th1} + R_{th2}) \\ &= 4.2 \times (33 + 7) \\ &= 168\,^\circ\mathrm{C}\end{aligned}$$

The winding temperature will be

$$\begin{aligned}T_{mot} &= T_{amb} + \Delta T \\ &= 20 + 168 \\ &= 188\,^\circ\mathrm{C}\end{aligned}$$

which is more than 100 °C hotter than the allowed 85 °C. The motor will overheat in this stalled condition and the windings will burn!

For the motor operating at the peak power point

$$\begin{aligned}\Delta T &= P_{dis}(R_{th1} + R_{th2}) \\ &= 1.13 \times (33 + 7) \\ &= 45\,^\circ\mathrm{C}\end{aligned}$$

The winding temperature will be

$$\begin{aligned}T_{mot} &= T_{amb} + \Delta T \\ &= 20 + 45 \\ &= 65\,^\circ\mathrm{C}\end{aligned}$$

which is more than 20 °C cooler than the allowed 85 °C, so the motor will be fine.

3.2.2.4 Examples of DC motor types

DC motors are probably the commonest of all the actuators used by biomechatronic engineers. This is because they are extremely efficient in converting electric power to mechanical power and there is a wide variety of types from which to choose.

A conventional low-cost iron core permanent magnet (PM) DC motor is shown in Figure 3.16. It can be seen that the armature (rotor) consists of a laminated iron core wound with insulated copper coils that are connected to a multi-contact commutator. The stator consists of two ferrite magnets within a cylindrical housing. End plates containing journal bearings hold the rotor between the magnets while allowing free rotation. Attached to one of the end plates is a brush assembly with metal brushes in contact with the commutator.

Modern high-performance motors made by companies like Maxon, Portescap or Faulhaber are mostly ironless as this reduces the inertia and increases the space available for coil windings. As can be seen from Figure 3.17, the rotor windings are

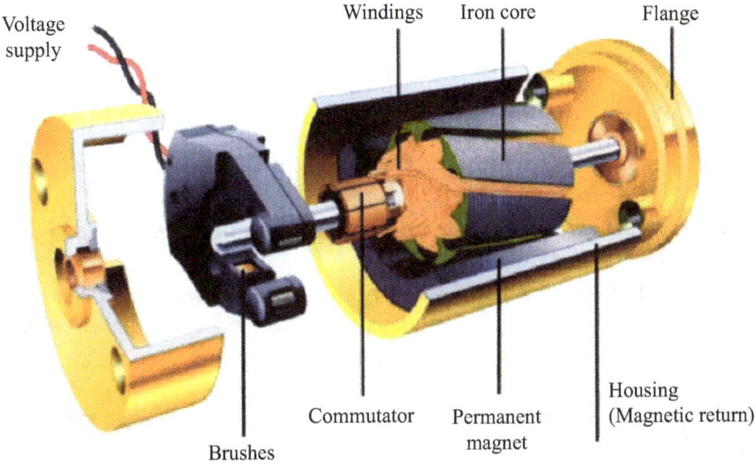

Figure 3.16 Typical permanent magnet DC motor, creative commons

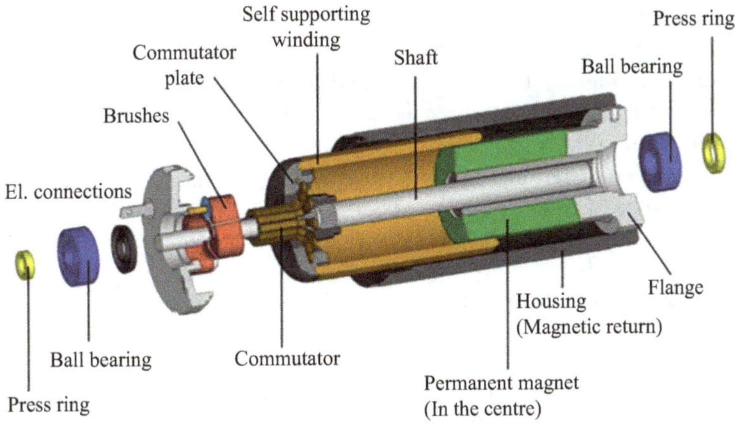

Figure 3.17 Construction of an ironless DC motor, creative commons

wound in a skewed, or honeycomb fashion to form a self-supporting hollow cylinder with the stator magnet in the centre. This construction places the windings close to the outside of the motor for improved heat dissipation.

Mechanical commutation has always been one of the major disadvantages of DC motors. This results in wear and hence limits the operational life of the motor; it generates metal or graphite particles and electromagnetic interference. Graphite brushes are generally used in larger motors for stop-start operation and if the motor controller uses pulse width modulation (PWM). Precious metal brushes ensure a highly constant and low contact resistance and so are good for low power continuous operation or for use in DC tachometers.

3.2.2.5 Selecting a DC motor

Selecting a DC motor for a particular application can be a rather involved process if undertaken comprehensively. However, it is not difficult to obtain a reasonable idea of the performance requirements from which to make a reasonable choice.

The following considerations can be important:

- Constant current operation of a DC motor produces constant output torque regardless of speed.
- Given a constant load (torque requirement), the motor speed is determined solely by the applied voltage.
- Power is the product of speed and torque. Maximum power is produced at an operating point that is defined by half the no-load speed and half the stall torque. A motor will seldom be operated at maximum output due to thermal considerations.
- As a rule of thumb, DC motors are typically operated at 70% to 90% of the no-load speed and from 10% to 30% of the stall torque. This is the region of maximum efficiency.
- For DC motors operated at a constant voltage, the higher the torque output, the lower the speed will be.
- Other factors to consider are size, environmental factors, weight and required life.

3.2.2.6 Example

As an example, consider the motor required to drive the turbine for a CPAP air pump. The nominal torque required is 3.5 mNm at a speed of 5000 rpm. A DC supply voltage of 20 V is available.

As a first step, determine the output power. Determine the conversion factor for rotational speed from rpm (n) to rad/s (ω), $2\pi/60 = 0.1047$

$$\begin{aligned} P_o &= 0.1047 n\tau \\ &= 0.1047 \times 5000 \times 3.5 \times 10^{-3} \\ &= 1.8 \text{ W} \end{aligned}$$

The motor should be rated at least 1.5 to 2 times the desired output power in relation to the maximum available at the nominal voltage. A motor with a maximum output power of between 2.7 and 3.6 W should be sufficient.

Table 3.2 Maxon RE 16 motor specifications

No load speed @ 24 V n_o (rpm)	7250
No load current I_o (mA)	3.11
Terminal resistance R (Ω)	42.8
Torque constant K_m (mNm/A)	31.4
Speed Constant K_e (rpm/V)	304
Output power P_o (W)	3.2
Maximum winding temperature, $T_{\max}$ (°C)	85
Thermal resistance: housing ambient, R_{th1} (°C/W)	30
Thermal resistance: winding housing, R_{th2} (°C/W)	8.5

Referring to the Maxon catalogue, the smallest motor to achieve this power rating is the RE 16 series (16 mm diameter × 40 mm long). The 24 V version has the closest operating voltage to the 20 V available. Some of the motor specifications are reproduced in Table 3.2.

As a first approximation, the no-load speed can be found by taking the specified value and scaling it by the ratio of the voltages

$$\begin{aligned} n_{20} &= \frac{20}{24} n_o \\ &= \frac{20}{24} \times 7250 \\ &= 6041 \text{ rpm} \end{aligned}$$

Since the desired speed is 5000 rpm, this represents 82% of the no-load speed, which falls well within the 70% to 90% range, and so the motor should operate at close to optimum efficiency.

The current through the motor is the sum of the no-load current and the load current

$$\begin{aligned} I &= \frac{\tau}{K_m} + I_o \\ &= \frac{3.5}{31.4} + 3.11 \times 10^{-3} \\ &= 114.6 \times 10^{-3} \text{A} \end{aligned}$$

The volt drop across the motor windings is the product of the motor current and the terminal resistance. The reduction in motor speed is equal to the product of this volt drop and the speed constant; therefore, the motor speed is

$$\begin{aligned} n &= n_{20} - IRK_e \\ &= 6041 - 114.6 \times 10^{-3} \times 42.8 \times 304 \\ &= 4550 \text{ rpm} \end{aligned}$$

This is reasonably close to the desired value; therefore, the motor will probably be suitable.

One final check is to determine whether the motor will be able to accommodate the heat rise in the coils. The power dissipated in the motor is

$$
\begin{aligned}
P_{dis} &= I^2R \\
&= (114.6 \times 10^{-3})^2 \times 42.8 \\
&= 0.6\ \text{W}
\end{aligned}
$$

The heat rise under steady state conditions is determined by the product of the dissipated power and the thermal resistance from the rotor coils to the air

$$
\begin{aligned}
\Delta T &= P_{dis}(R_{\text{th1}} + R_{\text{th2}}) \\
&= 0.6 \times (30 + 8.5) \\
&= 23.1\ ^\circ\text{C}
\end{aligned}
$$

Therefore, the motor winding temperature will be the sum of the ambient plus the temperature rise

$$
\begin{aligned}
T_{\text{mot}} &= T_{\text{amb}} + \Delta T \\
&= 20 + 23.1 \\
&= 43.1\ ^\circ\text{C}
\end{aligned}
$$

This is well below the 85 °C maximum operating temperature for the motor and so the motor will be capable of running indefinitely under these conditions (MicroMo 2008).

3.2.2.7 Powering DC motors

The simplest form of motor control is open loop. The voltage, or sometimes the current, is set to some preset value determined by the required output speed given a specific load torque. The voltage can be provided by a general-purpose linear or switched-mode regulator as discussed in Chapter 2, but it is more usual to develop specific electronics for motor control. Linear regulators are very inefficient, particularly if the motor speed needs to be controlled over a wide range. Therefore, most motor controllers use PWM to provide the required average DC voltage as shown in Figure 3.18.

PWM is achieved by connecting the DC supply to the motor at regular intervals to provide the required mean voltage. Because the switching rate is very high, the motor inductance and its rotational inertia ensure that it runs at a constant speed. The relationship between the switching frequency, f (Hz), the period, T (s) and the duty cycle $\eta_{\%}$ is as follows:

$$T = \frac{1}{f} \tag{3.36}$$

$$\eta_{\%} = 100\frac{t}{T} \tag{3.37}$$

where t (s) is the on period as shown in the figure.

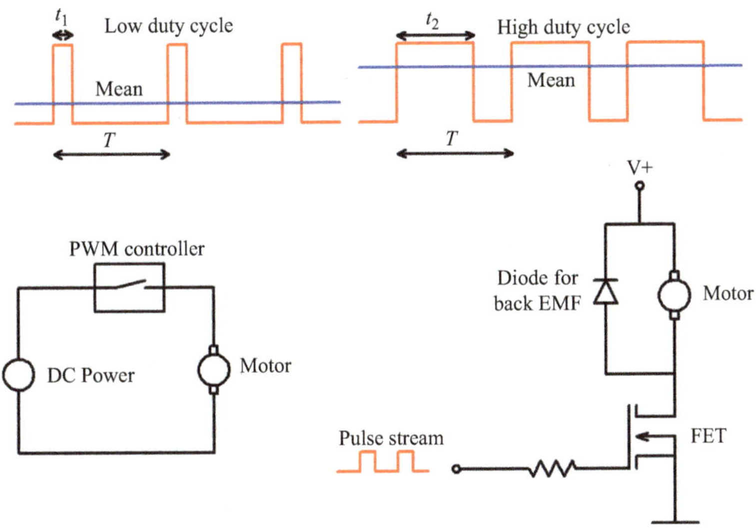

Figure 3.18 Pulse width modulation of a DC motor

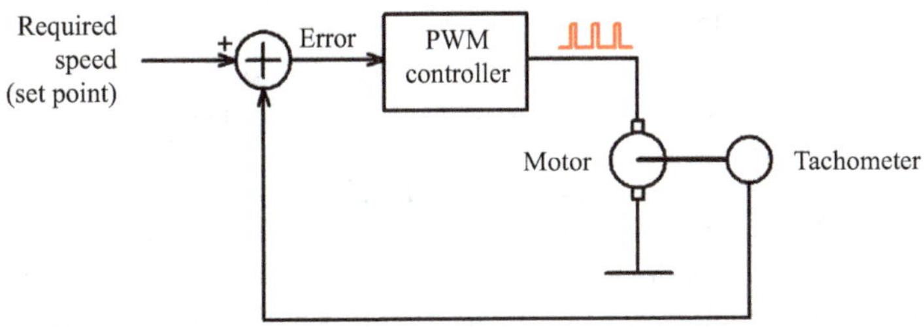

Figure 3.19 PWM motor speed control

Because of the high switching rate required, typically more than 1 kHz PWM-controllers do not use physical switches or relays, but power transistors as shown in the partial schematic.

To maintain a constant speed, or a specific motion profile that is torque dependent, closed loop or feedback control is generally used. This involves sensing the motor speed using one of the sensors discussed in Volume 1, Chapter 2 and using that to control the duty cycle of the PWM signal so as to maintain the required speed. This principle is illustrated in Figure 3.19.

Simple PWM controllers cannot reverse the direction of rotation of a motor unless a dual rail supply is used. This is often inconvenient, particularly if the motor is large and requires a high current, so a more common alternative is to use a switching configuration that can reverse the polarity of the motor supply. This configuration, known as an H-bridge for obvious reasons is shown in Figure 3.20.

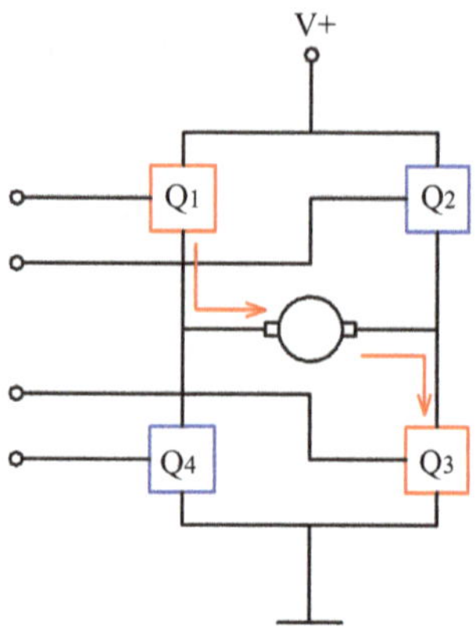

Figure 3.20 H-bridge control of a motor

The H-bridge uses four power transistors that are independently controlled to ensure that the appropriate pulse width is applied to the motor with the correct polarity. If transistors Q_1 and Q_3 are on simultaneously, then current will flow through the motor in the direction shown in the figure. However, if Q_2 and Q_4 are on, then the current flow is reversed, and the motor will turn in the opposite direction. Care needs to be taken to ensure that Q_1 and Q_2 or Q_3 and Q_4 are never conducting simultaneously, or they will provide a direct path to earth with catastrophic consequences.

Motor controllers can be built up using discrete components, but several sophisticated ICs are available that will probably perform the function much better. A good example is the National Semiconductor LMD18201 shown in Figure 3.21. This H-bridge can handle 3 A continuous and 6 A peak at a motor voltage of up to 55 V. In addition, it has several protection circuits including thermal and overcurrent.

As shown in the figure, it requires few external parts and is easily interfaced to an external microcontroller.

3.2.3 Brushless DC motors

Modern brushless DC (BLDC) motors are very similar in construction to permanent magnet AC synchronous motors. As shown in Figure 3.22, the brushless motor comprises a permanent magnet rotor, a stator winding and a number of hall or optical sensors to detect the rotor angle. These control current flow into the stator windings to turn the rotor.

The simple cylindrical three-phase motor shown in Figure 3.23 is commutated by a separate hall-effect device for each of the three stator phases. The changing position of the permanent magnet rotor is sensed by the hall device as the polarity of the passing rotor pole changes. This hall signal is amplified so that the stator coils are driven with the proper current.

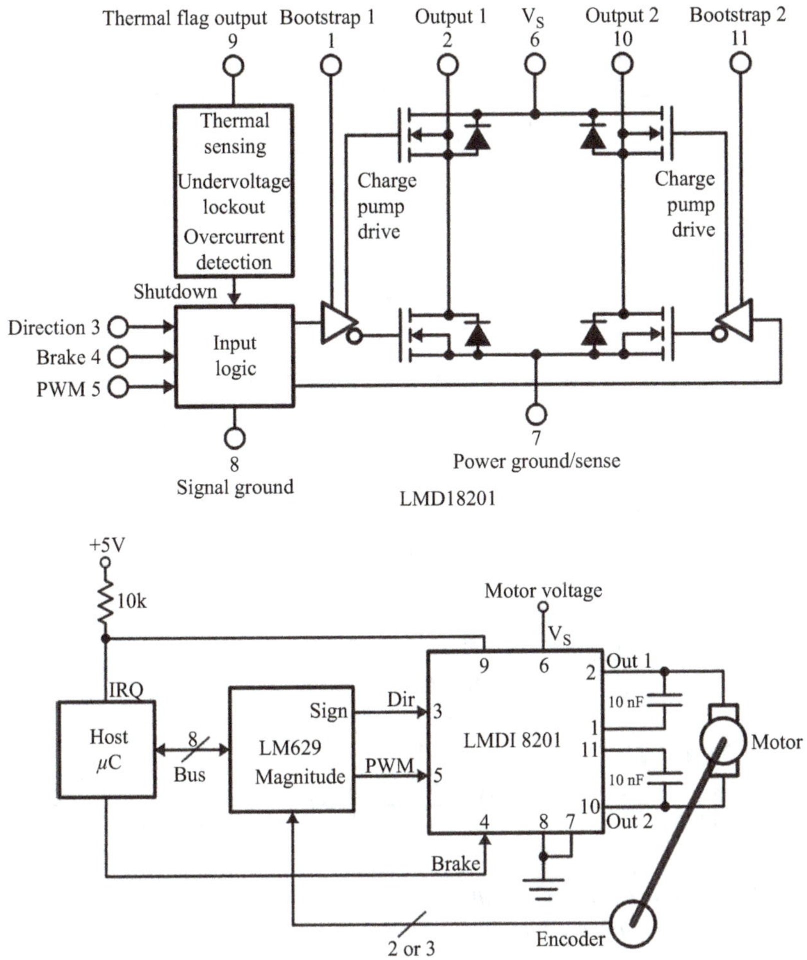

Figure 3.21 National LMD18201 H-bridge and application, reproduced courtesy of National Semiconductor

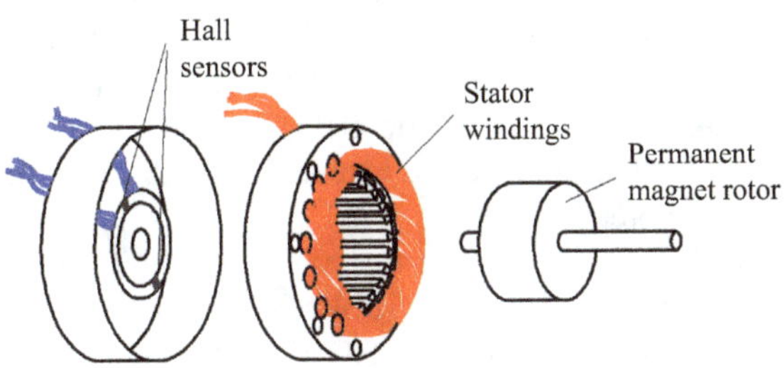

Figure 3.22 Exploded view of a brushless DC motor [adapted from Kuphaldt (2008)]

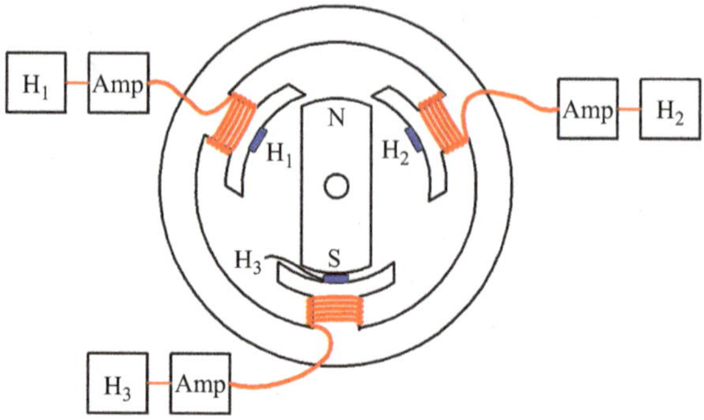

Figure 3.23 Simple BLDC motor [adapted from Kuphaldt (2008)]

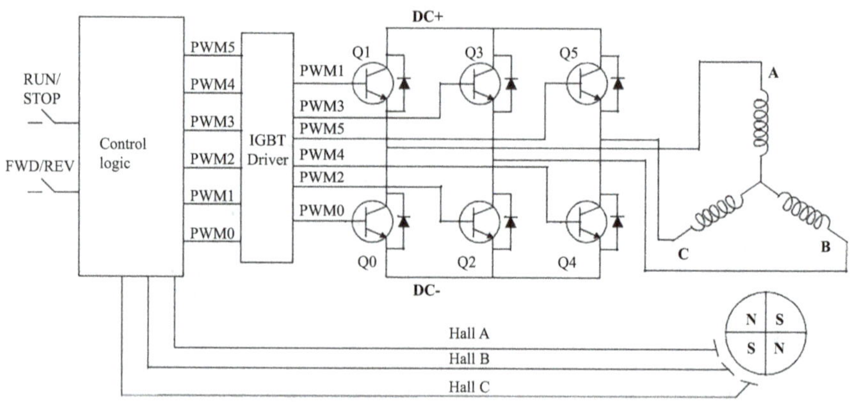

Figure 3.24 Controller for a 3-phase brushless DC motor (Yedamale 2003), reproduced courtesy of Microchip Technology

Most commercial BLDC motors have three-phase windings connected in a star topology as illustrated in Figure 3.24. A motor with this topology is driven by energising two phases at a time. The key to efficient operation is to sense the rotor position and then energise the phases to produce the most torque in the desired direction. The appropriate stator is energised when the rotor is 120° from alignment with the corresponding stator's magnetic field, and deactivated when the rotor reaches 60° from alignment. This process is repeated for the next circuit, and so on, until a full cycle is complete.

The hall sensors have digital outputs that are high for 180° of rotor rotation and then low for the remainder. They are physically separated by 60° so that an unambiguous digital code is produced for every 60° of rotation as shown in Figure 3.25.

Note that each drive phase consists of one motor winding driven high, one driven low and the last one left floating. A necessary precaution with this kind of driver is that both the high and low side drivers must never be actuated

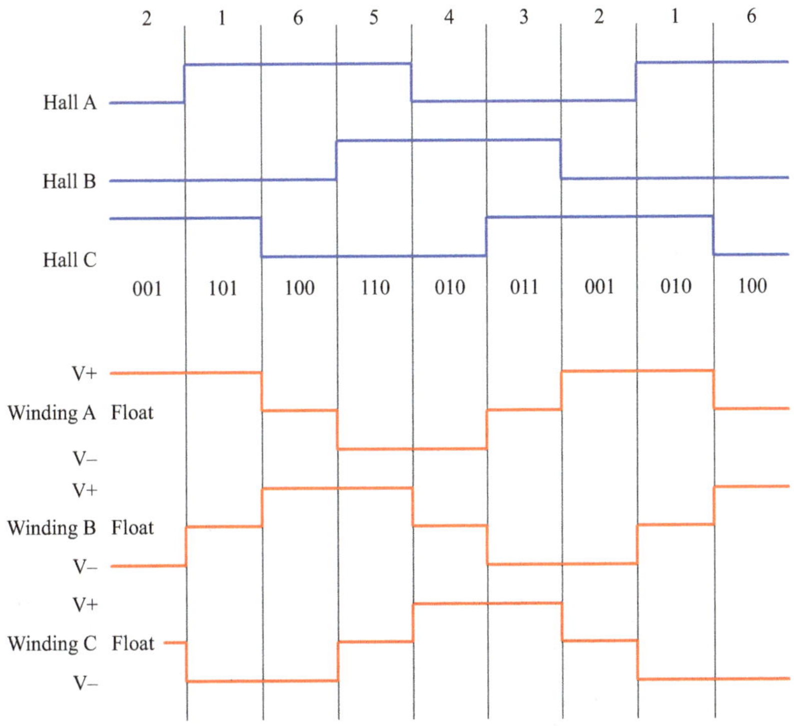

Figure 3.25 Hall sensor and drive timing [adapted from Brown (2002)]

simultaneously. This is often achieved using "dead time" control in which extra time is allowed for one driver to turn off before the next one is activated. In this configuration, there is a floating phase period between high and low drive periods so that the problem will never occur.

The torque-speed characteristics of BLDC motors are different to those of DC motors. There are two torque parameters, the stall torque, τ_s, and the rated torque, τ_r. During continuous operation, the motor can be loaded up to the rated torque which remains constant for a speed range up to the rated speed as shown in Figure 3.26. The motor can be run faster than this, up to about 150% of the rated speed, but the torque drops. In addition, the motor can generate torques higher than the rated value, but these must be limited in duration otherwise the motor can overheat.

It is possible to control a BLDC without hall sensor feedback. If the back EMF of each of the windings is monitored, it will transition through zero at about the same time that the hall sensor state would change. This transition can be used to control the commutation as discussed above.

Though BLDC motors are much more complicated than normal brushed types, they are often used in demanding applications because of their improved reliability and compactness. Table 3.3 provides a comparison between the main features of BLDC and brushed DC motors.

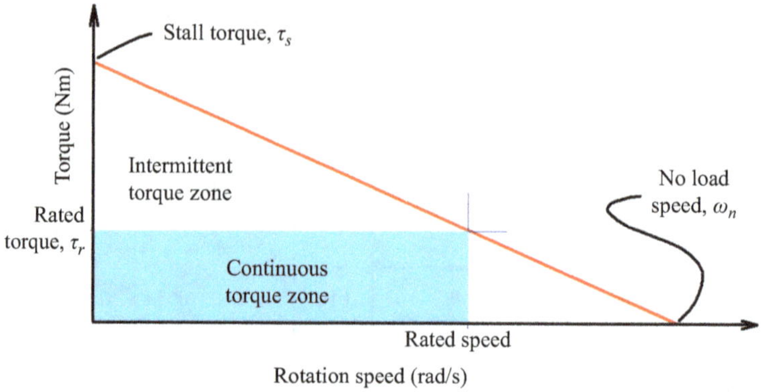

Figure 3.26 Torque-speed characteristics for a brushless DC motor

Table 3.3 Comparison between a brushless DC motor and a brushed DC motor (Yedamale 2003)

Feature	BLDC Motor	Brushed DC Motor
Commutation	Electronic commutation – hall sensors	Mechanical commutation - brushes
Maintenance	Less required due to the absence of brushes	Periodic maintenance
Life	Longer	Shorter
Speed/Torque	Flat – operation at all speeds with rated load	Moderately flat – at higher speeds brush friction increases, reducing output torque
Efficiency	High – no voltage drop across brushes	Moderate
Output Power/ Frame Size	High – good thermal characteristics because BLDC has windings on the outside in contact with the frame	Moderate to low – heat dissipation across the air gap
Rotor Inertia	Low because the rotor is only a permanent magnet	Higher inertia for iron rotor motors, lower for ironless types
Speed range	Higher – no mechanical limitation imposed by brushes	Lower – mechanical limitations of brushes
EMI	Low	High – brush arcing causes EMI
Cost	Higher – permanent magnet rotor expensive to manufacture	Low
Control	Complex and expensive	Simple and inexpensive
Control requirements.	A controller is always required to run the motor, though the same controller can also be used for speed control	No controller is required for fixed speed, a controller is required for variable speed operation

3.2.3.1 Selecting a BLDC motor

The parameters that govern motor selection for a particular application are (Yedamale 2003):

- Peak torque required for the application,
- RMS torque required,
- The range of operating speeds required.

Peak Torque: This can be thought of as the same as the stall torque and can be determined by summing the required load torque, τ_l, the inertial torque, τ_j, and the torque required to overcome friction, τ_f. Other factors also contribute to the peak torque requirements. They include windage loss that is contributed by the resistance of the air in the gap. These additional factors are difficult to calculate, so it is easier to add a safety factor of 0.2 when doing the calculation.

$$\tau_s = 1.2\left(\tau_l + \tau_j + \tau_f\right) \tag{3.38}$$

The inertial torque, τ_j, is the torque required to accelerate the load from standstill to operational speed. It is the product of the load plus rotor inertias, J_{l+m} (kg m^2), and the required acceleration, α (m/s^2).

$$\tau_j = J_{l+m}\alpha \tag{3.39}$$

The load and frictional torques are mostly determined by the mechanical system coupled to the motor shaft.

Root Mean Square Torque: The RMS torque can be thought of as the average torque requirements during normal operation. This depends on the duty cycle of the system, the load torque and required accelerations etc. In an application where the acceleration time is t_a (s), the run time is t_r (s) and the deceleration time is t_d (s), the RMS torque is

$$\tau_{\text{rms}} = \sqrt{\frac{\tau_s^2 t_a + \left(\tau_l + \tau_f\right)^2 t_r + \left(\tau_j - \tau_l - \tau_f\right)^2 t_d}{t_a + t_r + t_d}} \tag{3.40}$$

A good example of this is a powered prosthetic arm that may be called upon to generate high peak torques for short periods but will have a low RMS value. It should therefore be possible to under-specify the motor to minimise its mass, with the knowledge that the duty cycle will be low.

Speed range: The motor speed required to drive the application is determined by that application. In an air pump for a CPAP application, speed variations are not very frequent, and the maximum speed required will be very close to the average. However, in the prosthetic arm example, the peak speed requirement will be much higher than the average.

3.2.3.2 BLDC motor types

Motors can, of course, be manufactured in all shapes and sizes to fulfil particular torque and speed requirements of specific applications. These include long pencil-thin motors

Figure 3.27 Examples of some BLDC motors from various manufacturers

through flat pancake motors examples of which are shown in Figure 3.27. Some of the more common DLDC motors are those used in hard-disc drives, CD/DVD drives and electric fans.

A good example of one of the latest generations of high-power-density motors is the EC16 from Maxon. It is 16 mm diameter × 56 mm long, can rotate at up to 60,000 rpm and produces a mechanical output power of 40 W.

Power densities higher than this are available from the electric motors used to power radio control aircraft. These typically rotate at about 12,000 rpm and provide output powers of up to 300 W in a package less than 30 mm in diameter and only 40 mm long.

3.2.3.3 Biomechatronic applications

Brushless DC motors are the actuators of choice for many biomechatronic applications. They are available in a wide range of sizes with associated gearheads for almost any speed-torque requirement, which makes them suitable for driving pulsatile artificial hearts, other pumps and prosthetic limbs.

3.2.4 Stepper motors

Stepper motors are DC motors comprising a permanent magnet or variable reluctance core that can rotate in precise angular increments in either direction and can sustain a strong holding torque at zero speed. These motors receive digitally controlled pulses correctly phased that cause them to step by an incremental angle that is typically between 1.8° and 30° per step. Micro-stepping circuitry can be designed to provide more than 10,000 steps per revolution (Alciatore and Histand 2003). There are three basic types of stepper motor, variable reluctance (VR), hybrid and permanent magnet (PM) types.

Hybrid and PM steppers are bidirectional and can be operated over a wide speed range, from incremental to speeds as high as 1800 rpm, and speeds can be fixed or variable. The mechanical parameters including shaft and mounting configurations as well as gearboxes and pulley arrangements can be specified to meet particular requirements.

A commercial stepper motor has many poles that provide many equilibrium positions on the rotor. In the case of the PM variety, the stator consists of wound poles and the rotor poles are permanent magnets. Exciting different stator winding combinations move and hold the rotor at different positions. The VR motor has a ferromagnetic rotor that relies on the fact that the equilibrium points are in positions of minimum magnetic reluctance. These motors generally have a lower inertia than the magnetic variety, but they lack a holding torque unless energised. Hybrids are a combination of the two, with a core like a VR type that encases a permanent magnet.

Hybrid motors are typically chosen for applications needing a fine step angle output of 1.8°. They are also more efficient than PM steppers but also carry a higher cost. PM steppers are normally available in step increments from 3.6° to 18° and are inherently more adaptable. They are frequently less expensive and enjoy broad acceptance in diverse medical equipment applications.

PM stepper motors typically range in size from 15 to 70 mm in diameter and with gear reductions can generate output torques of 15 Nm or more.

3.2.4.1 Operational principle

To understand how a stepper motor works, consider the diagram shown in Figure 3.28. It consists of four stator poles and a permanent magnet rotor. The magnetic field on opposite poles of the stator can be reversed. At Step 0, the motor is in equilibrium with the opposing end permanent magnetic fields holding the rotor. If the polarity of one pair of stator windings is reversed, the rotor is no longer in equilibrium and will start to rotate as shown until it reaches equilibrium at Step 1. The polarity of the opposite pair of stator windings is then reversed and the rotor rotates again until it reaches equilibrium at Step 2. This continues indefinitely.

The timing of the field reversals is important, particularly in multi-pole motors where the response can be under-damped, and it is easy to skip a step if the rate is too high. Damping can be increased, but even so the motor needs sufficient time to settle between steps.

3.2.4.2 Driving stepper motors

Stepper motors use two types of coil windings, unipolar or bipolar, which determine the kind of driver that is used. Unipolar coils incorporate bifilar windings with

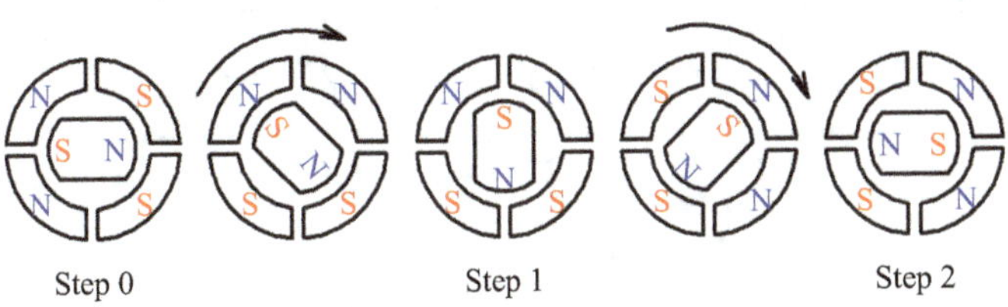

Figure 3.28 Stepper motor excitation sequence

one centre tap common to both coils. The centre tap is normally tied to positive voltage. Unipolar circuitry is more cost effective and is therefore used for high-volume applications. Bipolar coils contain one winding each with about the same number of turns found in unipolar coils, but with a larger wire gauge. This gives bipolar coils about 30% more torque, particularly at lower speeds. However, because it has no centre tap, a bipolar coil requires additional circuitry to switch the direction of the current. This is accomplished cost effectively with modern one-chip drivers.

Unipolar configurations use a single voltage rail but can still drive in both directions because of the way that the windings are configured, as seen in Figure 3.29. The conventional colour-codes for these stepper motors are indicated on the diagram.

To implement the appropriate magnetic configurations of the stator shown in Figure 3.28 each pole pair must be wound with a pair of complementary windings (bifilar windings) as illustrated in Figure 3.30. A useful result of this configuration is the ability to half-step, where only one pair of windings is excited. The switch sequences for full stepping and half stepping are listed in Table 3.4.

In the half-step mode, the resolution, or number of steps is twice that of the full step mode. In this case, it increases from 4 steps (90°) per revolution to 8 steps (45°) per revolution.

To improve the resolution, a process called micro-stepping can be used. Instead of operating the windings in a digital on/off mode, the current flow is

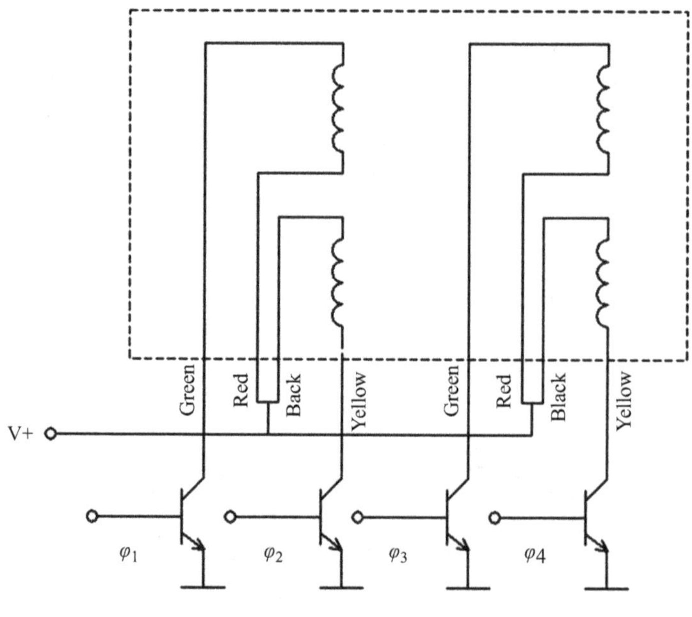

Figure 3.29 Schematic of a standard unipolar stepper motor drive configuration [adapted from Alciatore and Histand (2003)]

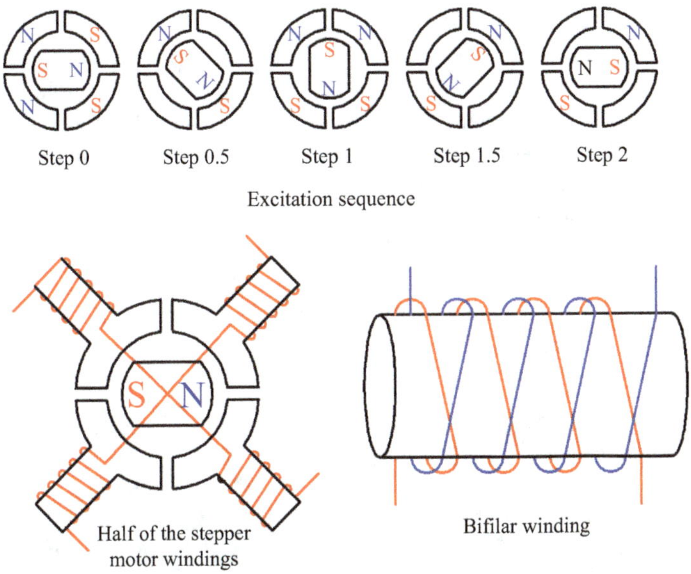

Figure 3.30 Operation of a unipolar stepper motor with bifilar windings

Table 3.4 Phase sequences for full and half-step drive

Step	φ_1	φ_2	φ_3	φ_4
1	On	Off	On	Off
2	On	Off	Off	On
3	Off	On	Off	On
4	Off	On	On	Off

Step	φ_1	φ_2	φ_3	φ_4
1	On	Off	On	Off
1.5	On	Off	Off	Off
2	On	Off	Off	On
2.5	Off	Off	Off	On
3	Off	On	Off	On
3.5	Off	On	Off	Off
4	Off	On	On	Off
4.5	Off	Off	On	Off

regulated to several different levels resulting in different magnetic equilibrium positions between the poles. In effect, discretised sine waves instead of square waves are applied to the poles with the result that a normal 200-step motor can be made to have 10,000 steps or more.

3.2.4.3 Real stepper motors

Figure 3.31 shows the photograph of a conventional stepper motor that has been dismantled. It is a hybrid configuration as it combines the variable reluctance toothed rotor with a permanent magnet core. In this case, the magnet is aligned along the axis of the motor so that one of the toothed sections is the south and the other is the north pole. The teeth in the two sections are misaligned by half a pitch.

In this configuration, no effective torque is generated by the magnetic field of the coils alone, and it is only in combination with the fixed magnetic field of the permanent magnet that a rotational torque is generated.

A simplified diagram showing the windings and the core of a typical stepper motor is shown in Figure 3.32.

Figure 3.31 Photograph of a dismantled stepper motor, reproduced courtesy of Arrow Electronics

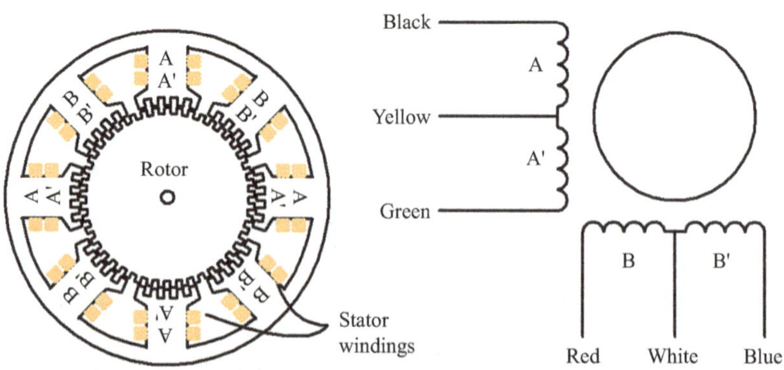

Figure 3.32 Typical stepper motor rotor and stator configuration

3.2.4.4 Stepper motor characteristics

Static characteristics: The stepper motor can be kept stationary by applying constant excitation to a single phase or a pair of phases. If an external torque is applied to the shaft, an angular displacement will occur. As the torque is increased, the angle will increase as will the resisting force until a maximum is reached. This relationship is known as the τ/θ characteristic of the motor. Once this is exceeded, the direction of the force will be reversed, and the rotor will be drawn to the next equilibrium point. The holding torque is defined as the maximum static torque that can be applied to the motor without causing continuous rotation.

The holding torque increases with current, and the relationship is known as the τ/I characteristic. This is typically very nonlinear for a variable reluctance motor, but reasonably linear for a hybrid, as shown in Figure 3.33.

Dynamic Characteristics: The torque-speed characteristics of stepper motors are nonlinear compared to DC motors. In addition, there are two different rotation modes as illustrated in Figure 3.34. The normal operational mode is the locked step mode within which the motor can be stopped, started and even reversed in the space of a single step. However, it is possible to continue to accelerate the motor by decreasing the step interval until the speed is too fast to allow for stopping and starting between steps. In this mode, the motor can only continue to travel in one direction and has to be decelerated into the lock-step region before it can be stopped and reversed.

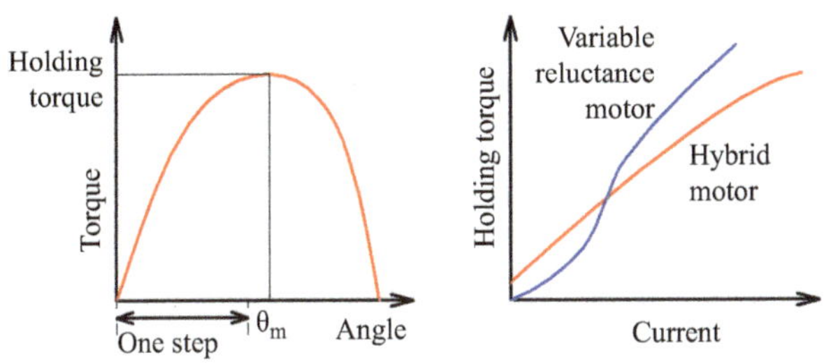

Figure 3.33 Static characteristics of a stepper motor

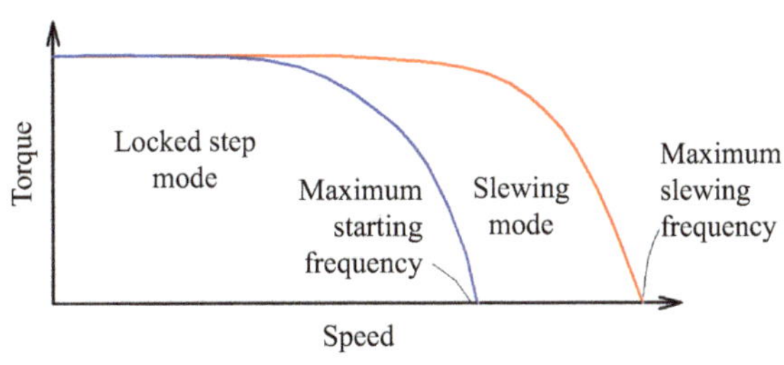

Figure 3.34 Stepper motor torque-speed curves [adapted from Alciatore and Histand (2003)]

At high speeds, the performance of stepper motors can often be analysed using the normal phasor analysis model that is used for synchronous AC motors.

Step response: At low speeds, each step is discernible and the overall behaviour is that of a series of step input transients. To determine their shape, the motor must be modelled as a set of nonlinear differential equations. For a variable reluctance motor, these equations are

$$\frac{di}{dt} = -\frac{1}{L}\left[R + \frac{dL(\theta)}{d\theta}\omega_r\right]i + \frac{1}{L}v \tag{3.41}$$

$$\frac{d\omega_r}{dt} = \frac{1}{J}\tau - \frac{1}{J}\tau_{load} \tag{3.42}$$

$$\frac{d\theta}{dt} = \omega_r \tag{3.43}$$

The motor response for a given step change on the input voltage can be obtained by solving these equations in conjunction with a number of initial conditions for the current, angle and angular rate. For a typical motor in isolation, the response is strongly under-damped, as illustrated in Figure 3.35 and for that reason it is often expedient to introduce some form of mechanical damping in the load.

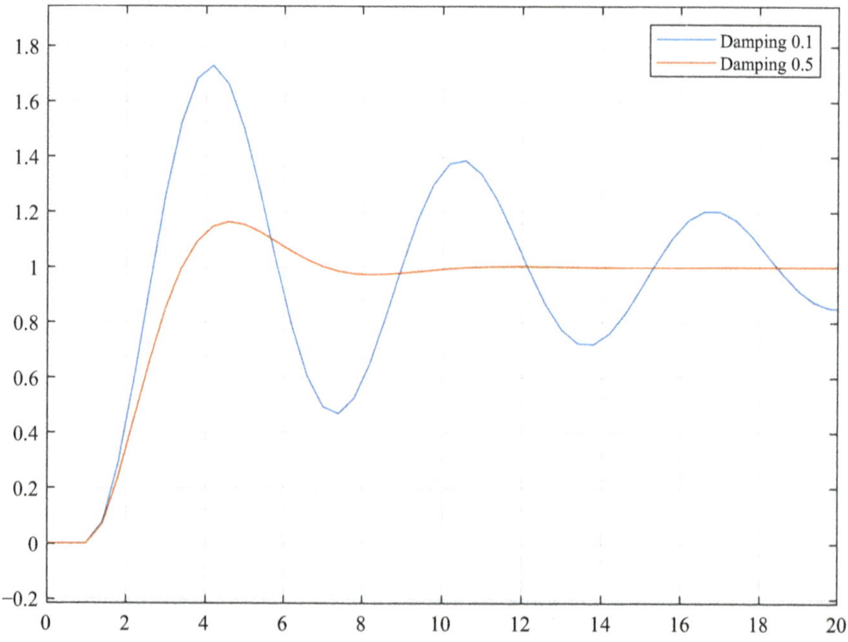

Figure 3.35 Step response without and with viscous inertial damping

Static Position Error: When a stepper motor is employed to drive a load, the equilibrium position will always be away from the ideal step position because the torque produced must balance the load torque. The difference between the actual equilibrium position and the ideal is defined as the static position error. It can be approximated by

$$\theta_{err} = \frac{\sin^{-1}\left(-\tau_l/\tau_{pk}\right)}{N_r} \tag{3.44}$$

where τ_l (Nm) is the load torque, τ_{pk} (Nm) is the holding torque and N_r is the number of steps per revolution.

Magnets: Major torque improvements can be made with stronger magnets, and upgrading ceramic magnets to rare-earth materials such as neodymium increases motor performance significantly. The strength of rare-earth magnets can be adjusted by varying the percentage of fillers used, typically boron, cobalt, or iron. Greater availability of rare-earth materials has recently driven rare-earth magnet prices to competitive levels. However, rare-earth magnets are more susceptible to damage from heat than their ceramic counterparts.

3.2.4.5 Stepper motor types

The advent of digital electronics and strong rare-earth magnets has been the primary driving force behind the proliferation of stepper motor types in a wide variety of applications.

The specifications of the Sanyo Denki 103-4902-0650 stepper motor shown in Figure 3.36 are listed as follows:

- Holding torque: 0.032 Nm,
- Step angle: 0.9°,
- Voltage: 6.6 V,
- Current/phase: 0.4 A,
- Ohm/phase: 16.5 Ω,
- Inductance/phase: 3.5 mH,
- Flange size: 39 mm^2 (1.54 in^2).

Figure 3.36 Sanyo Denki 103-4902-0650 Stepper Motor

Figure 3.37 Sanyo Denki 103H8222-0941 Stepper Motor

- Connection: 6 Lead,
- Rotor inertia: 0.009 kg m^2 ($\times 10^4$),
- Mass: 0.135 kg.

Specifications of the larger Sanyo Denki 103H8222-0941 stepper motor shown in Figure 3.37 are listed as follows:

- Holding torque: 4.13 Nm,
- Step angle: 1.8°,
- Voltage: 3.88 V,
- Current/phase: 4 A,
- Ohm/phase: 0.97 Ω,
- Inductance/phase: 3.60 mH,
- Flange size: 86 mm^2 (3.39 in^2),
- Connection: 6 Lead,
- Rotor inertia: 2.9 kg m^2 ($\times 10^4$),
- Mass: 2.5 kg.

3.2.4.6 Piezoelectric motors

As the name implies, piezoelectric motors rely on the change in shape of a piezoelectric material when an electric field is applied. As opposed to conventional piezoelectric actuators that rely on static strain to provide limited motion, these motors rely on a repeated cyclical process to provide unlimited motion. There are several different mechanisms used to achieve this, including resonant ultrasonic motors and non-resonant inchworm slip-stick motors.

With ultrasonic motors, travelling wave displacement around the stator contacts the rotor, and drags it around while inchworm types have a pair of actuators that operate sequentially in tandem to move the rotor. Slip-stick motors rely on the inertia of the rotor and differences between static and dynamic friction to move it (Spanner and Koc 2016). Two of these principles are shown in Figure 3.38.

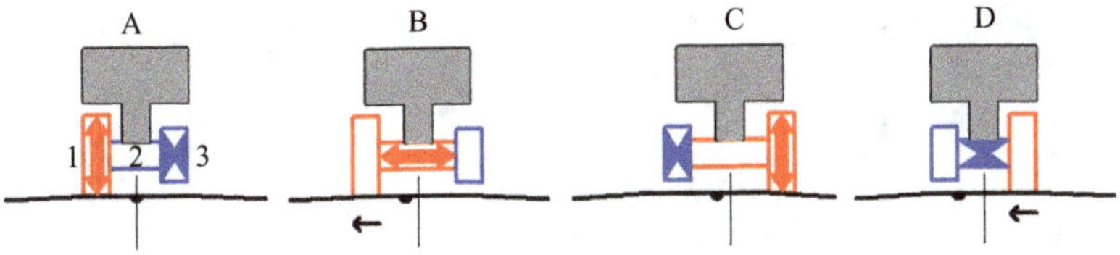

Piezoelectric element

	A	B	C	D
1	Expand	Hold	Release	—
2	—	Expand	Hold	Release
3	Release	—	Expand	Hold

(a)

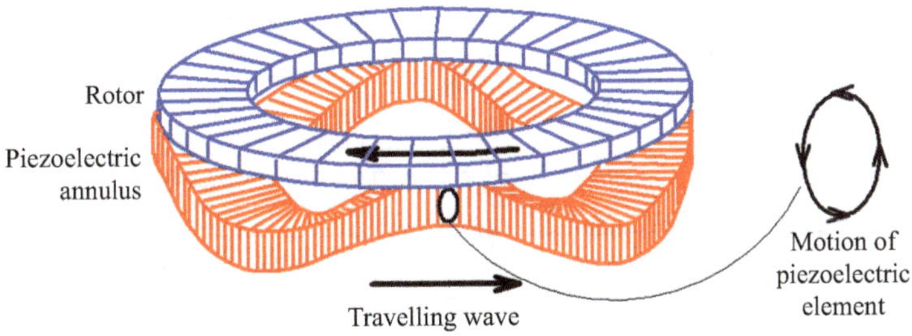

(b)

Figure 3.38 Working principles of travelling wave and inchworm piezoelectric motor types

Ultrasonic motors provide the following advantages compared to conventional electric motors:

- More efficient and so generate less heat for the same power output,
- Don't generate magnetic fields hence generate less interference,
- Very high positional accuracy compared to stepper motors,
- High holding force when turned off,
- Good dynamics and short response times,
- MRI compatible as they do not contain any ferromagnetic material,
- Rotor may be made of any material with a good friction coefficient,
- Potentially very low profile.

Specifications of a range of Tekceleo piezoelectric motors shown in Figure 3.39 are detailed in Table 3.5. These motors require a dedicated closed-loop position and speed controller and motor driver as adjacent pairs of elements of the piezoelectric annulus need to be provided with the correct quadrature voltage signal to produce the required travelling wave at the correct speed.

The torque-speed curves of these motors are similar to those of conventional stepper motors with little droop in the speed until the rated torque is reached, after which the speed drops off to zero at the maximum torque.

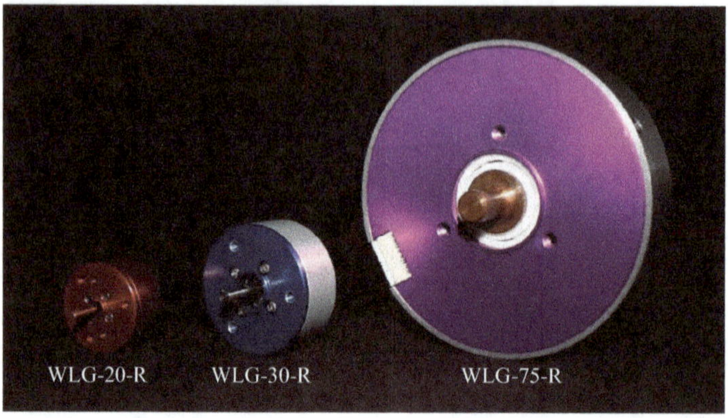

Figure 3.39 Range of travelling wave piezoelectric motors made by Tekceleo, reproduced with permission

Table 3.5 Specifications of a range of piezoelectric motors made by Tekceleo

Model	Rated power (W)	Maximum torque (mNm)	Rated torque (mNm)	Maximum speed (rpm)	Minimum speed (rpm)	Encoder precision (°)	Diameter (mm)
WLG-75-R	12	1200	750	175	1	0.0156	76
WLG-30-R	1.3	100	75	250	1	0.0450	30
WLG-20-R	0.55	30	20	380	1	0.0625	20

3.2.4.7 Biomechatronic applications

Peristaltic pumps provide medical facilities with accurate and repeatable pumping performance by using microprocessor-controlled stepping motors. They are used to pump blood for heart-lung machines and kidney dialysis units. By varying the diameter of the tubing or the step rate of the motor, the pump volume and rate can be easily changed and controlled (see Volume 2, Chapter 4). Various fluids can be processed using the same pump by simply changing the tubes and reprogramming the pump parameters. Nothing but the tube touches the fluid, thus eliminating the risk of cross-contamination between the pump and the fluid.

3.2.5 Linear actuators

There are many different methods of producing controlled linear motion, or linear positioning. Conventional means include air cylinders, ball screws, cables and pulleys, or even stacks of piezoelectric elements, but probably the most common is a stepping motor in which the shaft is replaced by a screw. The centre part of the rotor is tapped to convert it to a "nut" that is threaded onto a screw. By preventing the screw turning (anti-rotation), it can move axially to give the linear motion required. The increment, that is the linear progression corresponding to each step of the motor, is given by the screw pitch divided by the number of motor steps. This is selected to suit the actuator characteristics required. The longer the pitch, the quicker the advance but the lower the force of the device. Conversely, a fine pitch limits the advance rate but increases the force available.

Typical actuators of this type include the Portescap 26DAM series that provide linear force up to 33 N, linear step resolution of 0.025 mm, 0.05 mm and 0.1 mm and 3.4 W of power for fast response times and precise positioning capability. These linear actuators are available in captive and non-captive versions (see Figure 3.40), with unipolar or bipolar coil construction and industry-standard frame sizes.

Torque and efficiency calculations for this actuator type can be found in the section on translation screws towards the end of this chapter.

For higher performance applications that encompass longer life and smoother running, a ball-screw race usually replaces the simple "nut", as shown in Figure 3.41. In this case, the threaded rod rotates in place producing translation in the ball-screw race.

Figure 3.40 Non-captive stepper motor-driven linear actuator

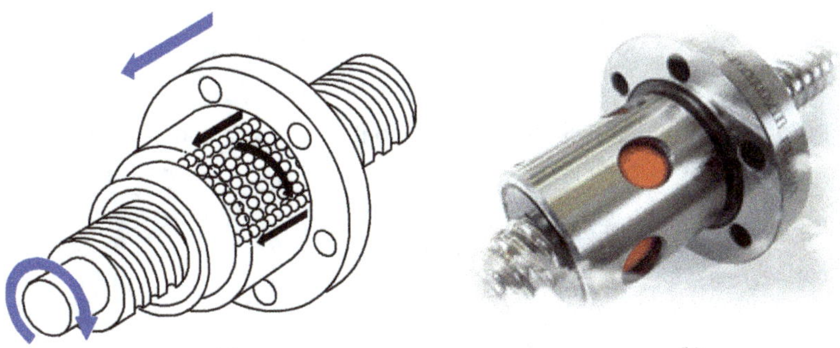

(a) (b)

Figure 3.41 Ball screw drive (a) drawing showing the flow of the bearings and (b) a photograph

(a) (b)

Figure 3.42 Worm gear drive actuator (a) photograph and (b) a schematic showing the internal mechanism

Another common drive mechanism uses a worm-gear to drive the linear actuator as shown in Figure 3.42.

3.2.5.1 Piezoelectric actuators

The piezoelectric effect is the generation of an electric charge by a crystalline material if it is stressed. It is caused by the stress-induced deformation of individual molecules within the lattice which exhibit changes in their charge distribution (also known as polarisation, not to be confused with the polarisation of light).

Natural materials such as quartz (SiO_2) as well as some synthetic ceramics and plastics exhibit this property. The latter are formed by poling (artificial polarisation), a process of aligning the dipole fields of the molecules in the lattice in the presence of a strong electric field, and then "freezing" them in place.

The piezoelectric effect is reciprocal insofar as applying a voltage will cause a strain in the crystal lattice and applying a strain to the lattice will generate a voltage across the faces. In each case, conductive electrodes on opposite faces are used to either apply the voltage or measure it.

The polarisation vector, P, of a piezoelectric material is the sum of the three orthogonal polarisation vectors (Fraden 1996)

$$P = P_{xx} + P_{yy} + P_{zz}. \tag{3.45}$$

Each of the individual vectors is related to the axial stress along the orthogonal axes

$$\begin{aligned} P_{xx} &= d_{11}\sigma_{xx} + d_{12}\sigma_{yy} + d_{13}\sigma_{zz} \\ P_{yy} &= d_{21}\sigma_{xx} + d_{22}\sigma_{yy} + d_{23}\sigma_{zz} \\ P_{zz} &= d_{31}\sigma_{xx} + d_{32}\sigma_{yy} + d_{33}\sigma_{zz} \end{aligned} \tag{3.46}$$

Where the constants d_{mn} (C/N) are the piezoelectric coefficients along the orthogonal axes of the material.

The force, F_z (N), produced by a piezoelectric material is proportional to the charge, Q (C). In the z-direction, this can be written as

$$F_z = \frac{Q_z}{d_{33}} \tag{3.47}$$

Because the crystal with its electrodes becomes a capacitor with capacitance C (F), the charge can be determined in terms of the applied voltage, V as follows

$$Q_z = CV \tag{3.48}$$

In turn, the capacitance is proportional to the electrode area, A (m^2) and inversely proportional to the thickness of the material, t (m)

$$C = \varepsilon\varepsilon_o \frac{A}{t} \tag{3.49}$$

where ε is the dielectric constant of the material and $\varepsilon_o = 8.8542\times10^{-12}$ C^2/Nm^2 is the permittivity of free space.

Substituting (3.48) and (3.49) into (3.47) gives the force in the z-direction as a function of the applied voltage

$$F_z = \frac{\varepsilon\varepsilon_o A}{d_{33} t} V \tag{3.50}$$

The displacement can be approximated from the generated force and the elastic modulus (Young's modulus), E (N/m^2) of the material. The relationship between the stress (F/A) and the strain ($\Delta t/t$) is

$$\frac{F_z}{A} = E\frac{\Delta t}{t} \tag{3.51}$$

Therefore

$$\frac{\Delta t}{t} = \frac{\varepsilon\varepsilon_o}{E d_{33} t} V \tag{3.52}$$

The magnitudes of the piezoelectric constant and Young's modulus for a number of common materials are given in Table 3.6.

Table 3.6 Properties of some piezoelectric materials

	Piezoelectric constant d **(C/N)**	**Relative dielectric constant**	**Young's modulus** E **(N/m^2)**
$BaTiO_3$	$d_{33} = 78\times10^{-12}$	1250 @ 20 °C	110×10^9
PZT	$d_{31} = 180\times10^{-12}$ $d_{33} = 110\times10^{-12}$	1700	83×10^9
PVDF	$d_{31} = 23\times10^{-12}$ $d_{33} = -33\times10^{-12}$	5-10	2×10^9
Quartz	2.3×10^{-12}	3.8	77×10^9

Figure 3.43 Piezoelectric actuators. Reproduced, with permission, from physikinstrumente (2008).

Piezoelectric actuators shown in Figure 3.43 are solid-state ceramic actuators that use this effect to convert electrical energy directly into linear displacement with an almost unlimited resolution. They rely on piezoelectric elements stacked in series with contacts sandwiched between each element. When a high voltage is applied to the elements, their individual lengths increase in the direction of the applied electric field by a small amount, resulting in a reasonably large total displacement (up to about 0.2% of the length of the actuator). Typical maximum displacements are between 5 μm and 200 μm with high force (up to 80 kN), high operational frequency and a long lifetime. Unfortunately, these transducers do suffer from creep as well as hysteresis and are therefore mostly operated in conjunction with an accurate displacement transducer in a closed-loop configuration.

For longer strokes, the same principle can be applied as is used in the piezoelectric motors discussed earlier in this chapter.

Other actuators operate on the bimorph principle. In this case, a slab of piezoelectric material is bonded to a metal backing. When a voltage is applied to the material, it will expand forcing the composite structure to bend in the same way as a bimetallic strip bends. Alternatively, slabs of material can be bonded to either side of a thin metal film to form a sandwich with the potential applied to one slab being of the opposite polarity to the potential applied to the other. These actuators can provide larger displacements than stacks, but the maximum force is reduced. The principle is used by ultrasound transducers (Brooker 2021).

The amount of deflection, Δx (m) at the tip of the cantilevered bimorph beam actuator shown in Figure 3.44 is

$$\Delta x = \frac{3Vd_{31}l^2}{4t^2} \tag{3.53}$$

where V (volts) is the applied voltage across the thickness of the material, d_{31} is the piezoelectric coefficient in the stretch (length) direction (see Chapter 2), l (m) is the length of the beam and t (m) is its thickness.

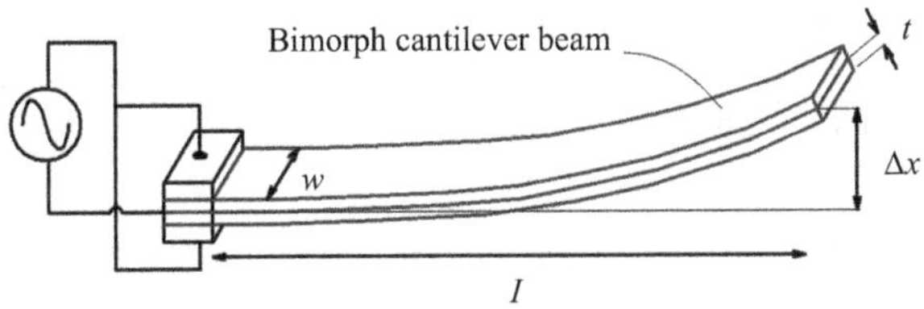

Figure 3.44 Configuration of a bimorph piezoelectric actuator

Similarly, the force, F (N), at the tip of the beam is

$$F = \frac{3VEwd_{31}t}{2l} \tag{3.54}$$

where E (N/m^2) is Young's modulus, and w (m) is the width of the beam

Worked Example 1:

Consider a typical low voltage actuator 50 mm high and 10 mm in diameter comprising 500 PZT elements each 100 μm high (including electrode thickness). A DC voltage of 20 V is applied to each element of the stack.

The force generated by each element can be determined from equation (3.50) and is proportional to the area. Therefore, calculate the area A (m^2)

$$A = \frac{\pi d^2}{4} = \frac{\pi \times (10 \times 10^{-3})^2}{4} = 78.6 \times 10^{-6}\text{m}^2$$

Now calculate the force per unit volt

$$\begin{aligned} F_z &= \frac{\varepsilon\varepsilon_o A}{d_{33}t} \\ &= \frac{8.85 \times 10^{-12} \times 1700 \times 78.6 \times 10^{-6}}{110 \times 10^{-12} \times 100 \times 10^{-6}} \\ &= 107.5 \text{ N/V} \end{aligned}$$

So, for an applied voltage of 20 V, the force is 2.15 kN, per element, and because the elements are in series, the total force is equal to the force of one element.

The strain can be determined from the force, the cross-sectional area and Young's modulus for the material

$$\begin{aligned} \frac{\Delta t}{t} &= \frac{F}{AE} \\ &= \frac{2.15 \times 10^3}{78.6 \times 10^{-6} \times 83 \times 10^9} \\ &= 329.6 \times 10^{-6} \end{aligned}$$

For a stack length of 50 mm, the change in length is $50 \times 10^{-3} \times 329.6 \times 10^{-6} =$ 16.5 μm

Note that the maximum applied voltage is governed by the electrical breakdown strength of the material that is between 1 kV and 2 kV per mm. In this example, the field strength is 20 V over 100 μm that is only 200 V/mm.

Worked Example 2:

Consider a bimorph cantilever 20 mm long consisting of two strips of 9 μm PVDF film. Calculate the displacement at the tip of the beam if 100 V is applied across the film.

$$\begin{aligned}\Delta x &= \frac{3Vd_{31}l^2}{4t^2} \\ &= \frac{3 \times 100 \times 23 \times 10^{-12} \times \left(20 \times 10^{-3}\right)^2}{4 \times \left(9 \times 10^{-6}\right)^2} \\ &= 8.5\ \text{mm}\end{aligned}$$

What is the force at the tip of the beam if the width of the beam is 10 mm?

$$\begin{aligned}F &= \frac{3VEwd_{31}t}{2l} \\ &= \frac{3 \times 100 \times 2 \times 10^{9} \times 10 \times 10^{-3} \times 23 \times 10^{-12} \times 9 \times 10^{-6}}{2 \times 20 \times 10^{-3}} \\ &= 31\ \mu\text{N}\end{aligned}$$

3.2.5.2 Bellows actuators

The metal bellows is essentially a hermetically sealed spring. When a pressure differential is applied across it, the bellows will compress and thus drive the actuator. When the pressure is removed, the bellows spring force will retract the actuator. In contrast to the conventional system that consists of the bellows and a pneumatic actuator, this construction provides a compact prime mover actuator as shown in Figure 3.45.

The relationship between the applied pressure, the displacement and the force for pneumatic and hydraulic actuators is described in more detail later in this chapter.

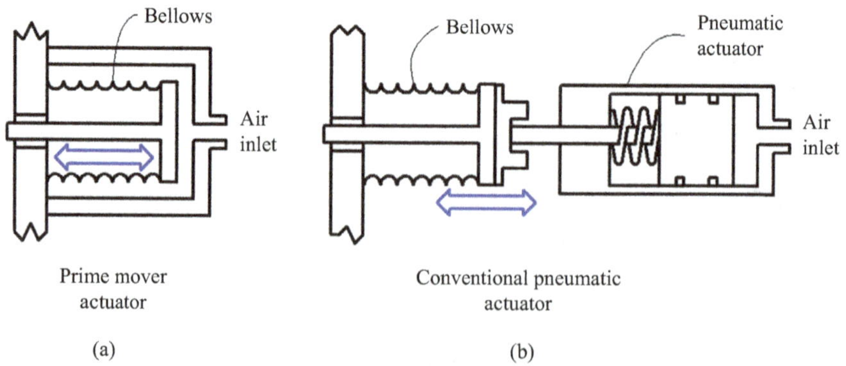

Figure 3.45 Pneumatic actuators

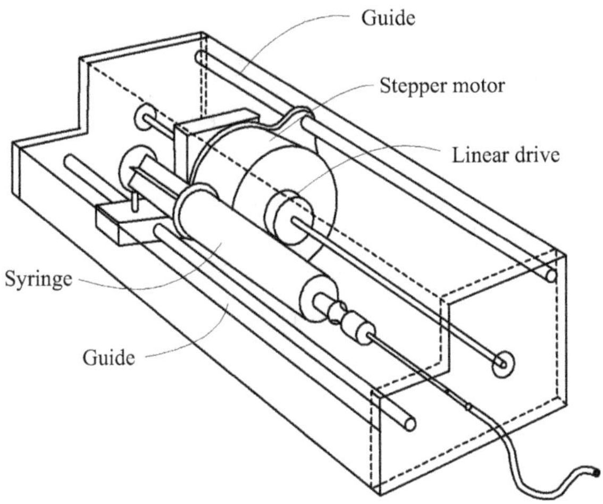

Figure 3.46 Precision syringe pump using a linear drive stepper motor

3.2.5.3 Biomechatronic applications

The stepper-motor-based syringe pump shown in Figure 3.46 automatically dispenses a precisely controlled amount of medication. Unlike an injection that is administered by a doctor in seconds, the microprocessor-controlled linear actuator can dispense medication over long periods of time at precise rates and volumes. Additionally, unlike an intravenous feeder that relies on gravity to dispense liquids, the syringe pump controls flow using pressure. The two major factors, flow rate and dispensed volume, are controlled by a stepper motor.

Electronic pipettes are also linear-stroke devices that utilise linear actuators to control the amount of fluid being dispensed. The shaft of the actuator is coupled to a pipette fitted with a piston. A microprocessor controls the step rate of the actuator that moves the shaft and piston in precise linear increments. Most modern electronic pipettes are portable handheld devices with built-in batteries and microcomputers.

3.2.6 Servo motors

A servomechanism, colloquially referred to as a servo is a device using error-sensing feedback to control the performance of a mechanism. Servo mechanisms often use a servo motor, but this is not necessary, as long as feedback is used to control some characteristic of the system. For example, an insulin pump driven by measured glycogen levels in the blood can be considered to be a servo mechanism.

This section will only be concerned with servo systems that use a servomotor to control position. These are of particular importance to active prostheses that are mostly powered by electric motors.

3.2.6.1 Radio control servos

The most common servo mechanisms are those which were originally developed for radio-controlled devices such as aircraft or cars, but which are now used for a

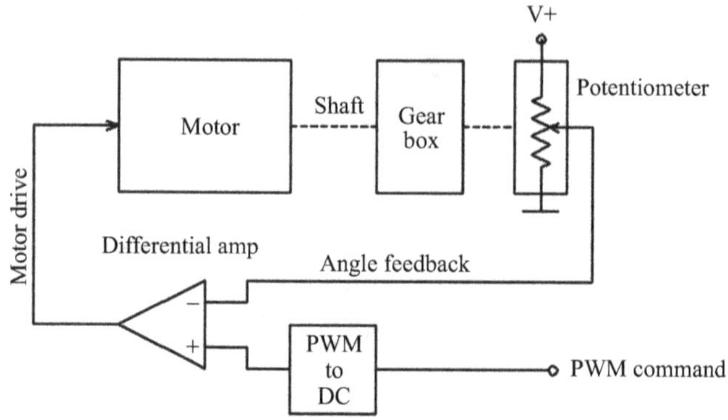

Figure 3.47 RC-type servo feedback mechanism

wide variety of precision position control applications. They consist of a DC motor mechanically linked to a potentiometer through reduction gears. Pulse-width modulated signals transmitted to the servo are translated into position commands by the internal electronics. The DC motor then drives the output shaft until feedback from the potentiometer reaches the commanded value. A schematic diagram illustrating the process is shown in Figure 3.47.

Due to their low cost, good reliability and the ease with which they can be controlled by microprocessors, RC servos are used in many mechatronic applications. Power and control are provided through 3 wires which are usually coloured as follows:

- Red: DC power (4.8–6 V),
- Black: Ground,
- Brown: PWM Position control.

The servo is provided with a pulse every 20 ms (or less) with a width that varies from 1.25 ms to 1.75 ms. These correspond to servo angles of 0° and 180° respectively. The neutral position that corresponds to an angle of 90° is provided by a pulse width of 1.5 ms. Physical limits to the angle of travel vary depending on the manufacturer but they are generally larger than the electrical limits.

Pulse width to voltage converter: The control pulse is fed to a pulse-width to voltage converter. This circuit charges a capacitor at a constant rate while the pulse is high. When the pulse goes low, the charge on the capacitor is output via a suitable high-impedance buffer amplifier. This produces a voltage related to the length of the applied pulse.

The circuit is tuned to produce a useful voltage over a 1 ms to 2 ms period. The output voltage is buffered to ensure that it does not decay significantly between control pulses, so the length of time between pulses is not critical. However, it should not exceed 20 ms for reliable operation.

Position sensor: The angular position of the servo motor output shaft is read by a sensor. This is normally a potentiometer that produces a voltage that is related to the absolute angle of the output shaft. The position sensor feeds one input of the error amplifier which compares the current position with the commanded position from the pulse width to voltage converter.

Error amplifier: The error amplifier is an operational amplifier with negative feedback. It will always try to minimise the difference between the inverting (negative) and non-inverting (positive) inputs by driving its output in the correct direction, as discussed in the section on op-amps in Chapter 5.

The output of the error amplifier is either a negative or positive voltage representing the difference between its inputs. The greater the difference between the two inputs, the greater the output voltage.

The error amplifier output has sufficient drive capability to power the motor. If the output is positive, the motor will turn in one direction, if negative the other. This allows the error amplifier to reduce the difference between its inputs (thus closing the negative feedback loop) and cause the servo to turn to the commanded position. It is essential that some form of filtering be applied to the loop; this can be electrical, or intrinsic to the mechanical configuration. Without it, the system response could be underdamped.

Different torques are provided by different RC servos, depending on their application. These range from about 0.6 kg-cm (58.8 mNm) to 24 kg-cm (2.35 Nm) for RC hobby types shown in Table 3.7 and Figure 3.48. They are determined by the motor size and its torque as well as the gear ratio of the gearbox.

As with most control systems, there is a tradeoff in performance among several parameters. In the case of these RC servos, the tradeoff is between the step response time (which should be fast) and the output torque (which should be high). Servo response is usually given as the time taken to step from 0° to 60°.

Table 3.7 Specifications of several Hitec servos

Model	Features	Weight (g)	Speed 0°–60° step (s)	Torque (kg cm)
HS50		6.1	0.09	0.6
HS-65MG	Metal gear, analogue	11.2	0.11	2.2
HS-82MG	Metal gear, analogue	19.0	0.10	3.4
HS-5645MG	Metal gear, analogue	60.0	0.18	12.1
HS-5955TG	Titanium gear, digital	62.0	0.15	24

Figure 3.48 Assorted servos, reproduced courtesy of Hitec

Futaba makes several high-speed, high-torque servos for robotic applications shown in Figures 3.49 and 3.50. These operate at 6 V and have the following specifications:

3.2.6.2 Professional servos

Hobby servos are obviously not suitable for biomechatronic applications where good reliability and long life are essential. In these applications, high-quality motors such as those made by Maxon and Faulhaber etc., should be used. In the past, a servo system would be assembled using a motor-gearbox combination with a separate encoder and power electronics but now integrated systems such as the micro drive system, described below, are starting to appear.

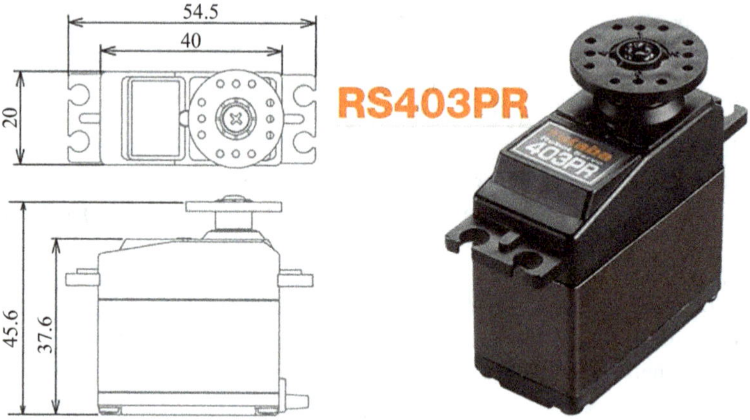

Figure 3.49 Futaba RS403PR digital coreless servo specifications

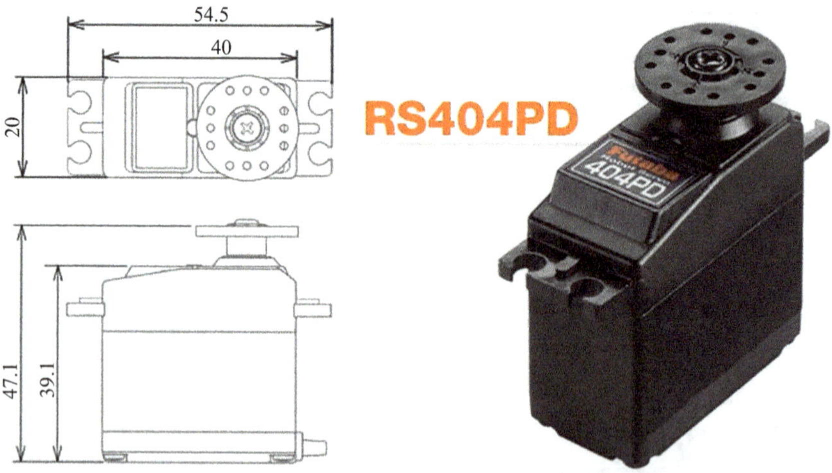

Figure 3.50 Futaba RS404PD digital iron core servo

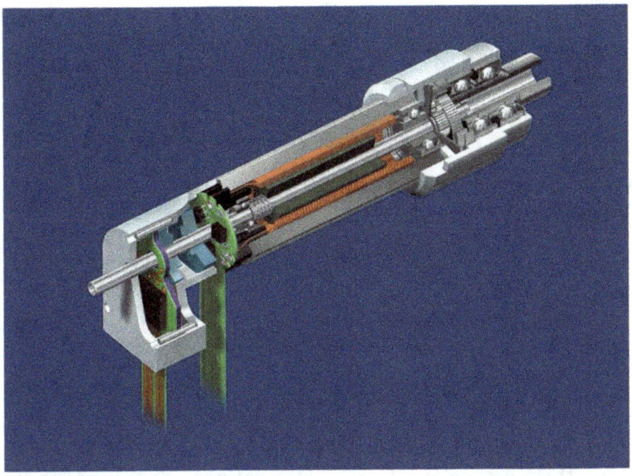

Figure 3.51 Maxon micro-drive servo motor, reproduced courtesy of Maxon

The micro-drive system shown in Figure 3.51 consists of a 6 mm BLDC motor, magnetic encoder and backlash-free harmonic drive gearhead. Supplying a constant 1.2 W, the motor can turn at up to 100,000 rpm and weighs only 2.8 g.

The backlash-free gearhead has a diameter of 8 mm and consists of a 1:160 reduction ratio harmonic drive and high-resolution 100-count incremental encoder. The positioning unit's shaft is hollow, and the drill-hole diameter is 0.65 mm.

3.2.7 AC motors

There are many different types of alternating current motors of which the most common are induction motors, hysteresis motors and synchronous motors. This section will outline the operation of induction motors as they are probably those that will be encountered most often in the biomechatronics field.

3.2.7.1 Induction motors

In an induction motor, the stator's magnetic field induces an alternating current into the rotor squirrel-cage conductors which constitute a shorted transformer secondary winding. This induced rotor current in turn creates a magnetic field. The rotating stator magnetic field interacts with this rotor field which attempts to align with it. The result is rotation of the squirrel-cage rotor. If there were no load torque, no bearing, windage, or other losses, the rotor would rotate at the synchronous speed. However, the slip between the rotor and the synchronous speed stator field causes the magnetic flux to cut through the rotor conductors and this develops torque. Thus, a loaded motor will slip in proportion to the mechanical load. If the rotor were to run at synchronous speed, there would be no stator flux cutting the rotor conductors, no current induced in the rotor, and hence no torque.

Motor Speed: The rotation rate of a stator rotating magnetic field is related to the number of pole-pairs per stator phase. In the full-speed case shown in Figure 3.52,

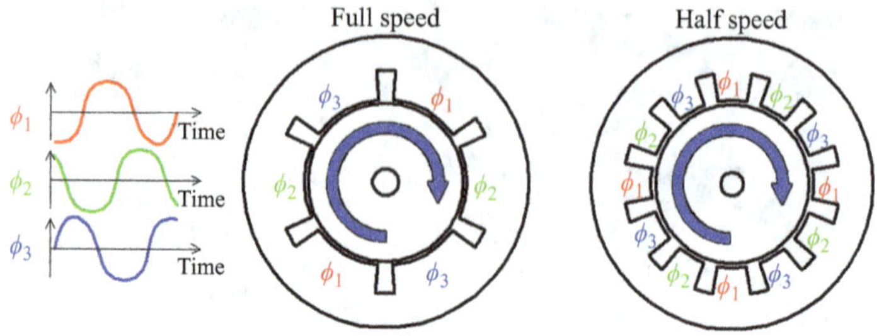

Figure 3.52 Three-phase induction motor with one and two pole-pairs per phase [adapted from Kuphaldt (2008)]

there are three pole-pairs and three phases. The magnetic field rotates once per sine wave cycle. For 50 Hz power, it rotates at 50 rotations per second, or 3000 rpm, this is the synchronous speed of the motor. Though the rotor of an induction motor never achieves this speed in practice, it is the upper limit of what is theoretically possible.

If the number of motor poles is doubled, the synchronous speed, N_s (rpm), is divided by two because the magnetic field rotates 180° in space for 360° of the electrical sine wave. The relationship is

$$N_s = \frac{60f}{n_p} \tag{3.55}$$

where f (Hz) is the mains frequency and n_p is the number of pole-pairs per phase.

Therefore, in Figure 3.52, the rotation rate for the half-speed option is

$$\begin{aligned} N_s &= \frac{60f}{n_p} \\ &= 60 \times 50/2 \\ &= 1500 \text{ rpm} \end{aligned}$$

because there are two pole-pairs per phase.

Torque: When power is first applied to the motor, the rotor is at rest, while the stator magnetic field rotates at the synchronous speed N_s. The stator field is cutting the rotor at the synchronous speed N_s and the current induced in the rotor shorted turns is a maximum, as is the frequency of the current. As the rotor speed increases, the rate at which stator flux cuts the rotor is the difference between synchronous speed N_s and actual rotor speed N, or $(N_s - N)$. The ratio of actual flux cutting the rotor to synchronous speed is defined as slip, s

$$s = \frac{N_s - N}{N_s} \tag{3.56}$$

The frequency of the current induced into the rotor conductors is only as high as the line frequency at motor start and decreases as the rotor approaches

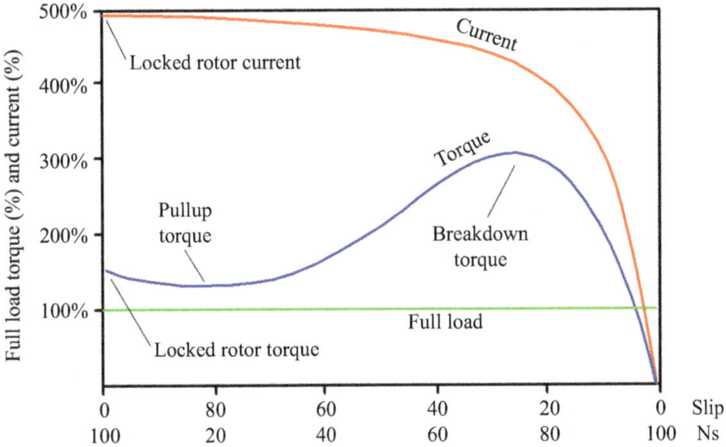

Figure 3.53 Torque and current vs slip for an induction motor [adapted from Kuphaldt (2008)]

synchronous speed. This rotor frequency, f_r (Hz) is

$$f_r = sf \tag{3.57}$$

Slip at 100% torque is typically 5% or less in induction motors. Thus, for $f = 50$ Hz line frequency, the frequency of the induced current in the rotor $f_r = 0.05 \times 50 = 2.5$ Hz.

Figure 3.53 shows the torque, speed and current relationships of a typical induction motor. It can be seen that the starting torque, known as locked rotor torque (LRT), is higher than 100% of the full load torque (FLT), the safe continuous torque rating. The LRT is about 175% of FLT for the example shown.

Starting current, known as locked rotor current (LRC), is 500% of full load current (FLC), the safe running current. The current is high because this is analogous to a shorted secondary on a transformer. As the rotor starts to rotate, the torque generally decreases slightly to a value known as the pull-up torque. This is the lowest value of torque ever encountered by the starting motor. As the rotor gains 80% of synchronous speed, torque increases from 175% up to 300% of the FLT. This breakdown torque is due to the larger than normal 20% slip. The current has decreased only slightly at this point but decreases rapidly beyond it. As the rotor accelerates to within a few percent of synchronous speed, both torque and current decrease sharply.

Slip will be only a few percent during normal operation. For a running motor, any portion of the torque curve below 100% rated torque is normal. The motor load determines the operating point on the torque curve. While the motor torque and current may exceed 100% for a few seconds during starting, continuous operation above 100% can damage the motor. Any motor torque load above the breakdown torque will stall the motor. The torque, slip, and current will approach zero for a "no mechanical torque" load condition (Kuphaldt 2008).

Efficiency: Large three-phase motors are more efficient than smaller 3-phase motors, and almost all single-phase motors. Large induction motor efficiency can be as high as 95% at full load. Efficiency for a lightly loaded or unloaded induction motor is poor because most of the current is involved with maintaining magnetising flux. As the torque load is increased, more current is consumed in generating torque, while the current associated with magnetising remains fixed. Efficiency at 75% FLT can be slightly higher than that at 100% FLT.

3.3 Hydraulic actuators

Hydraulic systems are designed to handle large loads by introducing high-pressure fluid into cylinders or sometimes rotary actuators. As shown in Figure 3.54, a typical hydraulic system comprises a pump to deliver high-pressure fluid, a pressure regulator to maintain a constant pressure in the rest of the system, valves to control flow rate and a distribution system consisting of several pipes. This infrastructure is all required to drive an actuator cylinder.

Hydraulic pumps are mostly powered by large AC electric motors. For large systems, pressures produced can range from 5 MPa to 20 MPa. The hydraulic fluid must be incompressible, have good lubrication qualities and corrosion resistance.

Most hydraulic pumps are defined as positive displacement because they provide a fixed volume of fluid with every cycle. These can include gear, vane and piston pumps amongst others. Because of the fixed volume flow rate, it is necessary to include a pressure relief valve (called a pressure regulator) to prevent the pressure from exceeding design limits for the components. The simplest pressure regulators are spring-ball types shown schematically in Figure 3.55. The threshold pressure is altered by adjusting the compression on the spring.

Hydraulic control valves can either be proportional, which allows any position between open and closed, or two-position open/closed only. The latter are usually described in terms of the number of ports and positions (similar to electrical

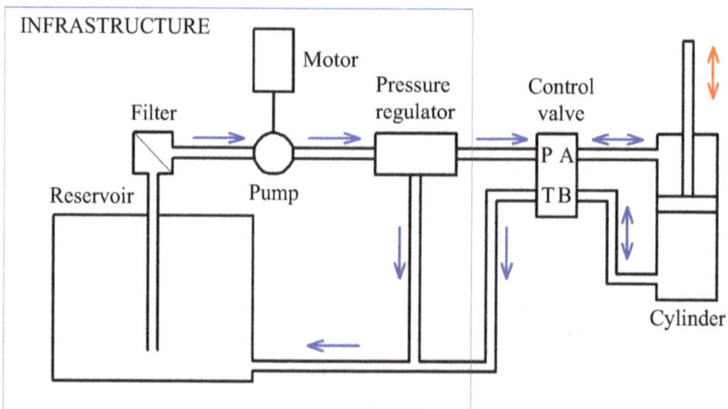

Figure 3.54 Components of an hydraulic system [adapted from Alciatore and Histand (2003)]

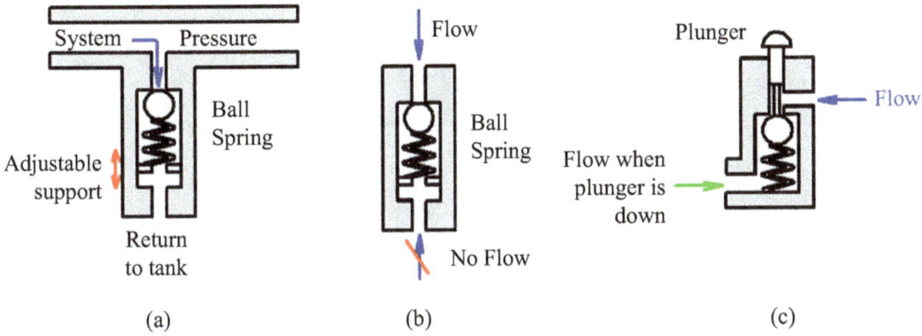

Figure 3.55 Hydraulic valves (a) pressure regulator, (b) check and poppet valves (c) [adapted from Alciatore and Histand (2003)]

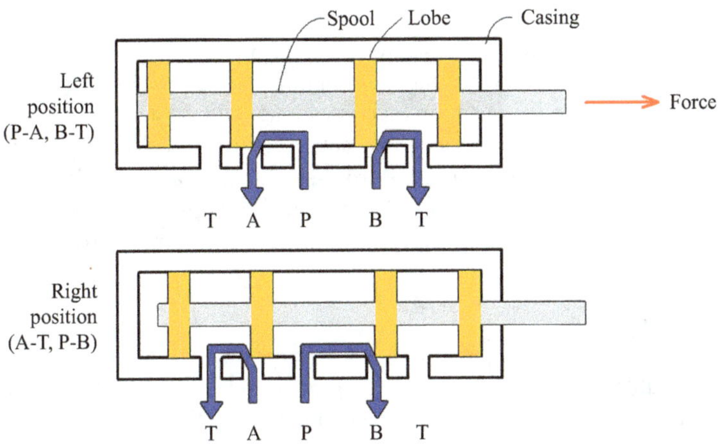

Figure 3.56 Hydraulic spool valve

switches). A four-port three-position valve is described as a 4/3 valve. Check valves allow flow in one direction only, while poppet valves are check valves that can be forced to allow reverse flow.

Spool valves are very common for controlling actuators as they can control multiple flows, and because the static hydraulic forces are balanced, actuators need only overcome the hydrodynamic forces associated with flow momentum. A linear force is applied by hand or using an electrical solenoid that switches inputs and outputs as illustrated in Figure 3.56.

Spool valves can be made proportional by moving the spool a distance proportional to some input command. Linear electrical actuators are usually used to control these valves. If the hydrodynamic pressures are very high, then direct electrical actuation becomes impractical and a smaller pilot hydraulic valve is used to control the larger valve.

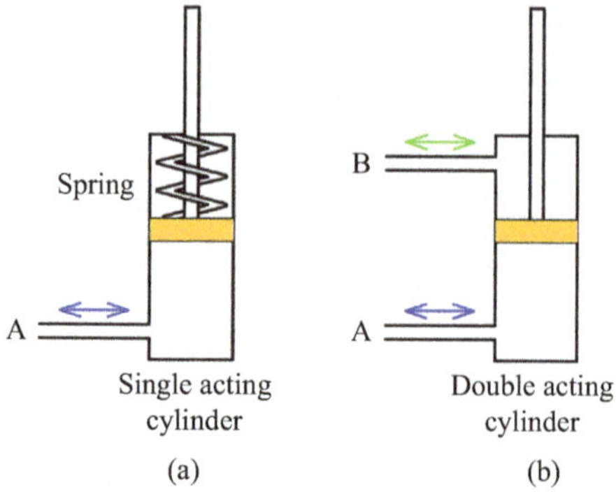

Figure 3.57 Hydraulic actuators (a) single acting and (b) double acting

The commonest hydraulic actuator is the simple cylinder with a piston driven by the pressurised fluid. These can be single acting, in which the cylinder is held in one position by the fluid pressure acting against a spring, or double acting where hydraulic pressure is used to drive the piston in both directions as illustrated in Figure 3.57.

The force available from a single-acting cylinder is the product of the fluid pressure, P_A (Pa = N/m^2), and the cross-sectional area of the piston, A_A (m^2), less the correcting force applied by the spring. The spring force is equal to the product of the spring constant, k (N/m), and the displacement from the neutral position, d (m)

$$F = P_A A_A - kd \tag{3.58}$$

In the double-acting cylinder case, the force is determined by the difference in the pressure between the A and B inlets and the area of the pistons. It can be seen that the piston area on the output side is reduced by the shaft area.

$$F = P_A A_A - P_B A_B \tag{3.59}$$

Hydraulic systems generate large forces from very compact actuators. They can also provide precise control at low speeds. However, their main drawbacks from a biomechatronic perspective include the large amount of infrastructure required, noisy operation and particularly the potential for fluid leaks. Hydraulic fluid is not very compatible with biological organisms, and because of the extremely high working pressure, fluid jets from pinhole-sized breaks are easily injected through the skin into the tissue below.

3.4 Pneumatic actuators

Pneumatic actuators are similar in principle to hydraulic systems, but they use compressed air as a working fluid. The components of a pneumatic system are similar to those of hydraulic systems in as far as they contain a compressor and a

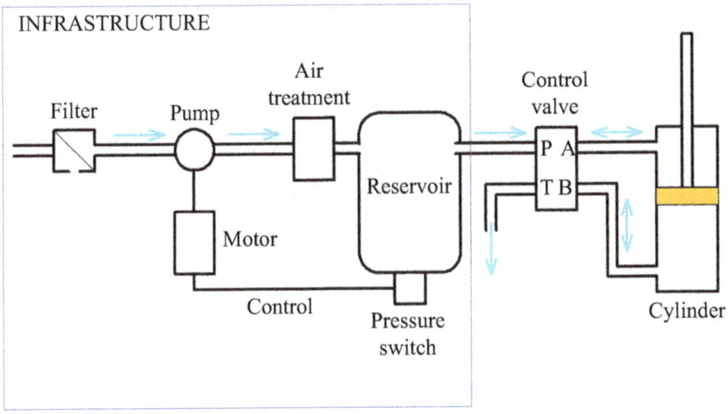

Figure 3.58 Components of a pneumatic system

reservoir, and this is fed to actuators through control valves as shown in Figure 3.58. Working pressures are limited to between 450 kPa and 1 MPa, and as these are much lower than those used by hydraulic systems, the forces available are much lower.

Filtered air is compressed by a motor-driven compressor after which it is cooled and dried before entering the reservoir. Unlike hydraulic systems, the reservoir is maintained at the working pressure, regulated by a pressure switch that controls the compressor motor.

Control valves and actuators operate in much the same way as they do in hydraulic systems, but instead of returning the fluid to a tank, it is vented into the air. Working with air makes them cleaner than hydraulics, but some form of lubrication is required to minimise wear on working surfaces.

Pneumatic actuators are ideal for providing low-force linear motion between two well-defined end points. However, because air is compressible, pneumatic systems are not good at providing accurate motion between the endpoints – particularly if the load is varying.

3.4.1 Pneumatic muscles

Pneumatic muscles also known as pneumatic artificial muscles (PAMs) are contractile or extensional devices that use pressurised air, but instead of a piston/cylinder construction these use rubber bladders surrounded by a scissor weave mesh. When the bladder is pressurised, its diameter increases, which forces the mesh to contract longitudinally with significant force. Because they only apply force in contraction, these actuators are generally grouped into pairs of agonists and antagonists similar to human muscles as shown in Figure 3.59.

PAMs were first developed under the name of McKibben artificial muscles in the 1950s for use in artificial limbs. These were commercialised in the 1980s as Rubbertuators (Wikipedia 2008).

They are lightweight because their main element is a thin membrane. This allows them to be directly connected to the structure they power, which is an

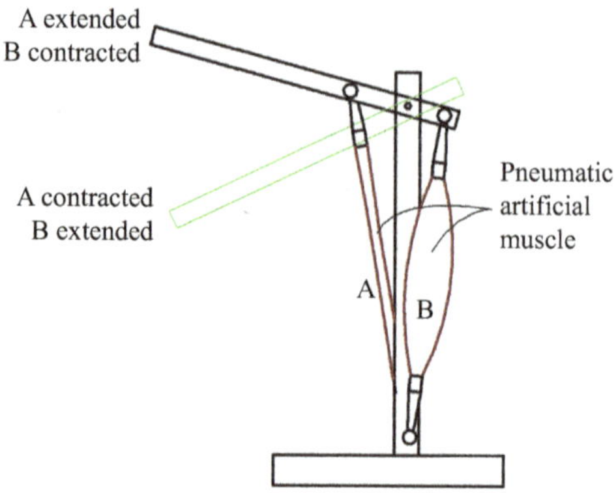

Figure 3.59 Pneumatic artificial muscles (PAMs) operating as an agonist–antagonist pair

advantage when considering the replacement of a defective muscle, or as part of a prosthetic limb. Another advantage of PAMs is their inherent compliant behaviour: when a force is exerted on the PAM, it allows some movement without increasing the force in the actuation. This is an important feature when the PAM is used as an actuator in a prosthesis that interacts with a human, or when delicate operations have to be carried out.

As a PAM contracts under constant pressure, the pulling force generated decreases. Therefore, the maximum possible force at a given pressure is obtained when the PAM is extended as far as possible. If the actuator is not taut when the air pressure is increased, it will not yield its full force. For a constant pressure, the relationship between the force and the length of the PAM is shown in Figure 3.60.

Data for several different PAMS manufactured by the Shadow Robot Company are summarised in Table 3.8 in terms of the percentage change in their lengths for various force and pressure combinations.

At a constant extension, the relationship between force and pressure is linear; therefore, some control of displacement against a spring-type load can be achieved by regulating the system pressure. The relationship between force and extension in PAMs mirrors what is seen in the length-tension relationship in biological muscle systems.

In summary, PAMs can be made with extremely high force-to-weight ratios as can be seen in Table 3.9.

Finally, because the air used to inflate these actuators is compressible, a PAM that uses long tubes must have a control system that can deal with a delay between the movement control signal and the effective muscle action. A PAM actuator system needs the infrastructure shown in Figure 3.58.

According to Ku and Bradbeer (2008), the response time of PAMS increases with applied load. It tests conducted using a 20 mm PAM, it took 0.7 s, 0.9 s and 1 s

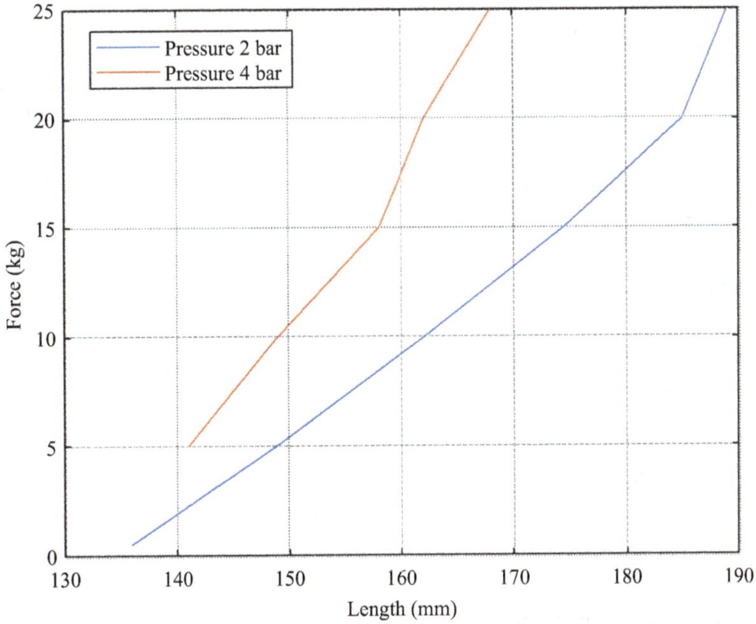

Figure 3.60 Relationship between force and length for a 210 mm long 20 mm diameter PAM at constant pressure

Table 3.8 Relationship between load, pressure and percentage decrease in length for various PAMs (Shadow_Robot_Company 2008)

6 mm PAM	**Load (kg)**				
Pressure (bar)	0.5	1	2	3.2	4.6
0	n/a	3%	2%	2%	1%
2	12%	10%	7%	5%	3%
4		20%	28%	17%	11%

20 mm PAM	**Load (kg)**					
Pressure (bar)	0.5	5	10	15	20	25
0	n/a	8%	6%	2%	1%	0%
2	35%	29%	23%	17%	12%	10%
4		33%	29%	25%	23%	20%

30 mm PAM	**Load (kg)**					
Pressure (bar)	0.5	10	20	35	50	70
0	n/a	2%	1%	0%	0%	0%
2	35%	33%	30%	27%	24%	21%
4		35%	33%	31%	29%	27%

Table 3.9 Specifications of some commercially available PAMs (Shadow_Robot_Company 2008)

Product	Diameter (mm)	Extended length (mm)	Pull at 3.5 bar (kg)	Maximum pull (kg)
	6	150	3	7
	20	210	12	20
	30	290	35	70

for 5, 10 and 12 kg of load respectively from the application of 4 bar of pressure to reach full contraction. These time constants are consistent with human muscle responses.

3.5 Shape memory alloy

Shape memory alloys (SMAs), of which the most common is Nitinol, are metals that exhibit two unique properties; these are pseudo-elasticity and the shape-memory effect. Though the effect was discovered in 1938, serious research only started in the 1960s. The most effective and widely used alloys include NiTi (Nickel–Titanium), CuZnAl, and CuAlNi.

3.5.1 Principle of operation

The properties described are made possible through a solid state-phase change. These are molecular rearrangements that occur in the SMA. A solid-state phase change is similar to a solid-liquid phase change in that a molecular rearrangement occurs, but the molecules remain closely packed so that the substance remains a solid. In most SMAs, a temperature change of only about 10 °C is necessary to initiate this phase change. The two phases that occur in SMAs are Martensite and Austenite.

Martensite is the relatively soft and easily deformed phase of SMAs that exists at low temperatures. The molecular structure in this phase is twinned as seen in Figure 3.61(b). Upon physical deformation, this phase takes on the deformed form, shown in Figure 3.61(c). Austenite, the stronger phase of SMAs, occurs at a higher temperature. The shape of the Austenite structure is cubic as illustrated in Figure 3.61(a). The un-deformed Martensite phase is the same size and shape as the cubic Austenite phase on a macroscopic scale, so that no change in size or shape is visible in SMAs until the Martensite is deformed (SMA 2008).

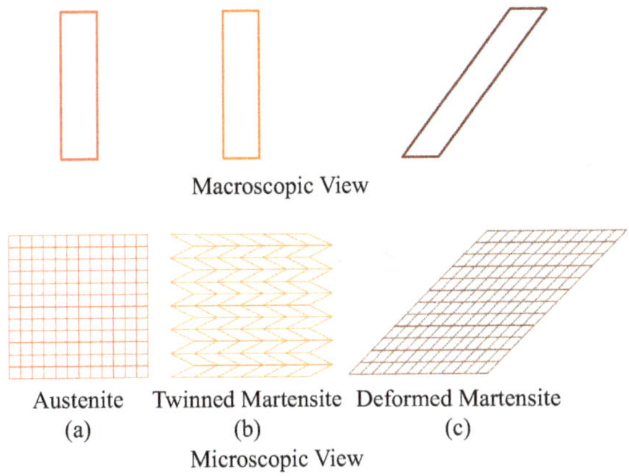

Figure 3.61 Macroscopic and microscopic views of the two phases of shape memory alloy [adapted from SMA 2008)]

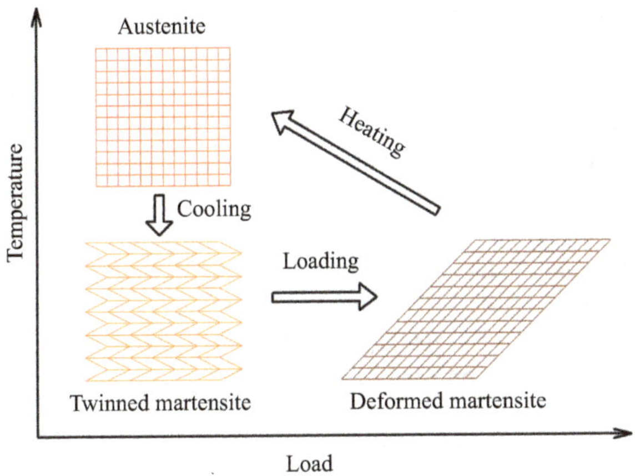

Figure 3.62 Diagram of the microscopic shape memory effect [adapted from SMA (2008)]

The temperatures at which each of these phases begins and finishes forming are a function of the applied load and the composition of the wire.

As illustrated in Figure 3.62, the shape memory effect is observed when the temperature of a piece of SMA is cooled to below the temperature M_f. At this stage, the alloy is completely composed of Martensite which can be easily deformed. After distorting the SMA the original shape can be recovered simply by heating the wire above the temperature A_f. The heat transferred to the wire is the power driving the molecular rearrangement of the alloy, similar to heat melting ice into water, but the alloy remains solid. The deformed Martensite is now transformed to the cubic Austenite phase, which is configured in the original shape of the wire.

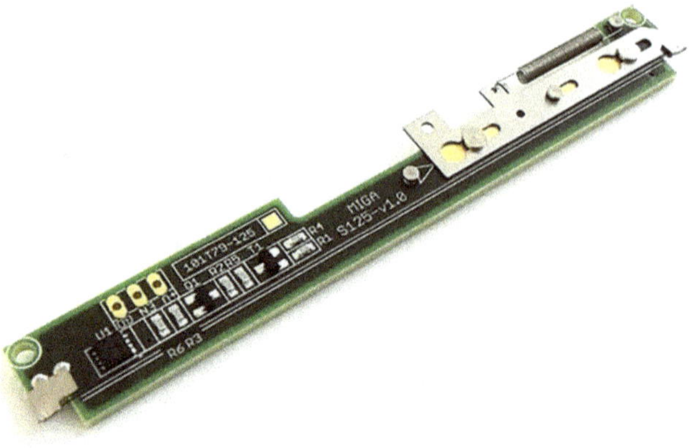

Figure 3.63 The S125-008 SMA actuator made by MIGA, reproduced with permission

Actuators made from SMA are manufactured by several companies. One example is the MIGA Motor Company (MIGA_Motors 2025) that makes a range of devices that can provide forces from 5 N from a unit weighing only 2.5 g up to 10,000 N for a unit weighing 30 g. Their latest actuator, S125-008, shown in Figure 3.63, has a stroke of 3.2 mm and a maximum force of 8800 N. It is only 94 mm long and weighs 4.75 g providing a force-to-weight ratio of nearly 900:1.

The SMA wire segments within the actuator are electrically heated by passing a current through them. When the temperature reaches 75 °C, the wires begin to contract, and are fully contracted by the time they reach 110 °C. If the temperature reaches 150 ° C, permanent damage to the SMA wires results. When power is removed, the wire must return to 60 °C before the actuator returns to the neutral position.

The S125 has an actuation time of less than 0.12 s when powered with 5 volts. With higher voltages, they can operate much more quickly, faster than 50 ms if the voltage rating for the actuator is exceeded, and it is powered with 24 V. Operation is simple with a three-wire interface V+, ground and enable. The actuator is energised whenever the Enable is raised 'high', and the onboard circuit allows the actuator to safely hold at the end of stroke for as long as is required. The operating voltage range in the range 3–12 volts is selected depending on the desired actuation speed.

Very accurate open-loop position control is not possible, as the actuator wires creep with age. Therefore, if accurate positioning is required, a separate sensor and closed-loop control are required.

Some of the main advantages of SMAs include:

- Bio-compatibility,
- Diverse fields of application,
- Good mechanical properties (strong, corrosion resistant),
- Strong contraction force at start of stroke,
- No inrush current or back EMF.

There are still some problems with SMAs that must be overcome before they can live up to their full potential. They are relatively expensive to manufacture compared with other materials such as steel and aluminium. Most SMAs have poor fatigue properties, with an equivalent steel component able to survive at least one hundred times as many cycles as an SMA would under the same cyclical loading conditions.

3.5.2 Biomechatronic applications

Because of their high force-to-weight ratio as well as the ease with which they can be configured for a specific space, SMAs have wide appeal as actuators in biomedical applications. These include actuators for finger joints, heart assist devices and vascular stents, some of which are discussed later in this book.

3.6 Mechanical amplification

As mentioned earlier in this chapter, individual actuators are often not well matched to the load they have to drive, in terms of displacement or, in some cases, speed or even torque. Mechanical methods of performing this matching process can be considered as mechanical amplification or mechanical impedance transformation. In an ideal situation, the actuator is matched specifically for the application, but in reality, off-the-shelf devices such as voice-coil actuators and motors are often used to drive biomechanical systems. In this process, the overall system efficiency is reduced somewhat, and less power is coupled to the output than could otherwise be obtained.

3.6.1 Linkages and levers

A link is a rigid body having two or more elements paired together and connected to other bodies for the purpose of transmitting force or motion (Ham *et al.* 1958). The term linkage refers, therefore, to a mechanism made up from links such as cranks, rods and levers that comprise elements moving in a linear fashion and in rotation.

The primary function of a link mechanism is to produce rotation, oscillation or reciprocating motion from the rotation of a crank, or the reverse. They may be used to convert

- Continuous rotation into continuous rotation with a constant or variable angular velocity ratio,
- Continuous rotation into oscillation or reciprocation (or the reverse),
- Oscillation into oscillation,
- Reciprocation into reciprocation.

Many mechanisms can be constructed from the simple constrained four-link structure consisting of four bar-shaped links and four turning pairs in an open or crossed configuration shown in Figure 3.64.

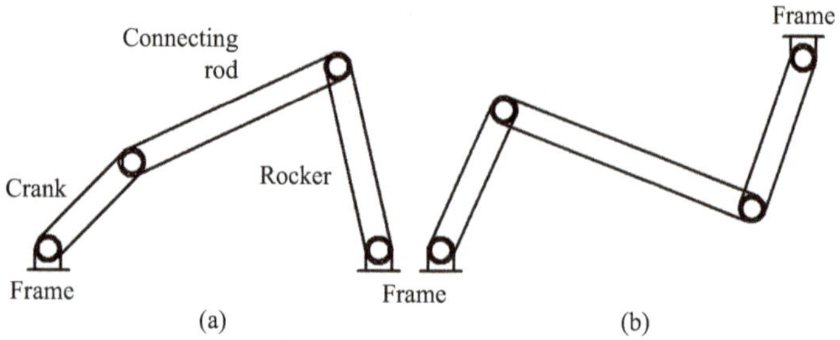

Figure 3.64 Open (a) and crossed (b) four-bar mechanisms

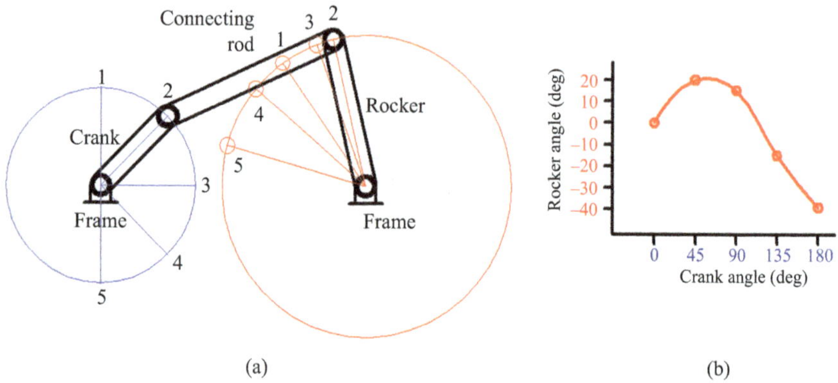

Figure 3.65 Determining the angle of the rocker arm (a) diagram of the linkages and (b) relationship between the angles

The fixed link may be an actual bar, but in most cases, it represents the frame of the machine. Generally, the other three links are referred to as the crank, the connecting rod (con rod) and the rocker or follower. Depending on the relative lengths of the links, the crank may make a complete rotation, or may oscillate, while the follower may oscillate, or in some cases, also rotate.

Considering the mechanism shown in Figure 3.64(a), the relative angle through which the rocker moves can be determined easily by finding the intersection of the circle from the tip of the crank arm and the locus of the tip of the rocker arm, as shown in Figure 3.65. In this example, the motion of the crank arm is limited to 180°, resulting in the irregular motion of the rocker arm through about 60°, as shown in the graph.

Care should be taken when designing these mechanisms that the lengths of the various elements are never such that the connecting rod and the rocker fall in a straight line. This is what is known as a dead point because any driving force transmitted to the rocker will be radial and hence cannot have any effect in rotating it.

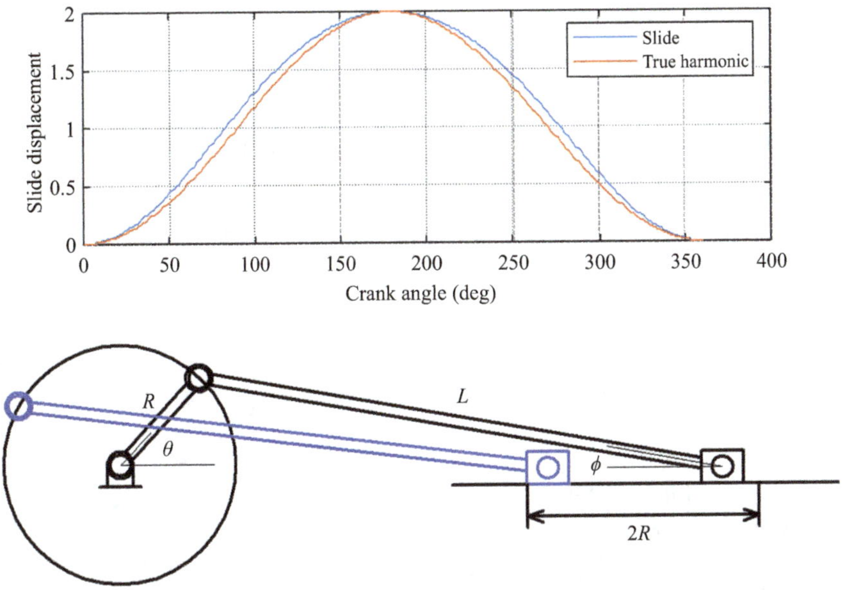

Figure 3.66 Displacement of a slider-crank mechanism with R = 1 and L = 4

One of the commonest applications of the slider-crank mechanism is to convert linear reciprocating motion of a piston into rotary motion, or vice versa. If the connecting rod has a finite length, the motion of the slider is not a simple harmonic motion because the connecting rod operates at an angle to the slider as shown in Figure 3.66.

Considering the figure, the displacement of the slider can be written in terms of the length of the crank, R (m), the length of the connecting rod L (m) and the two angles.

$$\begin{aligned} x &= R + L - R\cos\theta - L\cos\phi \\ &= R(1 - \cos\theta) + L(1 - \cos\phi) \\ &= R(1 - \cos\theta) + L\left[1 - \sqrt{1 - \sin^2\phi}\right] \\ &= R(1 - \cos\theta) + L\left[1 - \sqrt{1 - \left(\frac{R}{L}\right)^2 \sin^2\theta}\right] \end{aligned} \tag{3.60}$$

It is obvious that the second term of this expression describes the deviation of the displacement from simple harmonic motion, and that as the ratio R/L decreases – the length of the crank arm tends towards infinite – the difference reduces to zero.

Linkages can be used to facilitate quite complex motion of prostheses, particularly fingers, where a limited number of actuators must provide as much dexterity as possible. Some of these possibilities are considered in Volume 2, Chapter 5.

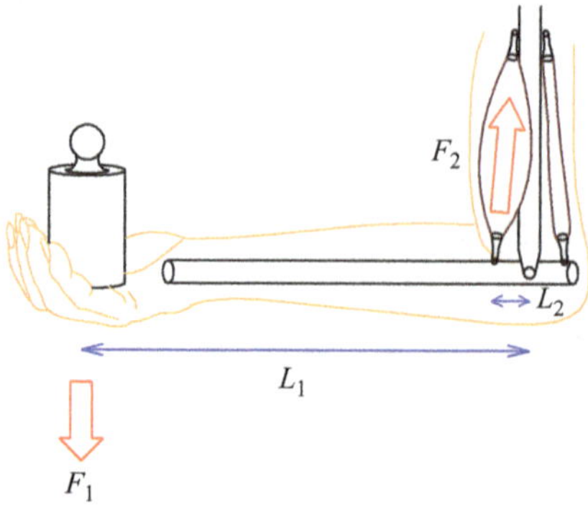

Figure 3.67 Prosthetic elbow actuated using PAMS

A lever, as illustrated in Figure 3.67, consists of a single bar with a fulcrum or pivot which attempts to match the force and displacement of the actuator with the requirements of the system. The equilibrium position of the mechanism can be determined by balancing the moments around the pivot point. However, in the case of the PAMs shown in the diagram, the analysis is complicated by the nonlinear force-to-length characteristics of the two actuator elements.

As a first approximation, it can be assumed that the triceps PAM is not contributing to the equation; therefore, in equilibrium the force relationship is

$$F_1L_1 - F_2L_2 = 0 \tag{3.61}$$

As the angle between the humerus and the ulna decreases from 90°, the moment is reduced proportional to the sine of the angle. For accurate analysis, a complete free-body diagram should be constructed for each activity, and these can be used to determine the required strength of the individual actuators.

Levers are commonly used in prosthetic limbs as they allow the high-force linear actuators such as those based on pneumatic, hydraulic, SMA and piezoelectric materials to be incorporated within the normal form of the limb. They are also very quiet compared to electric motor and gearbox-driven joints. Applications are discussed later in this chapter, and more comprehensively in Volume 2, Chapter 5.

3.6.2 Cams

Like linkages, cams are convenient, compact and simple mechanisms for converting rotary motion into other forms of motion. Cams consist of an element with a curved outline that rotates or oscillates to generate a predetermined motion to a second element called a follower. A common example of an ovoid cam with a roller follower is shown in Figure 3.68. In this example, the linear displacement of the follower is determined by the radius of the cam.

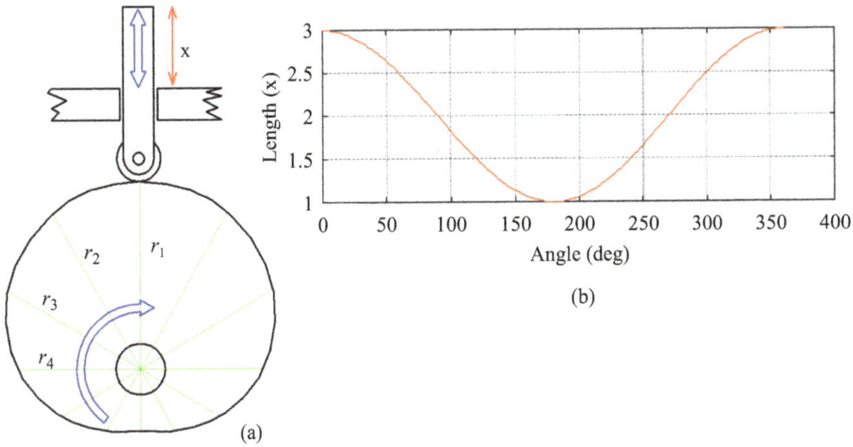

Figure 3.68 Diagram (a) cam mechanism with a roller follower and (b) the base curve generated by the cam

In this example, the cam mechanism has been developed to produce simple harmonic motion where the radius of the cam is

$$r = 2 + \cos\theta \tag{3.62}$$

The base curve in the figure shows the displacement of the follower. Its derivative and second derivative provide the velocity and acceleration respectively. In this example, these are well behaved because of the nature of the curve, but it is possible to produce cams where extremely high velocities and accelerations occur. This is generally undesirable, and care must be taken to ensure that the magnitude of the acceleration is not excessive. Another consideration is the pressure angle that is the deviation of the direction of the force of the cam compared to the motion vector of the follower. If the angle is large, the side thrust applied to the follower is also large and the follower can jam in its guides. This is limited by increasing the size of the cam while providing the same total displacement.

In addition to cams to produce simple harmonic motion, other profiles include straight lines and parabolic curves. The parabolic curve provides the smoothest motion (constant acceleration, and therefore a constant force to produce that acceleration) and the least power for operation. The end sections of these profiles are usually modified to limit the maximum acceleration, and hence limit the force required to move the follower (Ham *et al.* 1958).

Cam mechanisms are fairly common in biomechatronic applications. They can be used to generate specific motion profiles for prosthetic limbs or to drive pump diaphragms in pulsatile artificial hearts, for example.

3.6.3 Gears and belt drives

Gears and belt drives provide a means of matching the characteristics of a motor to the load using circular elements of different diameters. Considering the case where all of the power from the motor is transmitted to the load, the available load torque, τ_L (Nm) is

$$\tau_L = \frac{\omega_m}{\omega_L}\tau_m = \frac{r_L}{r_m}\tau_m \tag{3.63}$$

where ω_m (rad/s) is the speed of the motor, ω_L (rad/s) is the speed of the load and τ_m (Nm) is the motor torque.

The gear (or belt drive) ratio ω_L/ω_m is determined by the ratio of the two pitch circle radii r_m/r_L, which in turn determines the load speed as a function of the motor speed. The reciprocal of this gives the load torque as a function of the motor torque; therefore, decreasing the output speed results in increasing available torque, and vice versa.

In Figure 3.69, where r_1 = 40 mm is the pitch circle radius of the driven element, and the load is driven by the element with radius r_2 = 20 mm, the output speed will be double that of the input, and the available torque will be halved.

The design of gears must consider strength under static and dynamic loads, quietness and smoothness of operation, backlash, lubrication, temperature and heat dissipation as well as overall lifetime (Black and Adams 1968). Most motor manufacturers provide a wide range of gearheads designed for their motors, so it is seldom necessary to design gearing systems from scratch.

Two common gearhead designs are spur and planetary. In general, spur gearheads are simpler and less expensive than planetary units and perform well in

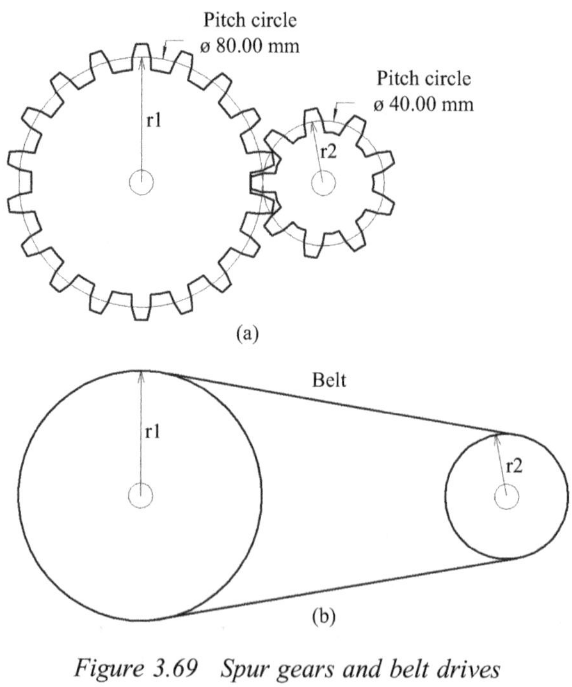

Figure 3.69 Spur gears and belt drives

low-torque applications. The torque capacity of spur types is limited because each gear in the train bears the entire torsional load. In contrast, planetary gearheads share the load over multiple planet gears. As shown in Figure 3.70, the input shaft drives a central sun gear that, in turn, drives the planet gears. Each of the planet gears simultaneously delivers torque to a rotating carrier plate (not shown) coupled to a geared output shaft.

The gear ratio, N, for a stationary ring gear, is determined by the internal diameter of the ring gear, ϕ_r (mm) and the outer diameter of the sun gear, ϕ_s (mm),

$$N = \frac{\phi_r}{\phi_s} + 1 \tag{3.64}$$

The formula holds if the respective diameters are replaced by the number of teeth in each case.

It is obvious that a 2:1 ratio cannot be achieved as it would require that the diameter of the sun and the ring gears be the same, and that is not possible. Typical ratios are from 3:1 to 7:1 as shown in Figure 3.71.

Because the size of the gears determines their load-carrying capacity, it is important to maintain a good balance between the diameters. Figure 3.72 shows the

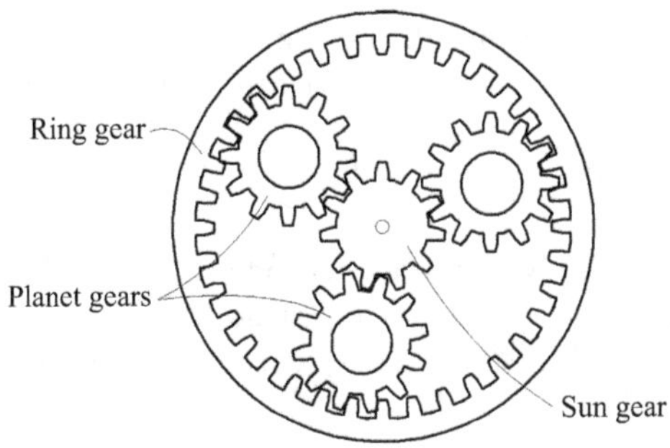

Figure 3.70 Gear configuration for a planetary gearhead

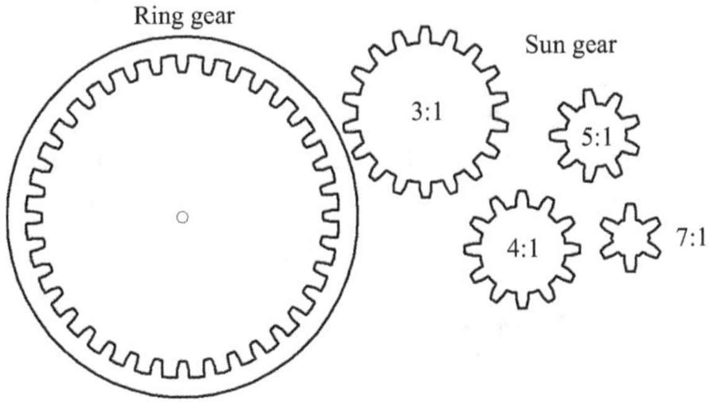

Figure 3.71 Size of sun gear compared to ring gear for different planetary gear ratios

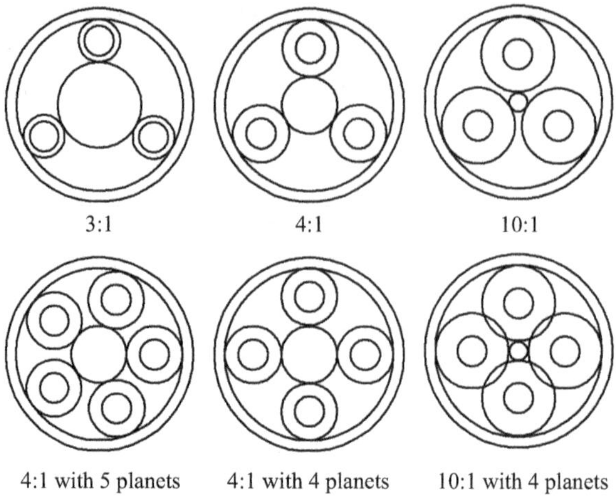

Figure 3.72 Different configurations for a planetary gearhead

gears in a 3:1, 4:1, and 10:1 system. For a 3:1 ratio, the sun gear is large, and the planets are small. In this case, the planets have thin walls and therefore limit the space for bearings and carrier pins, hence limiting the load torque. The 4:1 ratio is well balanced with sun and planets about the same size, while the 5:1 and 6:1 ratios (not shown) still yield fairly good balanced gear sizes. With higher ratios approaching 10:1, the small sun-gear becomes a strong limiting factor for the transferable torque.

Adding planets increases the torque transfer capacity of the arrangement. It can be seen in Figure 3.72 that, with lower ratios, additional planet gears can be used; but for higher ratios, such as the 10:1 shown, multiple gears would cause interference (Anthony 2008).

The material from which the gears are built is also a consideration in regard to torque capability. Sintered nickel-steel is often used because the sintering process produces gears that are able to run at closer tolerances. Additionally, because the material is porous, these gears hold lubricant better than steel units do. However, cut-steel gears tend to be more durable and are therefore a better choice for higher-torque applications. That notwithstanding, good lubrication is important regardless of gear material, especially at high speeds and loads. Planetary gearheads have an advantage because oil flying outward from the sun gear is captured by the planet gears and the carrier plate. Spur gearheads, on the other hand, fling lubricant off and away from the gears. This is one reason planetary gearheads have higher speed ratings.

Backlash is a measure of positional accuracy usually specified in arcminutes. For example, a typical spur gearhead has about 10 arcmin (0.1666°) of backlash, whereas its planetary counterpart may be rated better than half of that.

Reduction ratios for both spur and planetary gearheads range from near unity up to several thousand to one. Spur gearheads, with a single geared input shaft

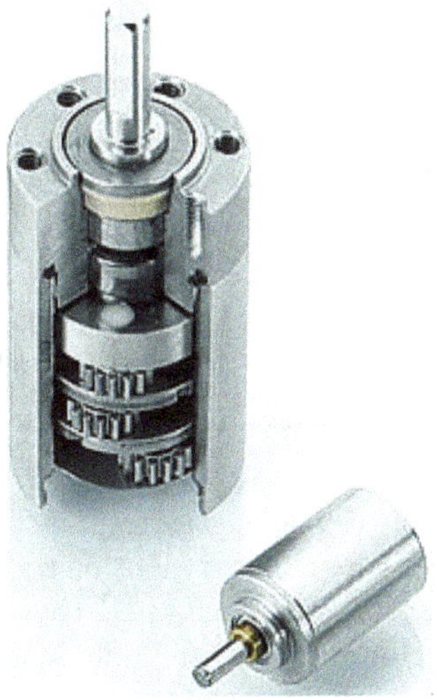

(a)

(b)

Figure 3.73 Stacked planetary gearhead

coupled to a geared output shaft (single stage), provide up to about a 6:1 reduction. As discussed previously, planetary units can reach 10:1 in a single stage. For higher ratios and proportionally greater output torque, multiple stages or gear sets are stacked together axially as shown in Figure 3.73. Increasing the number of stages increases the reduction ratio and output torque but also increases overall length as well as lowering the mechanical efficiency. A typical single-stage spur gearhead is about 90% efficient while a two-stage device is about 85% efficient. Planetary gearheads are slightly more efficient at approximately 97% and 94% respectively for one and two-stage units. Obviously, the efficiency decreases further as additional stages are included.

MicroMo Electronics offers a wide range of gearheads with diameters from 6 mm to 44 mm (to suit motor diameters). Reduction ratios are available from 3.7:1 to 23.104:1. Maximum input speeds range from 3500 rpm to 8000 rpm and continuous load torques are available from 25 mNm to 16 Nm.

As can be seen in Table 3.10, planetary gearheads are superior to spurs on all counts except mechanical noise and cost. Selection of the appropriate gearhead for a mechatronic application should take all these factors into account.

Belt drives are commonly used over longer centre distances than those used for gears. Conventional flat or V belts slip slightly so the driven speed is less than the ratio of pulley diameters. However, there is no slip in toothed belts.

The maximum amount of power that can be transmitted by a belt drive is determined by belt stretch or excessive slippage on the smaller pulley. This is in turn determined by the angle of contact, the belt tension and the coefficient of friction between the belt and the pulley.

Table 3.10 Comparison between planetary and spur gear characteristics

Factor	Planetary	Spur
Backlash	Lower	Higher
Efficiency	Higher	Lower
Load Capacity	Higher	Lower
Operating Speed	Higher	Lower
Mechanical Noise	Higher	Lower
Size	Smaller	Larger
Cost	Higher	Lower

In a toothed belt, the sheaves have axial grooves that engage teeth on the belt. The belts usually include a number of thin steel cables that carry tension under load which permits a light drive to operate at high speed (up to 100 m/s). These drives are compact, light, quiet and low maintenance. However, they are more sensitive to misalignment than flat or V belts (Black and Adams 1968).

3.6.4 Translation screw devices

Translation screws are commonly used compact devices to convert rotary motion produced by a motor into linear motion against large forces. In most translation screws, the screw rotates in its bearings and the nut moves axially as shown in Figure 3.74. However, in some devices the nut rotates while the screw moves axially with no rotation.

Consider that the nut of a screw is moved against an axial load by the rotation of the screw illustrated in Figure 3.75. The load on the nut will be transferred to the screw as a distributed load on the surface of the threads in contact. To facilitate analysis, it can be assumed that the load is concentrated at point o on the mean circumference, πd (m), of the thread.

Under static conditions, the direction of the load on the thread must be normal to the thread surface, along the vector ao as shown in the figure. As the screw rotates so that the nut is moved against an external force Q (N), the line of action ao will be rotated through the angle of friction ϕ (rad) to bo.

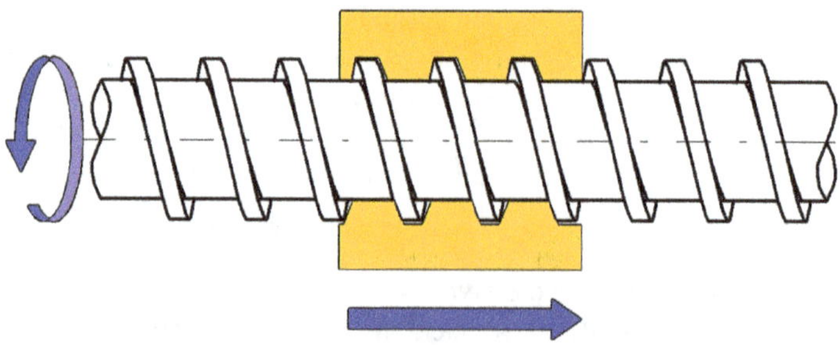

Figure 3.74 Translation screw operation

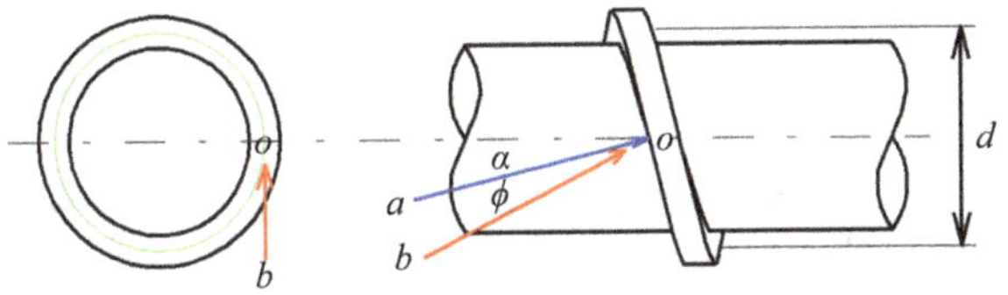

Figure 3.75 Forces on a square threaded translation screw

For equilibrium of forces, the component of *bo* parallel to the axis of the screw is

$$Q = bo \cos(\alpha + \phi) \tag{3.65}$$

where α (rad) is the lead angle of the screw and ϕ (rad) is the friction angle.

The component of *bo* at right angles to the axis of the screw is

$$F = bo \sin(\alpha + \phi) \tag{3.66}$$

Equating F in terms of Q

$$F = Q \tan(\alpha + \phi) \tag{3.67}$$

The torque, τ (Nm), is

$$\tau = F\frac{d}{2} = \frac{Qd}{2} \tan(\alpha + \phi) \tag{3.68}$$

To determine the efficiency of the screw, consider the required torque, τ_o (Nm) in the absence of any friction. Equation (3.68) can be rewritten as

$$\tau_o = \frac{Qd}{2} \tan \alpha \tag{3.69}$$

The efficiency is

$$e = \frac{\tau_o}{\tau} = \frac{\tan \alpha}{\tan(\alpha + \phi)} \tag{3.70}$$

The coefficient of thread friction $f = \tan \phi$ is about 0.1 for well machined, run in and well-lubricated threads. This increases to about 0.125 for average workmanship and quality. Static friction can be approximated as 1.33 times the dynamic friction (Black and Adams 1968).

Application of this technology in linear actuators was discussed briefly earlier in this chapter.

To improve the overall efficiency of translation screw devices, steel balls can be introduced between properly formed threads on the screw and nut. This ball-screw mechanism substitutes rolling friction for sliding friction. When the nut is translated by rotation of the screw, the balls run out of the thread of the nut and are returned to the other end via a recirculating groove. Efficiencies of higher than 90% can be achieved.

3.7 Twisted string actuators

As shown in Figure 3.76, a twisted string actuator consists of a pair of strings in parallel and connected to an electric motor. As the two strings twist, their length is reduced in a linear fashion and a contractile force is generated.

According to (Palli *et al.* 2015), if the finite stiffness of the strings is taken into account, it is reasonable to assume that twisted strings act as linear springs with the capability of resisting tensile forces only. The total length of the string L changes with respect to the unloaded length L_0 as a function of the fibre tension, F_i, and the string stiffness, K, normalised with respect to the length L_0.

$$F_i = \frac{K}{L_0}(L - L_0) = \frac{K}{L_0}\left(\sqrt{p^2 + (r\theta)^2} - L_0\right) \tag{3.71}$$

where r is the radius of the string, θ is the angular position of the motor and p is the resulting length of the transmission system. Each string acts as a spring with a deformation of $\sqrt{p^2 + (r\theta)^2} - L_0$ and so is proportional to the motor angle θ. It follows that the transmission length p is given by

$$p = \sqrt{L_0^2\left(1 + \frac{F_i}{K}\right)^2 - (r\theta)^2} \tag{3.72}$$

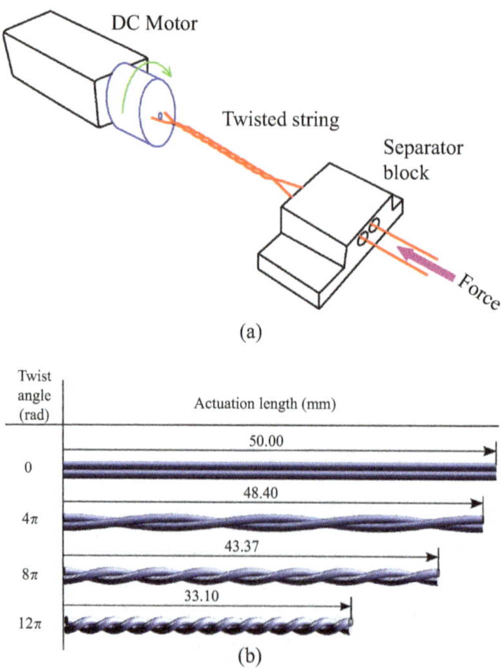

Figure 3.76 Schematic diagram (a) of a twisted string actuator and (b) twisting mechanism that shortens the string [adapted from Palli et al. (2015)]

Finally, the applied torque τ_L and the load force F_L can be derived from (3.71)

$$\tau_L(\theta, p) = 2\frac{\theta r^2 K}{L_0}\left(1 - \frac{L_0}{\sqrt{p^2 + (r\theta)^2}}\right) \tag{3.73}$$

$$F_L(\theta, p) = 2\frac{Kp}{L_0}\left(1 - \frac{L_0}{\sqrt{p^2 + (r\theta)^2}}\right) \tag{3.74}$$

3.8 Prosthetic hand actuation

Many attempts have been made to re-create human anatomy through mechanical means; most of these have met with limited success because the complexity of the human body makes it very difficult to duplicate even simple functions effectively. Most prosthetic hands in industrial production have been restricted to rugged two or three-finger grippers. They are used to execute relatively simple movements. Prosthetic hands for more delicate tasks have proven expensive and difficult to manufacture in the past due to the lack of available technology. However, in the past few decades, miniaturisation of electronics and mechanical systems has allowed good progress to be made, as discussed in Volume 2, Chapter 5.

In order to reproduce human extremities, there are a number of aspects that must be considered:

- The gripping force required to manipulate different objects (eggs, pens, tools),
- The motion capabilities of each joint of the hand,
- Actuator noise must be minimised,
- The ability to feel or touch objects (tactile senses),
- The method of controlling movement within the limb,
- Emulation of real human movement (smoothness, and speed of response).

Many different solutions have been proposed for these problems; some include using "muscles" controlled by air pressure, piezoelectric materials, or SMAs amongst others.

3.8.1 Shape memory alloy

SMAs mimic human muscles and tendons quite well. SMAs are strong and compact so that large groups of them can be used for prosthetic applications. In addition, the motion with which they contract and expand is very smooth, producing a life-like movement unavailable in other systems.

Creating human motion using SMA wires is a complex task, but the basics are straightforward, as illustrated in Figure 3.77. For example, to create a single direction of movement, a bias spring shown in the upper portion of the finger holds

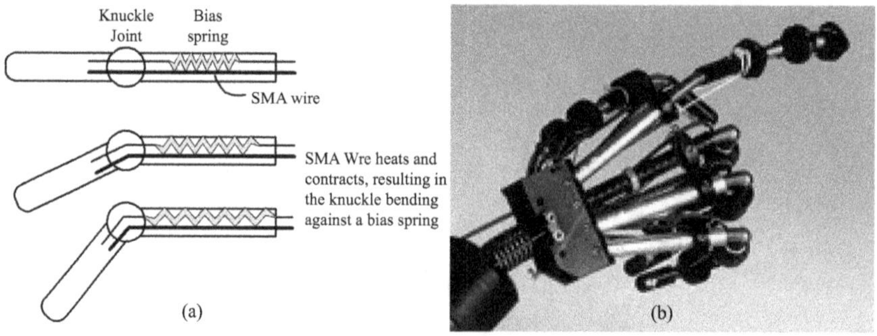

Figure 3.77 Using shape memory alloy to actuate a prosthetic hand

the finger straight, stretching the SMA wire. When the wire on the bottom portion of the finger is heated, it shortens, and the joint will be bent downwards. The heating takes place by passing an electric current through the wire.

There are still some challenges that must be overcome before prosthetic hands can become more commonplace. The first is generating the computer software used to control the artificial muscle systems within the robotic limbs. The second is creating sufficient movements to emulate human flexibility (i.e. being able to bend the joints as far as human beings can). The third problem is reproducing the speed and accuracy of human reflexes.

3.8.2 Electric motors

A combination of microelectronics and micromechanics has provided the means to produce prosthetic hands with separately controllable fingers and joints based on human hands. One example is the device developed by the German Aerospace Centre (DLR), in cooperation with the Harbin Institute of Technology (HIT). This prosthetic hand has evolved from the prototype (version I) to the more advanced version II as shown in Figure 3.78. They both use miniature actuators and high-speed bus technology.

Constructing a prosthetic hand with the strength and dexterity of a human hand requires at least four fingers: three fingers to allow the hand to grip conically shaped parts, and a thumb in opposition. Consequently, the DLR/HIT hand I consists of three fingers, each containing four joints with three degrees of freedom. The fourth finger, designed as a thumb, has four degrees of freedom.

Tiny but powerful motors and shaft encoders are fitted directly into the finger. Each finger joint also houses a contactless angle sensor and a torque sensor. Since both sensors provide extremely high-resolution feedback, a high-speed three-wire serial connection conveys sensed data back to FPGA-based controllers. For version I these are on a plug-in PCI card integrated in a standard PC, but for version II it has been possible to integrate all the processing into the hand.

The complexity of this prosthetic hand has its price with each finger requiring several separately controllable actuators. In the case of version I, there are twelve

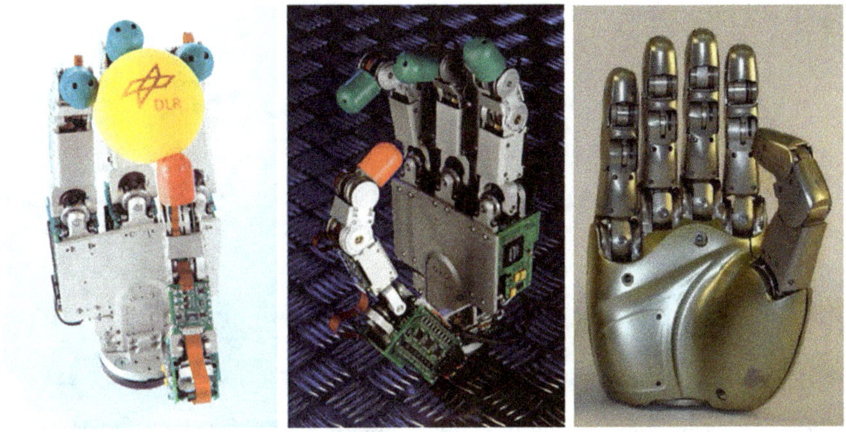

Figure 3.78 Evolution of the DLR/HIT prosthetic hand I and II that use miniature electric motors (MicroMo 2008), reproduced courtesy of Faulhaber

16 mm diameter Faulhaber analogue Hall sensor-controlled BLDC motors per hand. These are commercially available, high-performance motors. They can be connected with gear systems of the same diameter to form a single integrated unit. The motors produce an output power of 11 W and a maximum continuous torque of up to 2.6 mNm without gearing. However, for this application the motor's no-load speed of 29,900 rpm is reduced to 188 rpm through all-metal planetary gearheads with a ratio of 159:1. These simultaneously increase the torque to a maximum value of 450 mNm. Hall sensors provide rotor position to the controller and deliver the requisite feedback information with a resolution of at least 8 bits. The Hall sensors and motor form a compact 31 g unit with a length of only 28 mm and an outer diameter of 16 mm.

3.8.3 Pneumatic artificial muscles

The Shadow Hand is the closest robot hand to the human hand available. It provides 24 movements, allowing a direct mapping from a human hand to the prosthesis. In addition, it has integrated sensing and position control, allowing precise computer control.

The pneumatic version of the hand contains an integrated bank of 40 Air Muscles for actuation, some of which are visible in Figure 3.79. The muscles are compliant, which allows the hand to be used around soft or fragile objects, and because it can be fitted with touch-sensitive pads on its fingertips, it can be made sufficiently sensitive to detect an object the size of a small coin.

It is interesting to note that the latest iteration of the shadow hand is driven by electric motors, resulting in a power block that is much more compact than the Air Muscle actuated version shown here. This change in actuation method is discussed in more detail in Volume 2, Chapter 5.

Figure 3.79 The shadow hand in action (Shadow_Robot_Company 2008), reproduced with permission

References

Alciatore, D. and M. Histand (2003). *Introduction to Mechatronics and Measurement Systems*, 2nd Ed. Boston, MA: McGraw Hill.

Anthony, G. (2008). "The best balanced Planetary Ratio from a Torque Density Point of View." Retrieved December 2008 from http://www.motioncontrol.com/products/index.cfm/Balance-Planetary-Ratio-Torque-Density.

Black, P. and O. Adams (1968). *Machine Design*. Tokyo: McGraw-Hill Kogakusha, Ltd.

Brooker, G. (2021). *Sensors for Ranging and Imaging*, 2nd ed. Stevenage: The Institution of Engineering and Technology (IET).

Brown, W. (2002). "AN857: Brushless DC Motor Control Made Easy." Retrieved from http://www.microchip.com/.

Fraden, J. (1996). *Handbook of Modern Sensors*. New York: AIP Press, Springer-Verlag.

Ham, C., E. Crane and W. Rogers (1958). *Mechanics of Machinery*, 4th ed. New York: McGraw Hill Kogakusha.

Ku, K. and R. Bradbeer (2008). "Modelling Pneumatic Muscles as Hydraulic Muscles for Use as an Underwater Actuator". In: *Mechatronics and Machine Vision in Practice*. J. Billingsley and R. Bradbeer (eds). Berlin: Springer.

Kuphaldt, T. (2008). "Lessons in Electric Circuits: Volume II – -= AC." Retrieved July 2008 from http://www.ibiblio.org/kuphaldt/electricCircuits/AC/index.html.

MicroMo. (2008). "How to Select a DC Motor." Retrieved July 2008 from http://www.faulhaber-group.com/n390290/n.html.

MicroMo. (2008). "Micro-actuators Move Sensitive 3-Finger Hand." Retrieved June 2009 from http://www.micromo.com/n280126/n.html.

MIGA_Motors. (2025). Retrieved March 2025 from http://www.migamotors.com/Index.html.

Palli, G., L. Pan, M. Hosseini, L. Moriello and C. Melchiorri (2015). Feedback linearization of variable stiffness joints based on twisted string actuators. *IEEE International Conference on Robotics and Automation (ICRA)*. Seattle, WA, USA: 2742–2747.

physikinstrumente. (2008). “Piezo Actuators Overview.” Retrieved September 2009 from http://www.physikinstrumente.com/en/products/piezo/index.php.

Shadow_Robot_Company. (2008). Retrieved July 2008 from http://www.shadow-robot.com/.

SMA. (2008). “Shape Memory Alloys.” Retrieved July 2008 from http://www.cs.ualberta.ca/~database/MEMS/sma_mems/sma.html.

Spanner, K. and B. Koc (2016). “Piezoelectric Motors, an Overview.” *Actuators* 5 (1): 6.

Wikipedia. (2008). “Pneumatic Artificial Muscles.” Retrieved July 2008 from http://en.wikipedia.org/wiki/Pneumatic_artificial_muscles.

Yedamale, P. (2003). *AN885: Brushless DC (DLDC) Motor Fundamentals*. Microchip Technology.

Chapter 4
Sensory substitution, expansion and addition

4.1 Introduction

Sensory substitution is defined by Peter Meijer (Meijer 2006) as "the replacement of one sensory input (vision, hearing, touch, taste or smell) by another, while preserving some of the key functions of the original sense".

Due to the astonishing plasticity of the brain, sensory substitution devices can help the blind regain some sense of sight through touch or hearing, or help a deaf person regain some sense of hearing through vision or touch. It has been shown experimentally that if the brain is deprived of its main form of input, even for quite short periods, it will seek out other ways of gaining this information. Sensory substitution devices are designed to facilitate this process.

Extensions to substitution include sensory expansion – the widening of an existing sense such as the ability to detect electromagnetic signals outside the thermal and visual range, and sensory addition – the provision of a novel sense such as the ability to detect magnetic fields (Eagleman and Perrotta 2023).

In regard to vision replacement, direct cortical stimulation would appear to be the best interface as it does not interfere with other senses. However, the huge interfacing problems and invasive nature of the approach, together with the low-resolution results obtained until now, probably make it a long-term research item.

As an alternative, in the shorter term, when considering interference, a vibrotactile or electrotactile skin-stimulating system would seem to be a good choice because much of the skin plays only a subordinate role as a communication channel under normal conditions. Unfortunately, that same consideration might indirectly be its major disadvantage, because there is no strong evolutionary reason to expect the touch sensation to provide a high-bandwidth communication channel. This restriction does not just apply to the skin interface; it could be a limitation all the way up to the cognitive brain centres.

Experiments in the 1960s conducted by Paul Bach-Y-Rita using a 20 × 20 grid of 1 mm Teflon-tipped stimulators attached to the subject's back and activated by a video camera showed that if the subjects had control of the camera's pan and zoom facility, they were soon able to discern lines and edges. With some practice, they could even recognise depth, shadows and shapes, and even individual faces. This is an amazing result given the poor resolution of both the camera and the electrotactile stimulator. More important than this was the fact that the initial tactile sensations were soon forgotten, and perceived objects appeared to exist in three-dimensional space. However, this heightened perception capability only occurred if the subjects were moving the camera themselves, and measurements made using static cameras failed to produce any real perception. It was as if the perception was founded in the sensory-motor loop as a whole and not in the static transfer of an image to the array (Real and Araujo 2019).

It has been argued that reasonable resolution can be achieved by stimulating small areas of skin, but that the limited sensory bandwidth available would lead to severe information loss when stimulating a large matrix of skin positions. Experiments have yet to show that this is always the case, but it does suggest that using tactile substitution may not be the ultimate path to high-fidelity sensory substitution.

Knowing the importance of bandwidth in communicating data, in this case, detailed environmental information, an alternative would be to exploit the capabilities of the human hearing system. Although it cannot be claimed that this would produce a perfect solution, it is known that the human hearing system is capable of processing and interpreting extremely complicated and rapidly changing acoustic patterns, such as speech or music in a noisy environment. The available effective bandwidth, of the order of 10 kHz, corresponds to a channel capacity of many thousands of bits per second.

As with the tactile option, there may be restrictions imposed by the mechanics of the cochlea or with information encoding in the neural architecture. However, in spite of these uncertainties, the known capabilities of the human hearing system in learning and understanding complicated acoustical patterns provide a strong motivation for developing auditory sensory substitution systems.

4.2 Optical prosthetics – white canes, sonar and cameras

To date, the most successful and widely used visual prosthesis for the blind is the *white cane*. It can be used to detect obstacles on the ground, uneven surfaces, holes, steps and puddles. Even though the cane is inexpensive and is so light and small that it can be folded and tucked away in a pocket, only a small percentage (between 2% and 8%) of blind people actually use one. Some of the reasons for this poor uptake are considered in a recent study (Rasouli Kahaki *et al.* 2023). Even fewer use other aids, with only about 22,000 people worldwide using guide dogs at present, with fewer than 4000 dogs graduating from dog guide schools every year.

More advanced visual prosthetics can be divided into three major groups. First, there are the devices that use either ultrasound or a camera to sample the environment ahead of an individual and render the results into either a series of sounds or a tactile display. This process is known as sensory substitution because the sense of hearing or touch is substituted for that of sight. From this substitution, users are supposed to be able to discern the shape and proximity of objects in their path.

Sensory substitution prostheses based on ultrasound technology have been available for many years because that technology was really the only type that was portable and reasonably low cost. Since the beginning of this millennium, however, the advent of low-cost cameras based on charge-coupled devices (CCD) and more recently complementary metal oxide semiconductor (CMOS) arrays has spurred the development of camera-based systems.

4.3 Sonar-based systems

Sonar systems operate by transmitting a brief pulse of ultrasound from a transducer and then listening for an echo. As the speed of sound in air is reasonably constant (about 340 m/s), the round-trip time gives the range to the reflecting surface, and because the sound is radiated and detected over a narrow angular field, the beam-pattern of the transducer also gives a good idea of direction to the reflecting surface

For short-range applications, most ultrasonic transmitters are made either from thin wafers of piezoelectric material or use electrostatic transducers.

However, blind people can also use echolocation by emitting mouth clicks and interpreting the echoes that return. This ability allows them to identify objects' location, shape, size, distance, and even material without resorting to technological aids. Mouth clicks are brief (lasting about 3 milliseconds) and focused within a 60° cone by shaping the mouth. This is narrower than the pattern for typical speech (120°–180°). The clicks range in frequency from 2 to 4 kHz, providing the best auditory sensitivity. The repeat interval for clicks, known as the inter-click interval (ICI), depends on the environment and the echolocator's needs. Shorter ICIs allow for quicker updates on spatial information, while longer ICIs may be used in quieter environments or when focusing on distant objects. Expert echolocators instinctively adjust the intensity and frequency of their clicks based on an object's position (Thaler *et al.* 2017).

Blind echolocators often activate their primary visual cortex (V1), a brain region typically used for sight in sighted individuals. This phenomenon reflects neuroplasticity, where the brain repurposes visual areas for auditory spatial processing. Studies show that blind echolocators experience echoes as spatial representations akin to visual images.

Daniel Kish is a prominent expert who has been blind since infancy. He uses echolocation to engage in activities like mountain biking and teaches others how to develop this skill. Kish describes echolocation as constructing "images" from sound reflections (Wilson 2017).

4.3.1 Piezoelectric transducers

A typical piezoelectric transducer similar to the unit shown in Figure 4.1, such as the SensComp 40LT16, has the following characteristics (Senscomp):

- Centre frequency: 40 ± 1 kHz,
- Bandwidth: 2 kHz,
- Transmit sound pressure level (SPL): 120 dB minimum at 40 kHz; 0 dB re 0.0002 μbar per 10 Vrms at 30 cm,
- Maximum drive voltage: 20 Vrms,
- Beam angle (6 dB): 55°.

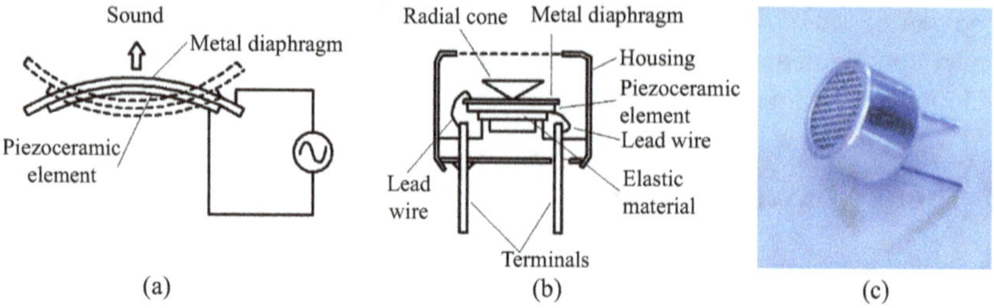

Figure 4.1 Principles and structure of a piezoelectric ultrasound transducer (Brooker 2021)

The matched receiver for this transducer is the 40LR16 that has similar characteristics:

- Centre frequency: 40 ± 1 kHz,
- Bandwidth: 2.5 kHz,
- Sensitivity: -65 dB at 40 kHz relative to 0 dB = 1 V/μbar.

Worked Example:

A short-range ultrasonic transceiver comprises a pair of SensComp piezoelectric transducers with specifications as listed above. Determine the acoustic power density 30 cm from the transmitter and calculate the received voltage level from a target with a target strength, $T = -10$ dB at a range of 5 m if the transmitter is driven by 20 V rms.

1 μbar = 0.1 Pa, therefore 0.0002 μbar = 20 μPa which is the standard reference level for SPL.

As the transmitter SPL in dB is listed for a drive signal of 10 Vrms and we want to operate at twice that, the new SPL needs to be calculated. The SPL increases with the square of the applied voltage; therefore

$$L_p = 120 + 20\log_{10}\frac{20}{10} = 126 \text{ dB}$$

The beam spread loss, H (dB) (the decrease in the transmitted power density as a function of range), obeys the inverse square law; therefore, it is

$$H = 20\log_{10}\frac{5}{0.3} = 24.4 \text{ dB}$$

The sonar range equation determines the echo power (SPL) back at the receiver, L_r (dB). It is determined by the transmit power, L_p (dB), the beam spread loss to the target and back to the receiver (dB), and the proportion of the power reflected by the target (dB).

$$L_r = L_p - 2H + T = 126 - 2 \times 24.4 - 10 = 67.2 \text{ dB}.$$

Next, we convert the received power (SPL) to the received pressure, P, relative to the reference pressure in μbar

$$P = 10^{L_r/20} = 10^{3.36} = 2291 \text{ relative to } 0.0002\ \mu\text{bar},$$

therefore

$$P = 2291 \times 0.0002 = 0.458\ \mu\text{bar}.$$

The sensitivity response of the receiver is specified to be −65 dBV/μbar, which equates to $10^{-65/20} = 5.62 \times 10^{-4}$ V/μbar; therefore, the output voltage will be $V_{out} = 0.458 \times 5.62 \times 10^{-4} = 257\ \mu$V.

4.3.2 *Electrostatic transducers*

Electrostatic transducers are typically larger than the piezoelectric variety as can be seen in Figure 4.2 and have narrower beamwidths. Because they are not resonant, they also exhibit a much wider bandwidth as noted in the abridged specifications for the SensComp 600 series reproduced below (Senscomp):

- Centre frequency: 50 kHz,
- Bandwidth: ≈ 40 kHz,
- Transmit SPL: 110 dB min at 50 kHz; 0 dB re 20 μPa at 1m (300 VACpp 150 VDC bias),
- Beam angle (6 dB): 15°.

In this case, the same transducer operates as a receiver with the following specifications:

- Sensitivity: −42 dB at 50 kHz relative to 0 dB = 1 V/μbar (for a 150 V DC bias)

Unfortunately, the wider bandwidth results in a much poorer receive sensitivity than that of the narrow band resonant variety. However, it results in superior range resolution.

Figure 4.2 SensComp 600 series transducer. The photograph is the property of SensComp and is reproduced with its permission

4.3.3 Atmospheric effects

As an acoustic signal propagates through the atmosphere, it attenuates. The main reason for this is the beam divergence that reduces the signal power density with the square of the distance travelled. The second mechanism is atmospheric absorption that converts some of the acoustic energy into heat. A reasonably simple model for the atmospheric attenuation, α_c (dB/m), which includes the effects of both temperature and humidity has been derived from measured data and is accurate for relative humidity greater than 30% (Knudsen and Harris 1950)

$$\alpha_c = \left(\frac{f}{1000}\right)^{3/2} \frac{0.283}{20 + \phi_t}, \tag{4.1}$$

where: f – frequency (Hz),

$\phi_t - \phi_{20}(1 + 0.067\Delta t)$,

ϕ_{20} – Relative humidity % at 20 °C,

Δt – Denotes the temperature difference from 20 °C.

Consider the attenuation, α_c (dB/m), determined using this model at a frequency of 30 kHz, for a temperature of 20 °C and a relative humidity of 30%

$$\alpha_c = \left(\frac{30 \times 10^3}{1000}\right)^{3/2} \frac{0.283}{20 + 30} = 0.93\text{dB/m}$$

A far more comprehensive model (Burnside 2004) has been used to generate the results shown in Figure 4.3. These results give an attenuation of 0.88 dB for a

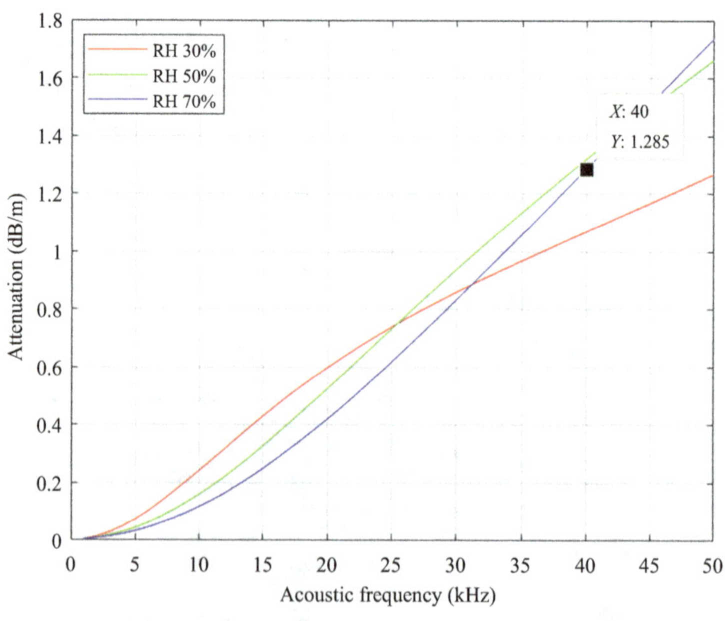

Figure 4.3 Sound attenuation as a function of frequency and relative humidity in air 20 °C, calculated from Burnside (2004)

frequency of 30 kHz, at 20 °C and a relative humidity of 30%. This result is in reasonably good agreement with the results from (4.1).

The maximum range that can be achieved by ultrasound sonar systems is determined by a combination of the beam divergence, the atmospheric attenuation and the target size. If the target is a fixed size, like a street sign, then the received power decreases with R^4, made up from an R^2 term on the way to the target and a second R^2 term on the way back. For a flat target like a wall, the received power decreases with the square of the range, R^2, because the reflecting area increases with R^2 which cancels one of the terms of the previous case. These losses in conjunction with an atmospheric attenuation of about 1 dB/m limit the maximum range to about 5 m. However, that is more than sufficient as an extension to the white cane.

The beamwidths of most ultrasound transducers are diffraction limited with the result that the larger the transducer diameter, the narrower the beam. As a rule of thumb, this relationship can be calculated as

$$\theta_{3dB} = \frac{25000}{df} \tag{4.2}$$

where θ_{3dB} (deg) is the half power beamwidth of the transducer, d (mm) is the diameter of the transducer and f (kHz) is the frequency.

The sizes of the transducers used, and hence their beamwidths, are determined by the application. For example, a unit mounted on the frame of a pair of eyeglasses would have to be much smaller diameter than one that is handheld or attached to a belt, and as a result its beamwidth would be much larger.

4.3.4 Some existing visual substitution systems

In the past fifty years, many sonar-based visual prostheses have been introduced, with the aim of improving mobility in terms of safety and speed. Some of the more successful devices are discussed in the following section.

4.3.4.1 Pathsounder

This is one of the earliest ultrasonic travel aids. Two ultrasonic transducers are mounted on a board that the user wears around the neck, at chest height. This unit provides only three discrete levels of feedback (series of clicks), coarsely indicating distances to an object.

4.3.4.2 Mowat sensor and derivatives

Several commercially available hand-held ultrasonic-based devices that inform the user of the distance to detected objects by means of tactile vibrations and/or sound have been produced over the years. The frequency of the vibration of these devices is typically inversely proportional to the distance between the sensor and the object.

The Nottingham Obstacle Detector (NOD) is an early hand-held sonar device that provides auditory feedback that indicates eight discrete levels of distance by different musical tones. The NOD has been commercially available since 1980 (Bissitt and Heyes 1980).

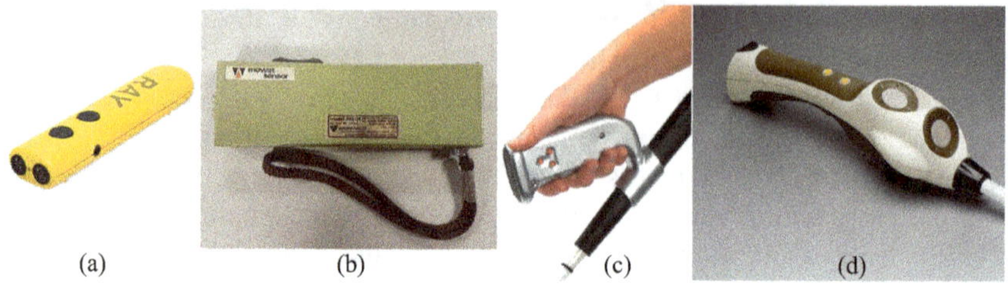

Figure 4.4 Photographs of hand-held ultrasonic prostheses: (a) Ray Ultrasound, (b) Mowat sensor, (c) K-sonar and (d) UltraCane, reproduced with permission from the various copyright holders including Vision Australia

Other sensors in this genre shown in Figure 4.4 include (a) the handheld Ray Ultrasound, (b) the original Mowat Sensor, (c) the K-Sonar shown attached to a cane and (d) the UltraCane built into the handle of a cane. The K-Sonar is one of the more successful prostheses which uses a wide band signal (one octave), probably based on frequency modulated continuous wave (FMCW) techniques to improve resolution and sensitivity. The multiple echoes are converted into "tone-complex" sounds which the user can learn to interpret for improved obstacle classification (K-Sonar 2008). Finally, the sleek UltraCane shown as the rightmost image uses two ultrasound sensors to probe the space in the direction of the cane and upward to detect targets at chest or head height. The feedback is delivered through two small vibrating buttons mounted in the handle.

4.3.4.3 Sonic pathfinder

This is a head-mounted pulse-echo sonar system comprising three receivers and two transmitters, controlled by a microprocessor. Developed in 1984 by Tony Heyes as an improvement to the NOD, it does not give information about surface texture, and normally the auditory display indicates only the nearest object, which is why it should be classified as an obstacle detector rather than as an imaging device. Heyes' approach is rather different from Kay's in that the Sonic Pathfinder deliberately supplies only the most relevant information needed by the user, whereas Kay strives for more information-rich sonar-based displays.

Heyes argues, for instance, that if some object moves away from the user, the user does not need to know this with regard to safe and efficient travel. In addition, this minimalist approach results in less confusion by minimising interference with hearing normal environmental sounds (e.g., traffic). The Sonic Pathfinder uses brief tones on a musical scale to denote distance, with pitch descending when approaching an object. Perception of objects on the left and on the right is further supported by tones for the corresponding ear. Two off-centre sonar beams are used to detect objects on the left or right to ensure that the user is aware of obstacles in their periphery.

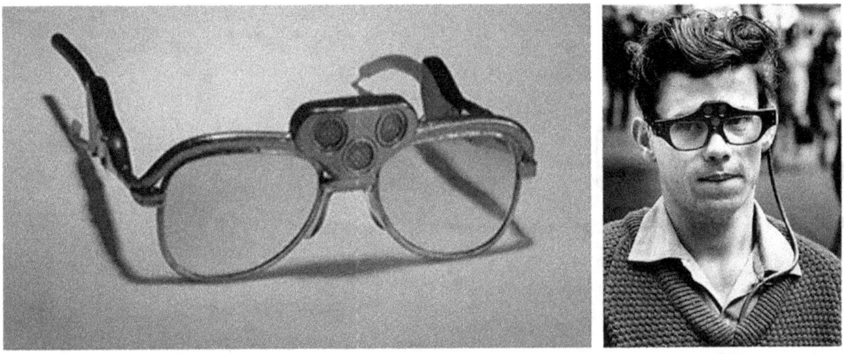

(a) (b)

Figure 4.5 Photograph of the original Binaural Sonic Aid (Kay 1974), reproduced courtesy Zabonne Ltd with permission

4.3.4.4 Binaural sonic aid (Sonicguide)

As can be seen in Figure 4.5, this device comes in the form of a pair of spectacle frames, with one ultrasonic wide-beam transmitter mounted between the spectacle lenses and one receiver on each side of the transmitter (Kay 1974). FMCW signals from the receivers are demodulated and presented separately to the left and right ear. The resulting *interaural amplitude difference* allows the user to determine the direction of an incident echo and thus of an obstacle.

In addition to providing wide field of view returns, the newer Trisensor is fitted with a narrow beam transmitter to provide information about the size and location of objects directly ahead. In both the Sonicguide and the Trisensor, distance to an object is encoded in the frequency of the demodulated low-frequency tone while surface texture is returned by the timbre of the audio signal.

Psychophysical studies conducted at short range (within arm reach) showed that users were able to judge the direction of a small cylinder to an accuracy of about 5° compared to 1° for a source of natural sound. In contrast, range estimation was improved, with subjects able to judge the distance to an object to an accuracy of 50 mm that was significantly better than their ability to judge the distance to a natural sound. At longer ranges (up to 5 m), users were able to judge direction to an accuracy of 6° and distance to within 300 mm (Warren and Strelow 1984).

4.3.4.5 NavBelt

This device was developed as part of a PhD thesis in 1989. It provided imaging and guidance operational modes. In the imaging mode, the device outputs a crude 120° wide view of obstacles ahead of the user. This was translated into a series of directional (stereo) audio cues that allowed the operator to determine which directions were blocked and which were clear. In the guidance mode, illustrated in Figure 4.6, the NavBelt knows the user's position and their ultimate destination

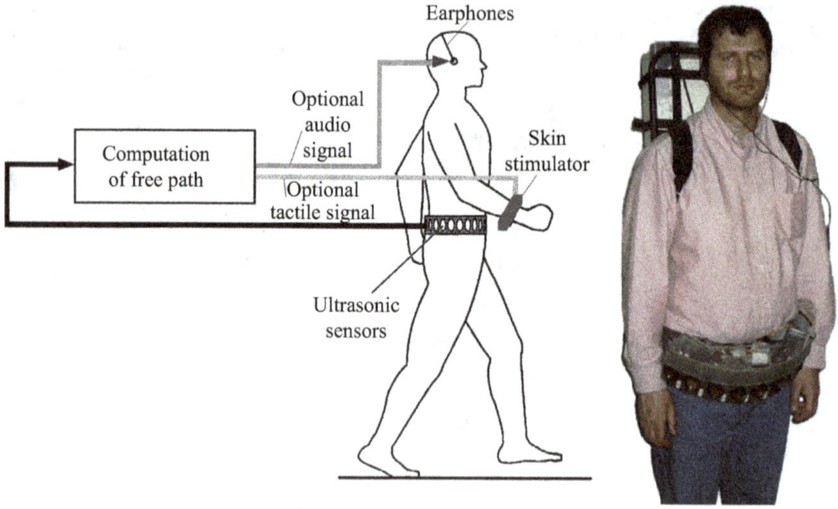

(a) (b)

Figure 4.6 Graphic showing the operational principle of the NavBelt and a photograph of the prototype (Borenstein and Ulrich 1997), reproduced with permission

from which its computer would calculate the free path and recommend the direction of travel (Borenstein and Ulrich 1997)

The sonar array used in the NavBelt uses Polaroid (now SensComp) transducers each with a beamwidth of 15° requiring that eight be used to cover the 120° for the situational awareness mode. Each of these sensors outputs the range to a single target within its beam, if there is one within range, to form a crude map of the obstructions around the user.

It is possible to process the phase information returned by individual elements of the array in much more sophisticated ways to improve the angular resolution of the sensor and to direct the beam.

As an introduction to the idea of beam steering, consider the array of ultrasound elements shown in Figure 4.7. In this case, each of the elements is radiating the same frequency signal, with the same phase in (a) but with a different phase – one that increases linearly across the array – in (b). If a line of constant phase is drawn for each of the elements, a linear phase front is produced parallel to the array in the first case and at an angle to the array in the second. This phase front defines the direction in which the beam is synthesised, or the direction from which a signal will be received.

In a conventional linear array, the power received by each element is the sum of the received powers scattered by the target from all the transmit elements. The voltage outputs of all N elements are summed without delays or phase shifts to give E_a (V) as shown in Figure 4.8. Since each element observes the same phase,

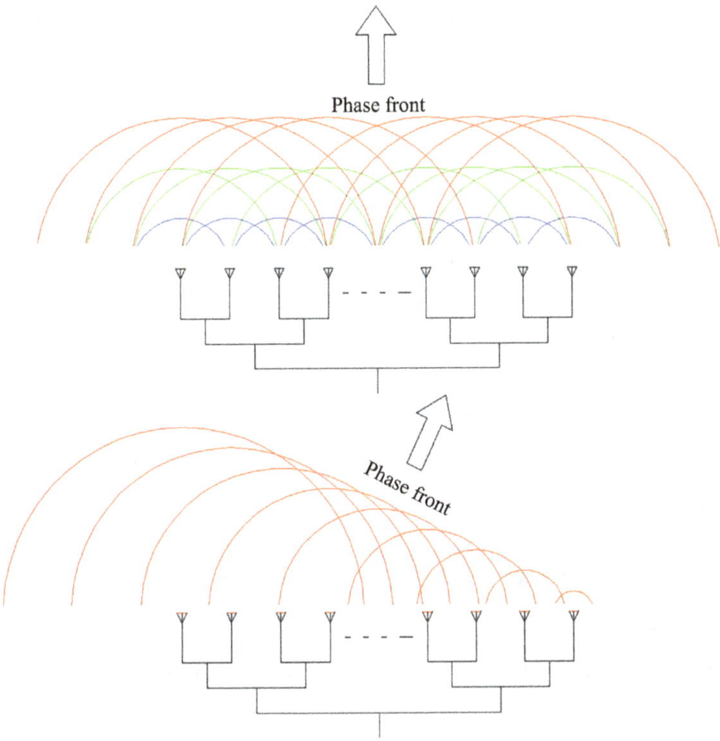

Figure 4.7 *Radiation from individual elements with (a) equal phase and (b) a linearly increasing phase shift combine to form a linear phase-front in a specific direction determined by the phase difference between successive elements (Brooker 2021)*

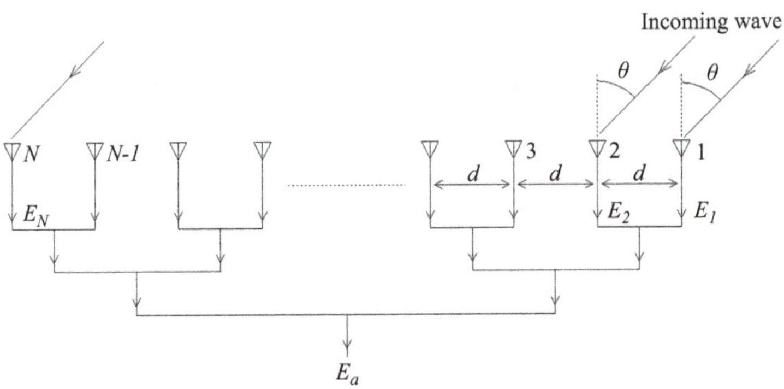

Figure 4.8 *The Array beam forming process is achieved by summing the signals from all the array elements in phase (Brooker 2021)*

ϕ_k(rad), the output after summing the signals from all N elements directly, with no phasing is given by

$$E_a = \sum_{k=1}^{N} \sin(\omega t + \phi_k) \tag{4.3}$$

It can be shown that the phase shift, ψ, between elements is constant and proportional to the spacing between the elements, d, and the sine of the angle of arrival, θ

$$\psi = \frac{2\pi d}{\lambda} \sin\,\theta \tag{4.4}$$

If the relative phase of each of the elements is referred to the centre of the array, then the equation for the sum can be written as

$$E_a = \sum_{k=1}^{N} \sin\left[\omega t + k\psi - \frac{N+1}{2}\psi\right] = \sin(\omega.t)\frac{\sin(N\psi/2)}{\sin(\psi/2)} \tag{4.5}$$

Substituting for ψ into equation (4.5) results in the classical equation that defines the voltage output of the one-dimensional (1D) linear phased array

$$E_a = \sin(\omega t).\frac{\sin\left\{\frac{N\pi d}{\lambda}\sin\theta\right\}}{\sin\left\{\frac{\pi d}{\lambda}\sin\theta\right\}} \tag{4.6}$$

By introducing linearly increasing phase shifts across the array, it is possible to steer the main lobe of the beam over a wide sector while still maintaining a reasonably narrow beam and low sidelobes as shown in Figure 4.9.

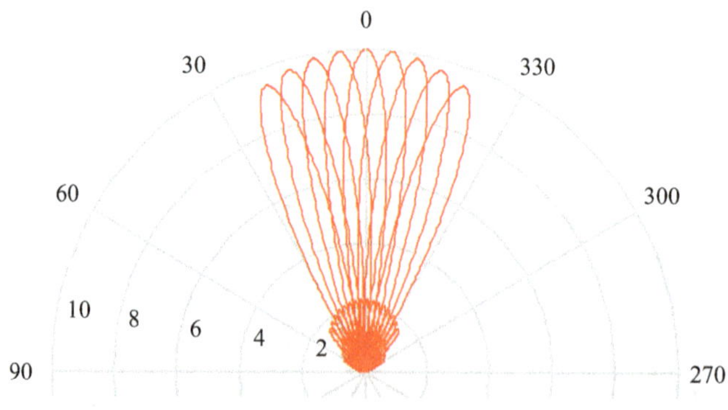

Figure 4.9 Beam scanning of a ten-element linear array with a 4 mm spacing between elements

```
% Phased Array
% phased_array.m
degrad = pi/180;
v = 340;                    % speed of sound in air (m/s)
f = 40.0e+03;               % frequency (Hz)
N = 10;                     % number of elements
lam = v/f;                  % wavelength (m)
dm = 4.0e-03;               % element spacing (m)

% generate angle theta
for ang = -20:5:20
       theta = (-pi/2:pi/1000:pi/2);
       theta0 = ang*degrad;        % beam direction
       % generate the phase shift per element
       phi = 2*pi*dm.*(sin(theta)-sin(theta0))./lam;
       % generate the amplitude factor
       amp_fac = sin(N*phi./2)./sin(phi./2);
       % generate the element factor
       el_fac = cos(theta);
       % intensity pattern abs(amplitude factor x element factor)
       int_pat = (abs(amp_fac.*el_fac));
       nn = find(int_pat<0);
       int_pat(nn) = 0;
       polar(theta,int_pat,'r');
       if ang == -20
              hold          % hold the plot to allow overlay
       end
end

% set image orientation
set(gca,'CameraUpVector',[1 0 0])
grid
```

In reality, the beam pattern of the array is modified by the shape of the beam of the individual elements. For that reason, small elements with wide beamwidths rather than the large elements shown in Figure 4.6 are used. This also allows the elements to be packed close together, typically $<\lambda/2$ apart, otherwise unwanted grating lobes appear on either side of the main lobe (Brooker 2021).

4.3.4.6 GuideCane

This is a fairly recently developed device that has been designed at the University of Michigan to help blind people navigate safely and quickly (Borenstein and Ulrich 1997). It consists of an array of ultrasonic sensors on a wheeled robotic vehicle that can navigate through a cluttered environment. The operational principles are shown in Figure 4.10.

The GuideCane is used much like a white cane though it is heavier and rolls on a pair of wheels. A computer-controlled steering servo directs the wheels with respect to the cane. An array of ultrasound sonar transducers mounted on a circular arc senses the environment to the front and sides. In addition, a flux gate compass and wheel encoders provide information about the direction and distance travelled. Finally, a miniature, thumb-controlled joystick is positioned on the handle of the cane.

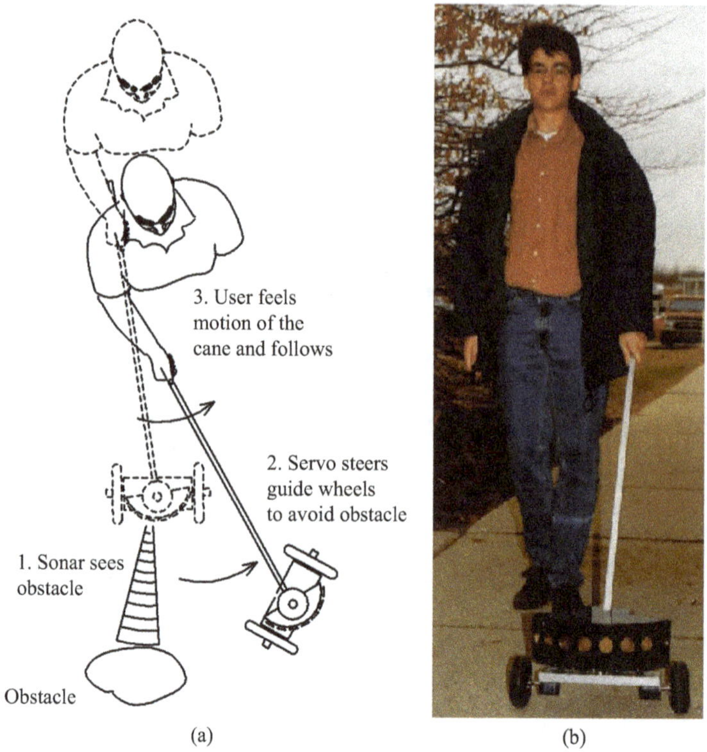

Figure 4.10 Schematic of the operational principles and photograph of the GuideCane prototype in Operation [adapted from Borenstein and Ulrich (1997)], reproduced with permission

The joystick commands the initial direction of travel, either straight ahead or to the left or right from the current direction, and the steering servo will turn as the GuideCane is pushed forward until it has reached that direction after which it will follow a straight path. The sonar array images over the 120° sector ahead looking for obstacles as it moves forward. This obstacle map is then used by the onboard computer to calculate the best direction to travel, even through a cluttered environment. If an obstacle is detected in the path, the steering servo drives the angle of the GuideCane wheels in a direction to avoid it. The operator feels the changed resistance on the cane even before the direction change is detected and intuitively follows the direction suggested with almost no hesitation.

The GuideCane solves the problem of stairs in two ways. When a drop off is reached, the wheels drop off the edge providing a fail-safe indication of the fall. In the other direction, the main forward-looking array sees the bottom step as an obstacle, but the forward-looking up-facing sensor measures a longer range. If the difference between the two ranges is about 300 mm, the object is treated as a flight of stairs. However, if the two distances are almost the same, then the object is treated as a wall and avoided.

GuideCane incorporates two additional functions. The first is a side-looking sonar for wall following and the second is GPS which can be used for navigation of preprogrammed routes. The latter are generated automatically during a "lead-through" run under the guidance of a sighted person.

For indoor navigation, dead reckoning based on compass readings and wheel-sensor odometry can be used for short distances, but the integrated error grows too large to achieve the required accuracy for distances further than a few metres. However, algorithms such as simultaneous localisation and mapping (SLAM) (Leonard and Durrant-Whyte 1991), a well-known algorithm used by the robotics fraternity, could be implemented to solve this problem.

4.3.5 Issues with sonar-based systems

No visual prostheses are without problems, and all of those here based on sonar discussed have some shortcomings:

- With most of the systems, the user must actively scan the environment to detect obstacles (no scanning is needed with the *Sonicguide*, but that device does not detect obstacles at floor level). This procedure is time-consuming and requires the user's constant activity and conscious effort.
- The user must perform additional measurements when an obstacle is detected, in order to determine the dimensions of the object. A path must then be planned around the obstacle. Again, a time-consuming, conscious effort that reduces walking speed.
- One problem with all blind aids based on acoustic feedback is their interference (called *masking*) with the blind person's ability to pick up environmental cues through hearing (Kay 1974, Lebedev and Sheiman 1980, Brabyn 1982). This is a serious concern as almost all blind people use acoustic information for situational awareness, obstacle detection and navigation.
- The GuideCane is large and rather unwieldy, it must be lifted up steps and roadside curbs. The wheels can easily wedge into cracks and holes in the pavement.

4.4 Laser-based systems

Introduced in 1973, the C-5 laser cane manufactured by Nurion-Raycal and shown in Figure 4.11 is based on optical triangulation with three collimated lasers and three photo-diodes as receivers (Benjamin *et al.* 1973). The three modules are aimed parallel to the cane, at 45° and perpendicular to it. It can detect obstacles at head-height, drop-offs in front of the user, and obstacles up to a range of 1.5 m or 3.5 m ahead of the user depending on the close/far setting.

As laser and lidar technology has developed over the past fifty years, lidar-based sensory aids have become far more sophisticated. A good example of the genre is the LidSonic (Busaeed *et al.* 2022) that uses a high-resolution scanned lidar to augment the outputs from a conventional sonar of the sort described earlier

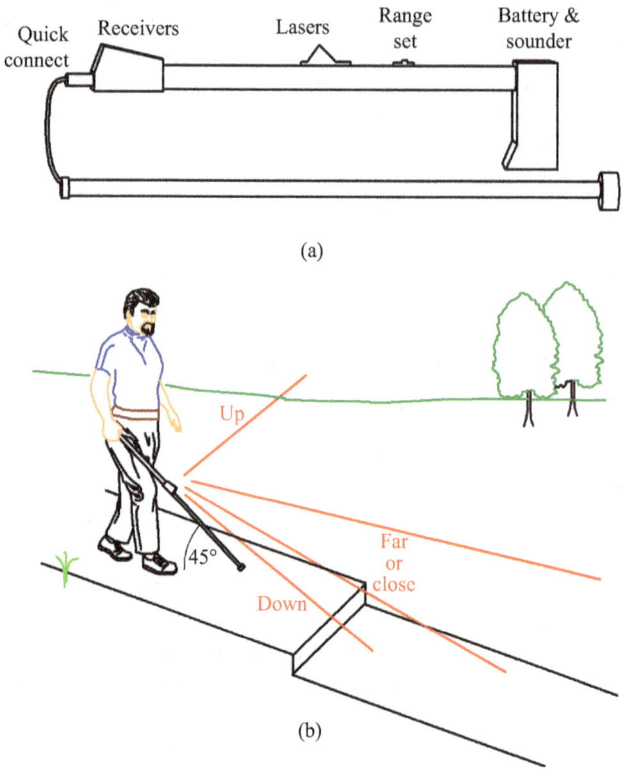

Figure 4.11 *Sketch (a) of the Prototype C-5 laser cane and (b) the application showing the directions of the laser beams, [adapted from Benjamin et al. (1973)]*

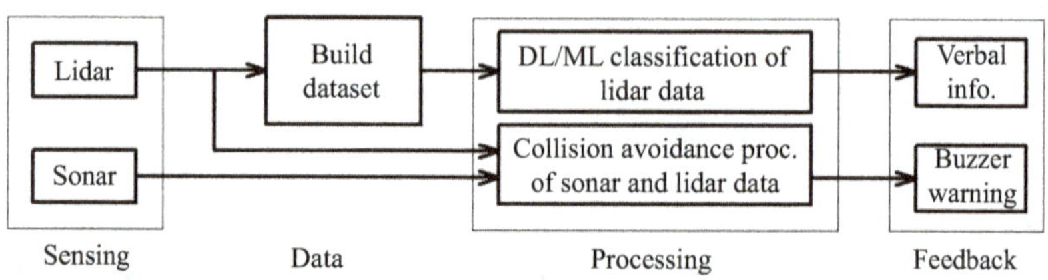

Figure 4.12 *Functional Overview of the LidSonic System, [adapted from Busaeed et al. (2022)]*

in this chapter. Figure 4.12 shows the high-level functional overview of the system. A Bluetooth link transfers both Lidar and Sonar data to a dataset on a smartphone where deep learning/machine learning (DL/ML) algorithms classify the individual targets in the scene, after which verbal feedback communicates this information to

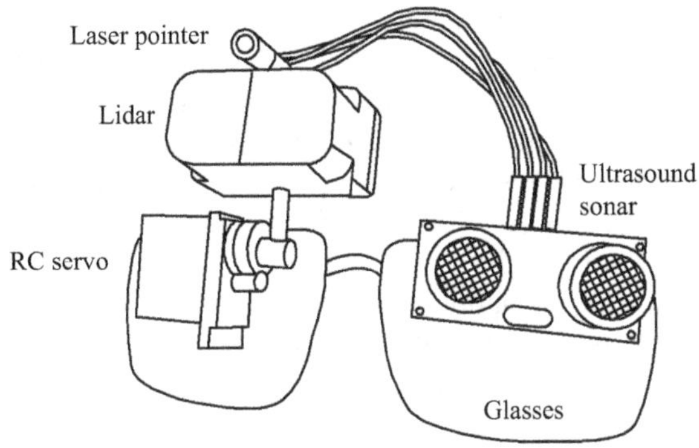

Figure 4.13 LidSonic System mounted on a glasses frame, [adapted from Busaeed et al. (2022)]

use user. During the operational mode, real-time data from both sensors is processed and a buzzer warns the user of any obstacles in their path.

The system shown in Figure 4.13 mounted on a pair of glasses, comprises a scanned TFmini-S Lidar that runs from 5 V and provides a new range measurement every millisecond over a range between 10 cm and 12 m. The accuracy out to 6 m is ±6 cm. When scanned by the RC servo, it produced high-resolution profiles of the ground and any obstacles in the beam. Its main disadvantage is that it cannot detect transparent objects such as glass doors. The HC SR04 ultrasound sonar is incorporated to compensate for this. It has a beamwidth of about 15° and a maximum range of about 3 m.

4.5 Visual substitution

Effective vision substitution modalities depend on the probing sensor, typically a sonar or a camera, and the substitution sense, generally tactile or auditory.

One of the major problems with auditory sensory substitution is the masking effect, in which sound cues normally used by a blind person are masked by feedback from the prosthesis. Under laboratory conditions, the interference of an auditory prosthesis with normal hearing is probably not an issue, and it is possible that it may be one that can be overcome outdoors provided the system does not block the hearing system completely. If the system only blocks subtle clues about the environment, normally perceived and used only by blind persons, but replaces them with much more accurate and reliable feedback, it may be an acceptable trade off.

4.5.1 Input from sonar

As the resolution of sonar prosthetics has improved, so has the requirement for higher and higher bandwidth perception. This detailed sensory environment has been championed by Professor Leslie Kay starting with the Sonicguide in 1974. More recently, he has improved the angular resolution of his Binaural Sensory Aid (BSA) system using a third narrow-beam transmitter for creating an additional monaural signal. This newer system was first called the Trisensor but is now known as the KASPA system (Kay's Advanced Spatial Perception Aid). In 1998, this system won Kay the Saatchi and Saatchi Innovation in Communication Award.

Kay is an advocate for supplying blind users with a rich sensory environment, and KASPA is an example of such a prosthesis that is shown in Figure 4.14. It represents object distance by pitch and surface texture through timbre (the sum of several harmonically related sounds). To achieve this, use is made of the high-resolution capability of wide bandwidth FMCW signals. The improved, but still modest, resolution positions Kay's system somewhere between obstacle detection and environmental imaging. The best angular resolution is about one degree in the horizontal plane (azimuth detection) for the central beam, which is quite good, but vertical resolution (elevation) is poor. This makes the "view" akin to using binoculars to view the world through a narrow horizontal slit.

It is difficult to settle on a good measure for equivalent image resolution as offered by sonar approaches, because sonar is not just based on mapping a

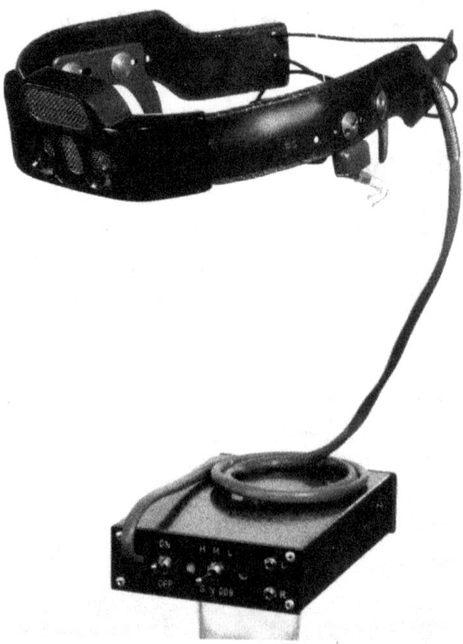

Figure 4.14 Photograph of one of the KASPA prototypes, reproduced courtesy Zabonne Ltd with permission

two-dimensional (2D) projected image to another sensory modality. It can be used to provide a much richer representation of the environment based on ultrasonic interference and acoustic delay patterns. In this regard, Kay's work does demonstrate some features (resolution, texture, parallax) that support its classification as an imaging device for vision substitution.

4.5.2 Input from cameras

Concerning the input from the environment, true visual, camera-based input has some major advantages over the use of sonar echo patterns. The possibilities of sonar have been investigated rather extensively, and its short operational range makes it impossible to perceive distant objects which are useful for guidance. Furthermore, visual input matches the most important information providers in our surroundings, such as road signs, books and television. From a technical perspective, it is difficult to obtain unambiguous high-resolution information using a scanning sonar, while any commercially available low-cost camera will perform the function well.

It should be noted that this only holds if the camera input is supplemented by depth-clues through changes in perspective, to resolve ambiguities in distance. Relative depth information can be obtained from changing the relative positions of the viewer and the environment in combination with some knowledge of the sizes of objects in the scene. An equivalent of binocular vision is no strict prerequisite, but there might be advantages in presenting the views from the individual eyes as sound patterns to their associated ears.

Some research groups have tried to simulate vision substitution as accurately as possible. For example, Capelle (Capelle *et al.* 1998) represents the centre of the visual field with more pixels, just as it is in the eye, with each pixel being represented by a particular frequency. Because this system cannot present the visual world continuously, adjacent pixels cannot always be represented by neighbouring sinusoids, which makes the representation counter-intuitive and potentially more difficult to interpret than a linear mapping.

4.5.2.1 The vOICe system

A different approach was followed by Peter Meijer (Meijer 2006). In order to increase the image resolution obtainable via an auditory representation and to decrease the required bandwidth, a time-multiplexed mapping is performed to distribute an image in time. The 2D spatial brightness map of a visual image is scanned and transformed into a 2D frequency and time map. It is thought that the subdivision of only one of the two spatial dimensions of the visual image into individual scan lines will be more amenable to a later mental integration than the subdivision in both dimensions into individual pixels.

The human brain is far superior to most, if not all, existing computer systems in rapidly extracting relevant information from blurred and noisy images. Therefore, no attempt is made by Meijer to filter any redundancy out of visual images that are being transformed into sound patterns. From a theoretical perspective, this means that the available bandwidth is not exploited optimally. However, by keeping the mapping as direct and simple as possible, the risk of accidentally filtering out important clues is reduced.

The principles used for spatial encoding of the visual signal to a sound sequence are shown graphically in Figure 4.15 for a grey scale image. Image processing has reduced the spatial resolution to only 110 pixels (10×11), and the intensity to only three grey-scale levels ($M=10$, $N=11$, $G=3$). The mapping translates the vertical position of each pixel into a frequency proportional to its position in the column with the amplitude proportional to the pixel brightness. The horizontal position maps to a delay after the start time (denoted by a click), or to an amplitude difference in a stereo system.

As the complete audio signal from each column is output simultaneously, and there are N columns, the sound is available for T/N seconds for a total scene update rate of T seconds. For each column, j, every pixel is used to excite an associated sinusoidal oscillator within the audible band, with the lowest frequencies at the bottom and the highest at the top. In addition, the oscillators form an orthogonal basis in which they are all integer multiples of some reference frequency. This ensures that all of the information is preserved in this transformation from geometrical to Hilbert space (Meijer 2006).

The M oscillator signals from column j are superimposed with the corresponding sound pattern presented to the ear for T/N seconds before the sound from column j+1 is output. This continues until the Nth column has been converted after T seconds, when the whole cycle is repeated for a new image. Images are separated by a synchronisation click which is essential to ensure that the subject can reorientate laterally. In addition, the relative amplitude of the sounds differs in the two ears to produce a stereo effect, so that the sound appears to come from the correct direction.

The spectrum of a sinusoidal signal with duration T/N seconds includes other frequency components. These can be determined by convolving the Fourier

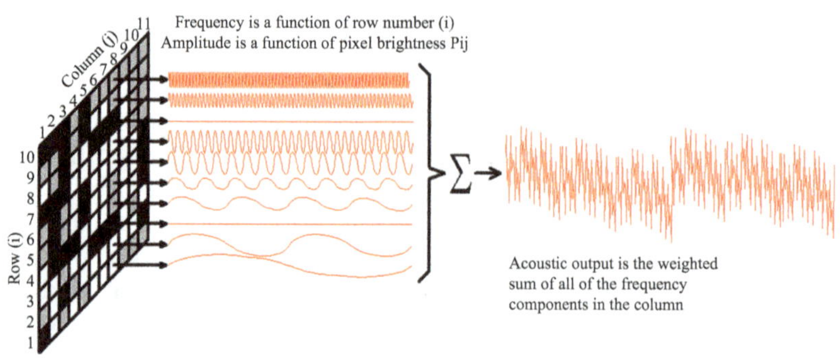

Figure 4.15 Conversion of an 10×11 grey scale image to sound [adapted from Meijer (2006)]

transform of the rectangular window with that of the sinusoidal signal. One important consideration is that even the lowest frequency is represented by a reasonable number of complete cycles during the observation time; otherwise, it would not be interpreted correctly by the ear.

This processing strategy has been implemented in a sensory substitution prosthesis called vOICe, with which, according to Meijer, it is possible to learn to sense instinctively how the features of a soundscape correspond to objects in the physical world. Pat Fletcher, a proficient user of the vOICe who could see until age 21, describes the sound images in her head as "ghostly" but real. At a meeting of the Cognitive Neuroscience Society in New York, researchers from Harvard Medical School announced that when they viewed the activity in the brains of two vOICe users (one blind at birth, the other who went blind later in life), it was in many respects like that of a sighted person while seeing. However, in tests undertaken by Alex Storer (Storer 2006), though his subjects' identification capabilities improved using vOICe for long periods, none felt that the experience was visual in nature.

For obvious reasons, not everyone has the inclination to walk around with a head-mounted camera and a laptop, so Meijer has modified his setup to work using a camera phone. Now, after downloading a simplified version of the software from www.seeingwithsound.com, practically anyone can point their camera phone at what they want to see and listen to what it looks like.

With the now ubiquitous availability of smart glasses, the vOICe system can be operated from any of a wide range of options such as the VISION-800 AR glasses shown in Figure 4.16.

So far, the research community has not yet found conclusive answers with respect to the potential of this approach for the blind, and the relevant limitations in

Figure 4.16 New generation of the vOICe software runs on smart glasses such as the VISION-800 AR glasses

human auditory perception and learning abilities for comprehension and development of visual interpretation remain largely anecdotal. In addition, the training effort is expected to be significant while involving perceptual recalibration for accurate sensorimotor feedback. In 2010, the US National Science Foundation funded the first controlled study to look at the benefits of the system while simultaneously attempting to find an optimal training protocol (Trivedi 2010).

One of the key research questions is to what extent the use of a sensory substitution system cannot only provide synthetic vision in a functional sense for extended situational awareness through active sensing, but also lead to visual sensations through forms of induced artificial synaesthesia.

Other researchers including Stefan Strahl of the UCL Ear Institute in London and Lucy Irving of the University of Brighton have been examining similar aspects of sensory substitution using the vOICe system, with most of their results available on the web.

Functional MRI results show that in novice users, the auditory cortex is involved in the interpretation of the soundscape, but after 10 - 15 hours of training, activity in the visual cortex and an area known as the occipital tactile-visual (LOtv) region increases. Fairly recent experiments on advanced users including Pat Fletcher involved the use of repetitive transcranial magnetic stimulation (rTMS) to close down the LOtv region. When this occurred they lost the ability to "see" and became confused and frightened (Trivedi 2010).

4.5.2.2 The sound of vision system

The Sound of Vision (SoV) project is much more recent than the vOICe and has attempted to tackle all of the challenges that limited the former's adoption including wearability, real-time operation, pervasiveness, utility and cost. To achieve these goals, the system would have to be operational both indoors and outside irrespective of illumination conditions. The system architecture comprises the following four steps:

- 3D data acquisition from the surrounding environment,
- 3D environment reconstruction and segmentation,
- Haptic and audio modelling of the processed 3D scene,
- Haptic and audio feedback to the user.

To achieve the 3D acquisition, two depth sensors are used, the PS1080 IR-based depth sensor from Occipital, Inc. San Francisco, California and the LI-OV580 stereo vision camera from Leopard Imaging Inc., Fremont California. To track the orientation of the head and cameras, the LPMS-CURS2 IMU from LP Research Inc. Tokyo is rigidly mounted to the headgear as shown in Figure 4.17.

Data capture can include each of the 3D acquisition sensors individually or in combination as well as using recordings captured earlier. This sensor fusion maximises the probability that good environmental data can be obtained irrespective of the situation.

Data fusion includes both colour and depth information to identify various structures such as doors, signs and written text. Essential for scene segmentation,

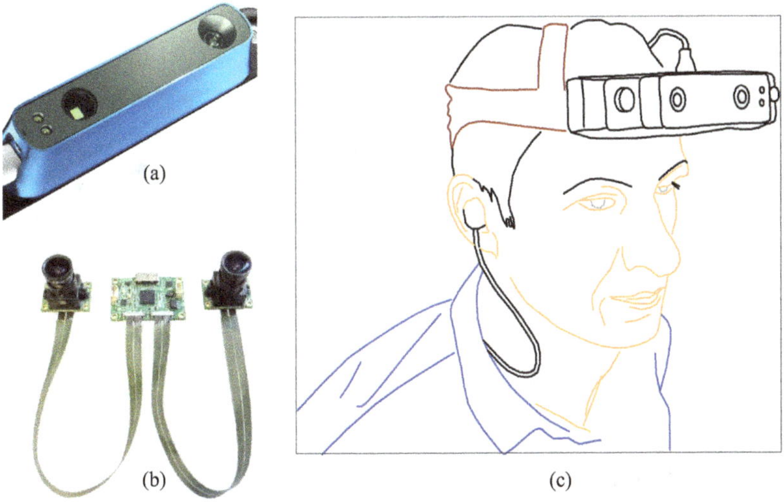

Figure 4.17 The Sound of Vision Headgear showing (a) Structure Sensor PS1080 from Occipital, (b) Stereo Camera LI-OV580 from Leopard Imaging and (c) the integrated system [adapted from Caraiman et al. (2019)]

processing needs to detect the ground (and the ceiling indoors) and the partitioning of the remaining 3D information as potential obstacles.

Navigation looks at the 3D content and updates the scene structure regularly to ensure efficient obstacle avoidance, while attempting to minimise the number and complexity of the messages transmitted to the user in such a way that interpretation is possible even while the user is walking. The user is provided with continuously updated feedback based on a labelled depth map of the scene, or with specific haptic/sound patterns encountered by an ever-expanding sphere. Additionally, user feedback is augmented with information regarding dangerous situations including the existence of negative obstacles such as holes or stairs leading downward or obstacles at head height. Outputs are encoded as acoustic tones or the vibrations of a custom-designed haptic belt (Caraiman *et al.* 2019).

The significance of this approach includes the following:

- perception of the environment and the ability to move independently without the use of predefined tags or external sensors,
- the system is pervasive in that it can operate indoors or outside irrespective of variations of illumination,
- a rich and naturalistic description of the environment based on original combinations of sound and vibration,
- a real-time description of the environment is provided and updated sufficiently fast to allow the user to walk comfortably,
- efficient training programs and tools based on virtual and augmented reality implemented together with feedback from trained instructors.

4.5.3 Cutaneous tactile displays

Electrotactile stimulation, sometimes called electrocutaneous stimulation, evokes tactile sensations within the skin at the location of the electrode by passing a local electric current through it. In contrast, vibrotactile stimulation evokes tactile sensations using mechanical vibration of the skin, typically at frequencies between 10 Hz and 500 Hz (Kaczmarek *et al.* 1991).

Because of the relative simplicity of the technology required, there are many systems, both commercial and experimental, that have been developed over the past half century to perform these functions. They range from single-element devices, through lines of transducers to two-dimensional arrays.

Single stimulation points can convey information through the skin using variations of intensity/amplitude, frequency or both. The source of information could be temporally varying, such as a simple auditory prosthesis that translates the envelope of inputs from a microphone directly into vibration intensity.

One-dimensional (1D) arrays consisting of two or more actuators in a line present spatial information more naturally. A good example of this is feedback from a prosthetic arm in which the elbow angle is mapped to the excitation of a linear array mounted elsewhere on the body. From a visual substitution perspective, a 1D array can map the outputs of an array of sonar sensors for devices such as the NavBelt.

Two-dimensional (2D) displays, also referred to as tactile vision substitution (TVS) devices, can display spatial information to the skin in a similar manner to the way that the lens of the eye presents spatial information to the retina. At its most primitive, each pixel from a camera is mapped directly to an individual vibrotactile or electrotactile element that is excited in proportion to the pixel intensity. Many researchers including Bach-y-Rita developed systems in the 1970s to study the skin's ability to interpret visual information. They found that subjects could immediately recognise vertical, horizontal and diagonal lines, and that with experience, users could recognise common objects and even people's faces (Kaczmarek *et al.* 1991).

The commercially available Optacon (optical to tactile converter) and its more modern derivatives convert the outlines of written letters and simple graphics to a vibrotactile array on the user's fingertip. Exceptional users can read ordinary printed text at up to 90 words per minute using such devices.

4.5.3.1 Limitations

Despite the fact that touch, like vision encodes information spatiotopically, there are some cognitive limitations constraining the bandwidth of any visual equivalent sensory substitution method. There is little evidence to suggest that extensive brain plasticity has been demonstrated in response to high rates of spatiotemporally varying information presented via the skin surface irrespective of whether that surface is the back, stomach, forehead, or tongue. The consensus is that the skin interface will be restricted to various special-purpose devices that enable specific activities such as navigation, the control of locomotion, pattern perception, etc (Spence 2014).

4.6 Auditory substitution

Because of the large percentage of people worldwide who suffer from disabling hearing loss, researchers are starting to build sophisticated sensory substitution for the deaf. The work done extends far beyond the original flashing light to indicate door knocking or phone ringing that has been around for decades. Leveraging advances in high-performance computing into a sound-to-touch sensory substitution device worn as a vest, users were able to feel the sonic world around them. The signal processing of this sound field tried to emulate the cochlea by separating the sound into several frequency components spatially separated across the skin. Unfortunately, conveying cochlea-like quality information would require a few thousand vibrotactile devices, which is completely impractical. However using judicious compression techniques allows the number of tactors to be reduced significantly, to the extent that the array can be fitted across a chest strap or wrist band (Novich and Eagleman 2015). As an example, a wrist strap acoustic to vibrotactile device was shown to elicit activity in both the somatosensory and auditory regions, with anecdotal evidence suggesting that users no longer perceive the vibrations but instead actually hear the sounds being reproduced (Malone *et al.* 2021).

Given the quality of this form of sensory substitution, its low cost and simplicity, it is a wonder that it hasn't proven to be used as an alternative to the incredibly invasive and side-effect prone Cochlear Implants (Eagleman and Perrotta 2023).

4.7 Other applications

4.7.1 Proprioception

Until the fairly recent process of reinnervation of both motor and sensory nerves, people with prosthetics had a great deal of difficulty learning how to use them due to the lack of proprioception. Proprioceptive feedback of the position of the limb joints in relation to each other and to gravity can be used to stimulate individual or arrays of vibrotactile devices to overcome this deficit to some extent.

4.7.2 Balance

As discussed in the section on the BrainPort device later in this chapter, head orientation was fed to the tongue array with the position of the electrotactile stimulation determined by the tilt angle. Interestingly, residual benefits persisted long after the device had been removed, as the brain learned to use the residual vestibular signals to best advantage (Eagleman and Perrotta 2023).

4.8 Sensory expansion

Expansion of the normal human sensory capability includes increasing the wavelength range of both visual and audio perception available as well as improving the minimum detectable signal level using audio or optical amplification.

4.8.1 Image intensifiers

Most image intensifiers use a microchannel plate (MCP) between the photocathode and the phosphor as illustrated in Figure 4.18. Early MCPs consisted of about a million hollow glass tubes fused together into a disc. These tubes are about 10 μm in diameter and 1 mm long with the inside walls of each coated with a secondary emitting material to act as electron multipliers (Lampton 1981). A single-stage tube with an MCP can produce optical gains of up to 5×10^4. A modern second generation MCP, contains between 2 and 6 million holes to provide better resolution.

In the configurations shown in Figure 4.19, the potentials applied to the MCP are intermediate to those of the anode and the cathode. The proximally focused intensifier is similar to a two-electrode intensifier in as far as the gaps between the photocathode and the MCP and the MCP and screen are sufficiently short to minimise electron dispersion. In the focused intensifier case, the electrons emitted by the photocathode are focused by an electric field onto the MCP.

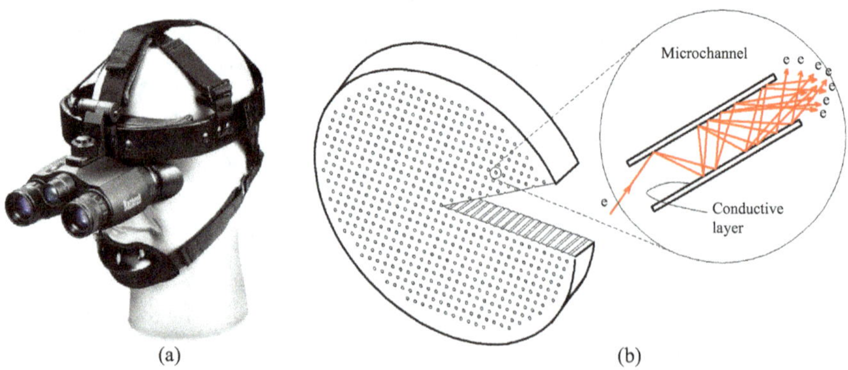

Figure 4.18 Microchannel plate (a) night vision goggles and (b) schematic principle of operation (Brooker 2021)

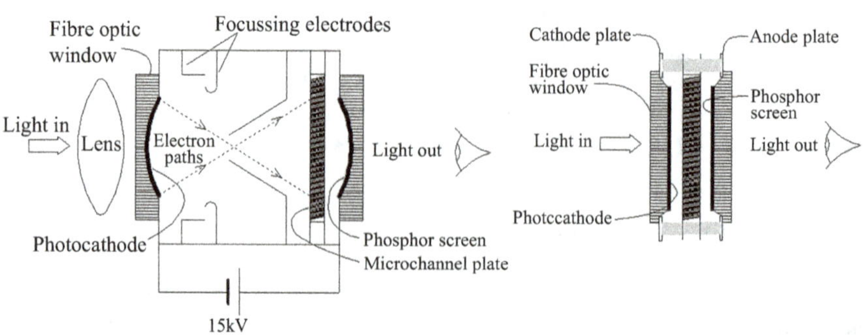

Figure 4.19 Second generation tubes using microchannel plates (Brooker 2021)

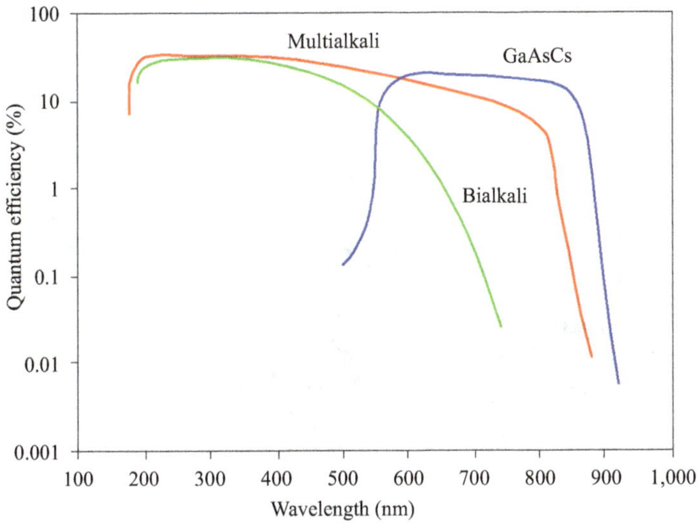

Figure 4.20 Third generation photocathode responses (Brooker 2021)

Third-generation tubes are similar in construction to second-generation devices but have a more sensitive GaAs or Multialkali photocathode, which extends into the near-infrared (NIR) band (450 to 950 nm) and beyond, as shown in Figure 4.20.

It is clear that the latest generation of image intensifiers improves both the minimum detectable signal as well as extends the wavelength range into the near infrared providing sensory extension.

4.9 Sensory addition

Sensory addition involves the introduction of completely new sensing modalities into the brain using various interfaces. One of the most common is the introduction of tiny neodymium magnets into the fingertips to allow the users to identify static or varying magnetic fields. The former is recorded as a small pressure while the latter is sensed as a vibration if the frequency is sufficiently low for detection.

More sophisticated and less invasive methods have used electromagnetic sensors and vibrotactile transducers incorporated into a wristband. This allows for the detection of higher frequency electromagnetic signals, and with the aid of some processing, the separation of different frequencies either spatially as occurs in the cochlea or as changes in the vibration frequency (Eagleman and Perrotta 2023).

The ability to detect the magnetic field around the Earth is another common application. In our laboratory, we have performed several experiments using wrist bands, ankle bracelets and haptic belts containing magnetometers and an array of vibrotactile transducers to indicate to the user the direction of magnetic north. The

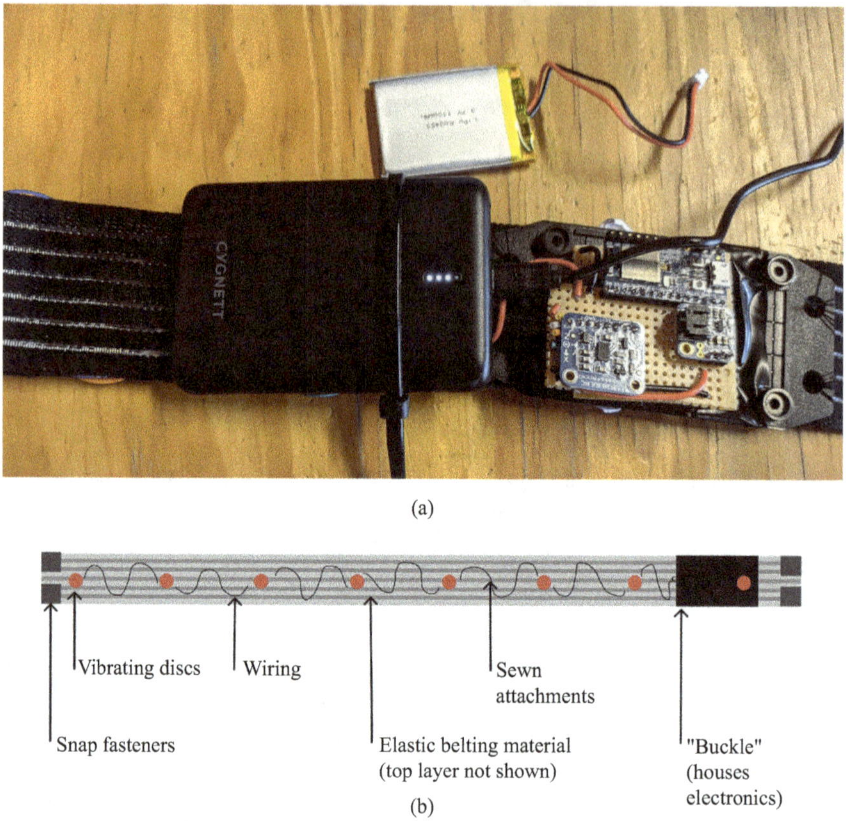

Figure 4.21 Photograph (a) of the extrasensory belt and (b) a schematic showing the placement of the vibrotactile discs [adapted from Player (2021)] reproduced with permission

objective of these experiments was to determine whether users could integrate the information directly into the brain without any cognitive overhead. Figure 4.21 shows the prototype haptic belt developed by Player (Player 2021). Tests conducted using the device showed that it could help with localisation while bush walking and maintaining a specific bearing during rowing.

4.10 The skin as an interface

Six types of tactile receptors (shown in Figure 4.22) have been identified and characterised in the skin, though all types are not available everywhere across the body with some present beneath hairy surfaces and some in hairless (glabrous) regions. These can be characterised by their response to a step change in the applied pressure to the skin. The response is measured by the change in the receptor's action potential firing rate.

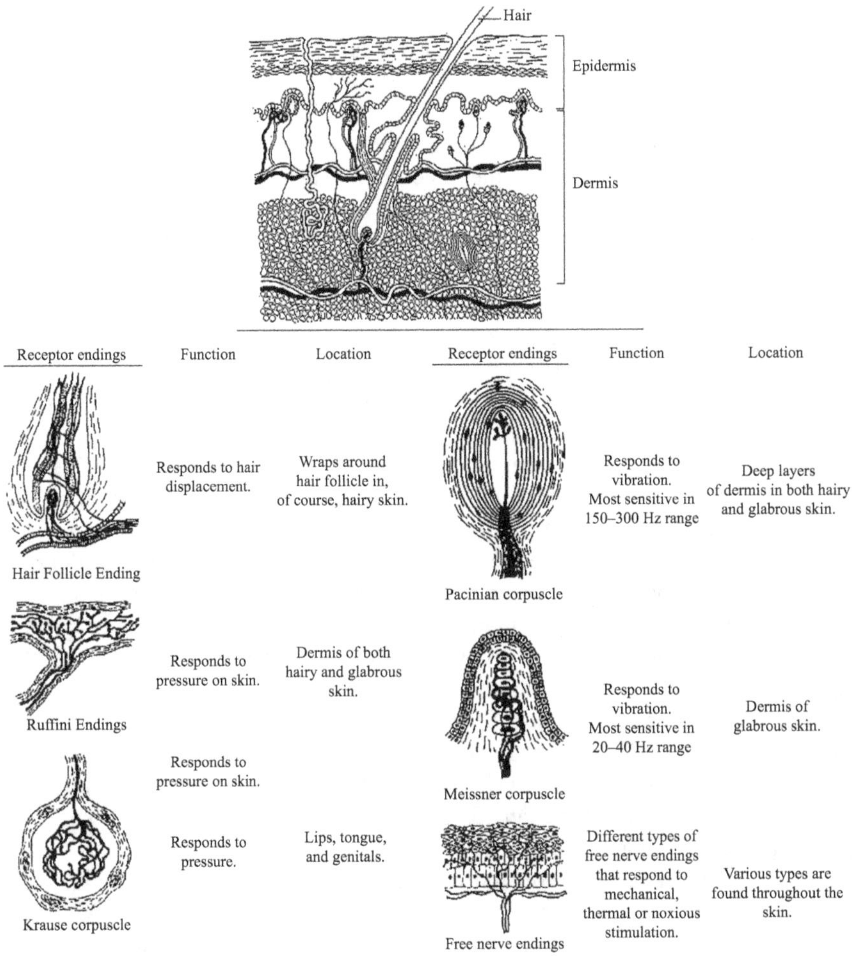

Figure 4.22 Anatomy of the skin and tactile receptors [adapted from Brodal (1969)]

Merkel cells innervate the human fingertip at a density of 100 elements/cm^2 and are mainly responsible for the detection and identification of spatial patterns. Meissner corpuscles are even more densely packed in the human fingertip (150 elements/cm^2) and are highly sensitive to dynamic skin deformation over a wide receptive field. These corpuscles poorly resolve spatial information but encode for skin motion by detecting low-frequency vibrations indicative of object slip. Pacinian corpuscles are fast-acting receptors with a broad receptive field, extremely sensitive to the smallest skin motion (in nanometres) and responsible for the perception of high-frequency stimuli. Their peak sensitivity occurs at stimulation frequencies of 200 to 300 Hz. Ruffini endings have an unknown function and have never been found in the skin of the fingertips (Pasquero 2006).

Temperature is sensed by two types of free nerve endings with one type measuring heat and the other type measuring cold. For temperatures above 45 °C and below 10 °C, temperature also stimulates pain endings.

Tests with fine wires and small vibrators have been used to quantify the sensitivity of different areas of skin. The static force required (Weinstein 1968) varies from 0.05 g on the lips to 0.63 g on the fingertips and belly, and up to 3.5 g on the sole of the foot. The fingertip threshold corresponds to a skin indentation of about 10 μm.

With regard to vibration, Geldard (1957) used a 1 cm^2 vibrator and found that the fingertips were the most sensitive of all the body locations by at least an order of magnitude. The abdomen was found to be 60 times less sensitive than the fingertips for 200 Hz vibrations.

The threshold amplitude for vibrotactile stimulation also increases after a strong conditioning stimulus with the reduction in sensitivity being proportional to the amplitude of the conditioning stimulus and its duration.

For vibrotactile and electrotactile displays, the spatial resolution of the skin is extremely important. The basic test is the two-point discrimination threshold (TPDT) and is summarised in Table 4.1. However, other methods to determine tactile spatial resolution include the determination of the minimum width of a deep groove which can be detected on an otherwise smooth surface. This turns out to be 0.87 mm on the fingertip, much smaller than TPDT. The orientation of grooves in a grating is another test that shows that the fingertip is much more sensitive than expected. In this case, the resolution is 0.84 mm.

The skin can also recognise a frictionless position shift of a stimulus ten times smaller than the TPDT can. This indicates that the skin's spatial resolution is much better for some tasks than the TPDT suggests. A good example is the ability of the fingertips to distinguish between different grades of sandpaper with very fine spatial frequencies (Kaczmarek *et al.* 1991).

The dynamic range of vibrotactile stimulation for a 0.78 mm^2 stimulator is about 40 dB if the maximum amplitude of the stimulation is limited to 0.5 mm.

Table 4.1 Static simultaneous two-point discrimination thresholds (Kaczmarek et al. 1991)

Location	**Static touch (mm)**	**Vibrotactile (mm)**	**Electrotactile (mm)**
Fingertip	3	2	<7
Palm	10	–	8
Forehead	17	–	–
Abdomen	36	–	10
Forearm	38	–	9
Back	39	11–18	5–10
Thigh	43	–	10
Upper arm	44	–	9
Calf	46	–	9

4.10.1 Vibrotactile displays

Most attempts at building vibrotactile displays have focused on devices that stimulate the fingertips because of their high tactile acuity. Some researchers have investigated other regions including the tongue and the lips which are also sensitive, or even the torso and the thighs which offer the advantage of a large available area. The relative proportion of the sensory cortex associated with these sections is graphically illustrated in Figure 4.23.

Most vibrotactile arrays operate by indenting the skin with arrays of pins that rise out of a surface to create a discrete representation of a texture or pattern as shown in Figure 4.24.

Other techniques include using vibration, heat/cold, blowing compressed air or changing shape. Actuators themselves include shape memory alloy (SMA), piezoelectric materials, motors and solenoids, pneumatic valves, rheological fluids and pistons. Some of these are considered in Figure 4.25.

Unfortunately, most of the tactile displays built to date fail to convey meaningful tactile information and to be practical at the same time. The devices are too big; they do not yield enough force or are constrained to too low bandwidth; they are limited to a small number of actuators with low-spatial density; they require constant maintenance or are simply too expensive and too complex. These failures are reflected by the relative scarcity of tactile displays that have made it to the commercial world.

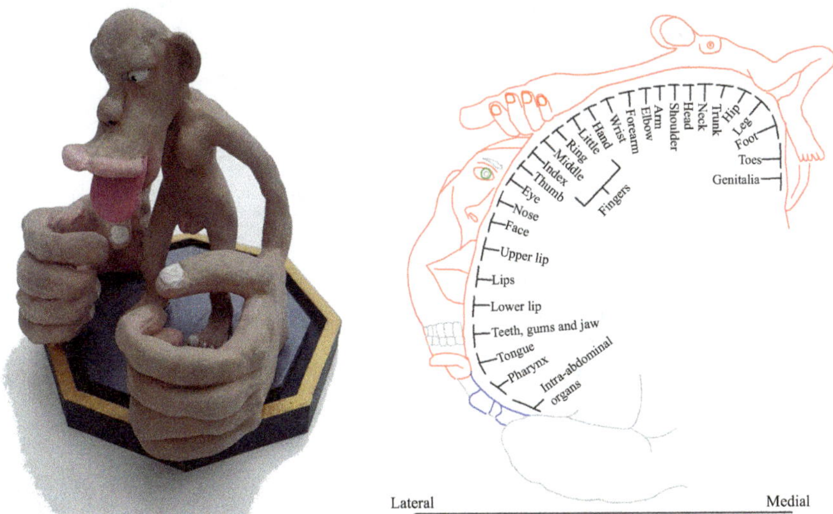

(a) (b)

Figure 4.23 Sensory homunculus (a) (made by Teja Digumarti) showing the relative proportions of the sensory cortex associated with touch as a volume and (b) as the length of the black dashes in the line

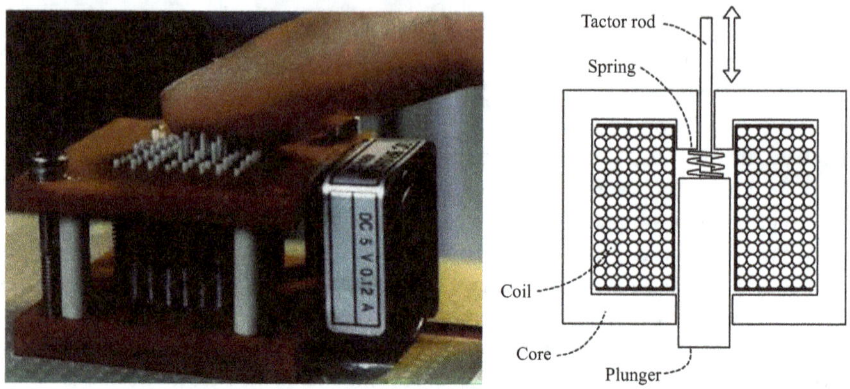

(a) (b)

Figure 4.24 Conventional vibrotactile display, reproduced courtesy Forschungszentrum Karlsruhe

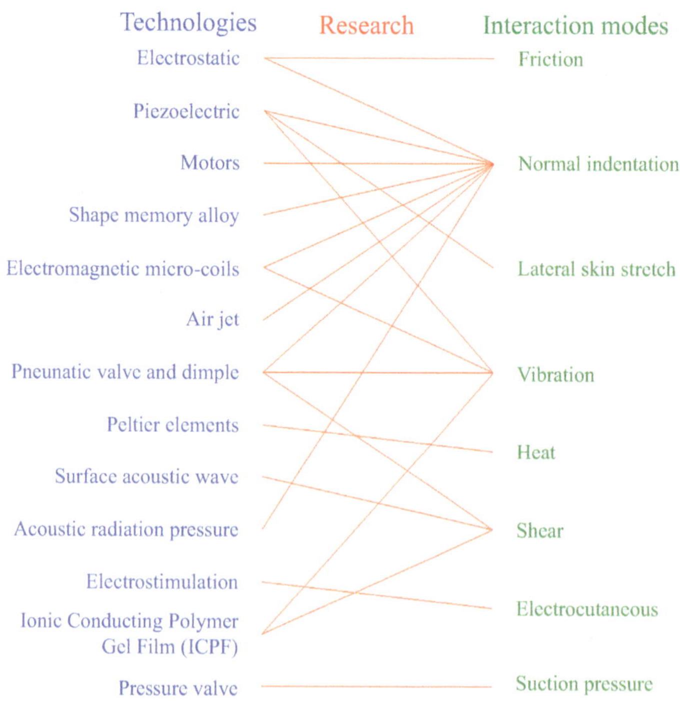

Figure 4.25 Tactile display technologies and interaction. [Adapted from Pasquero (2006)]

Frequency of vibration is a parameter that influences the quality and intensity of perception, with the sensitivity affected by factors like body position, skin temperature, and underlying tissue type (bone, fat, muscle, or a combination). Values between 50 and 300 Hz are generally used as being in the range where the tactile corpuscles are most sensitive. The use of oscillating pressure also adds new degrees of freedom to the design of vibrotactile stimuli. These include waveform shape, for example, sinusoidal or square, and amplitude modulations (at different modulation frequencies) of the carrier frequency.

Several features of vibrotactile stimuli can be modulated to convey information over this sensory channel (Cincotti *et al.* 2007). The list can be divided into two subsets. The first includes features related to physical perception and are:

- *Frequency*: the main spectral component of the periodic stimulus,
- *Intensity*: the amplitude of stimulation (measured either as force applied or as displacement produced),
- *Timbre*: the complexity of the stimulation waveform (i.e., the content of harmonics in the spectral representation),
- *Duty cycle*: the ratio of the durations of the on and off time on an elementary stimulation,
- *Spatial location*: the body part or the pattern of parts that are stimulated.

Features in the second subset are clearly perceived by an individual, but do not rely on any specific property of the receptors. They need to be interpreted on a cognitive level:

- *Rhythm*: the sequences of stimulation and pauses, with specific durations, that compose the current message, that is, a triplet of stimuli such as a Morse coded SOS,
- *Tempo*: the speed, due to longer or shorter duration of the whole message, given at a fixed rhythm,
- *Flutter*: an amplitude modulation of the stimulation carrier frequency that can either be perceived as an increase and decrease in the intensity if modulation is slower than 5 Hz or as "roughness" if modulation is faster than 10 Hz.

Braille, one of the earliest tactile display methods, was invented two hundred years ago as a means of giving blind people access to the written word. This was originally in the form of a 2×3 array of embossed dots on paper, but nowadays it is often displayed on computer peripherals called refreshable Braille displays that use 2×4 arrays. These displays have not changed significantly over the past 50 years with typical systems using cantilevered bimorph piezo actuators (reeds) supporting vertical pins at their free end, as shown in Figure 4.26. On actuation, the selected reeds bend upwards pushing the pins above the surrounding surface to generate the selected Braille character.

The individual cells as illustrated in the figure are simple and relatively inexpensive, but the overall cost to generate a line of Braille using 40 to 80 cells becomes significant (Levesque *et al.* 2005).

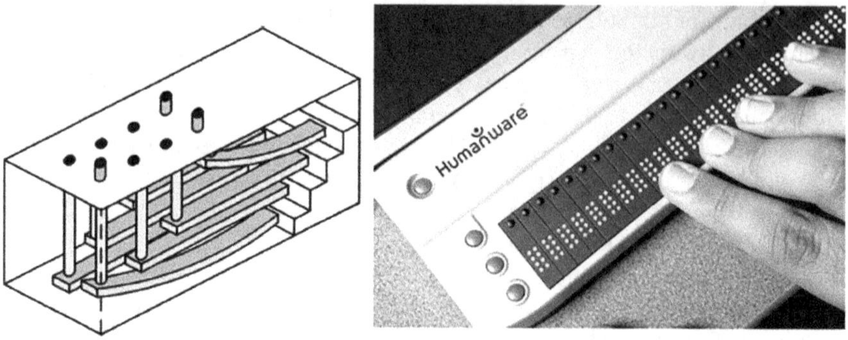

(a) (b)

Figure 4.26 Refreshable Braille display showing (a) a single cell and (b) a complete array made from an array of cells, reproduced courtesy HumanWare with permission

In contrast, an innovative braille module recently developed by Vijay Varada was awarded the Hackaday Prize for 2023. Using 3D printing and some clever electromechanics, these modules cost just over one dollar compared to \$100+ for commercial piezoelectric modules.

As an extension to the Braille units shown previously, one of the most successful applications of vibrotactile displays is the Optacon (OPtical TActile CONverter) manufactured by Telesensory Systems Inc. This device consists of a main electronics unit about the size of a large paperback connected by a cable to a small camera module. The camera module consists of 144 phototransistors arranged in a 24×6 configuration along with two small rollers for easy movement and two small lights for illumination. The electronics unit includes a vibrotactile display onto which the user places his finger and then tracks along a line on a printed page with the camera module that feeds the letter-sized images back to the display. The tactile array contains a 24×6 array of 0.08 mm diameter pins, each of which can be independently vibrated at about 230 Hz by a piezoelectric bimorph reed. The Optacon II, shown in Figure 4.27 provides a 20×5 array. As the user moves the camera, tactile images of print letters are felt moving across the array of rods under their finger (Efron 1977). Reading speed is limited to about 100 wpm compared to 250 wpm or more using Braille. But that was not the point, the Optacon offers access to innumerable sources of the written word not previously available to the blind – labels on jars, CD or DVD covers, computer screens (with a special lens attachment), ordinary newspapers and a host more. The Optacon was discontinued in 2000, and Telesensory Systems ceased its operations in 2005 (Stronks *et al.* 2015).

Several derivative products using similar principles have been developed to extend the capability of these vibrotactile systems because they have proven to be so useful to blind people. Devices like the videoTIM have a much larger tactile display and are therefore more amenable to representing simple graphics,

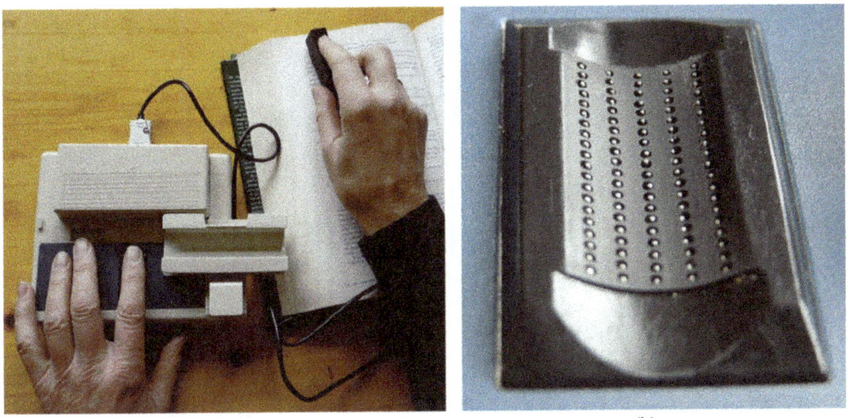

(a) (b)

Figure 4.27 Photos of the (a) Optacon II and (b) the vibrotactile module, reproduced courtesy Stichting Nationaal Blindenmuseum with permission

handwritten text and formulae. They are also able to read keys and messages on smartphone and computer displays if optical character recognition (OCR) software is not available.

Vibrotactile research at Johns Hopkins University has included the development of a 400-element array for fingertip stimulation. This consists of four planes of actuators slightly staggered so that the pins can all pass through a uniform array of holes in a fingertip-sized plate. This large number of actuators in a small area is required for research as it is of the same order as the number of tactile corpuscles on the fingertip.

The McGill Haptic lab has been working on the development of an up-to-now unexplored mode of interaction for tactile displays. Their researchers believe that lateral skin stretch is sufficient to give the impression of skin indentation. The objective is therefore to create a device capable of conveying meaningful tactile sensations by lateral skin stretch at the fingertip. The project is based on the simple realisation that it might be possible to compensate for both our limited knowledge of touch and the current shortcomings of haptic technology by making use of tactile illusions that are less complex to implement with state-of-the-art actuators.

This research has led to the development of a tactile display called the Stimulator of Tactile Receptors by Skin Stretch (STReSS). Various prototypes of the device have been built based on the principles shown in Figure 4.28 (Pasquero 2003), and these show potential.

STReSS is a computer-peripheral device that stimulates the fingertip by lateral skin stretch. It is composed of a miniature array of 100 (prototype #1) or 50 (prototype #2) bimorph piezoelectric actuators that induce time-varying

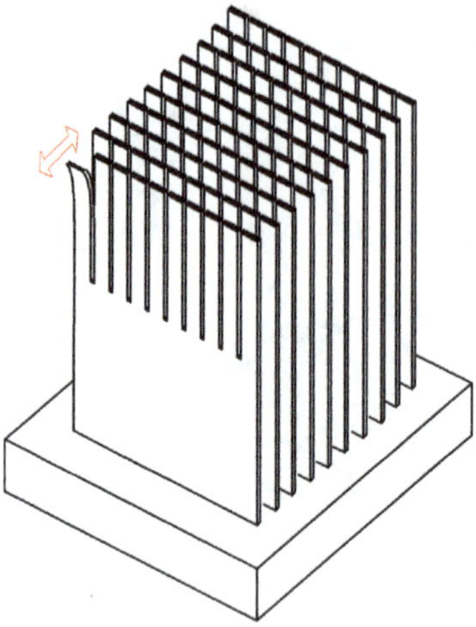

Figure 4.28 McGill University STReSS tactile display [adapted from Pasquero (2003)]

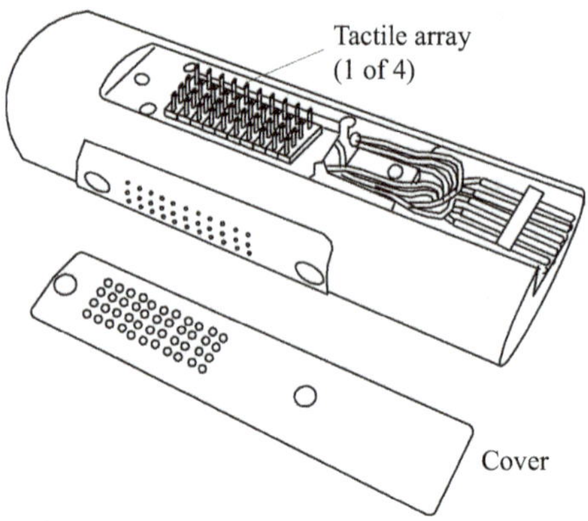

Figure 4.29 Vibrotactile haptic device [adapted from Becker et al. (2009)]

programmable strain fields at the fingertip to convey tactile information such as texture or small-scale shape. Tactile signals are generated on a personal computer and then fed to the display via a USB port.

Vibrotactile devices are also used for haptic feedback on handgrips as can be seen in Figure 4.29. Haptic devices are often used in tele-remote-control applications where visual and audio feedback is insufficient to generate the required "emersion".

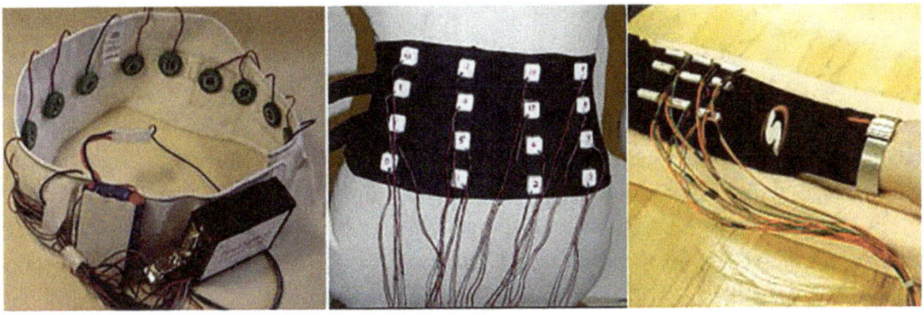

(a) (b) (c)

Figure 4.30 Vibrotactile devices for the torso and limbs, reproduced from Nagel et al. (2005) with permission

From a visual prosthetic perspective, such devices can replace the handle of a cane where they become sensory substitutes and convey visual information to the palm of the hand.

Larger tactile devices are also made for the torso or limbs where the density of sensory corpuscles is much lower. A number of these devices are shown in Figure 4.30. They are often made from micromotors driving an eccentric mass, similar to those commonly found in portable consumer equipment like smartphones. They are therefore very low cost. The vibration frequency is proportional to the applied voltage and can therefore be controlled. Unfortunately, the response time is poor.

Alternative designs are based on the construction methods for earphones where the diaphragm is replaced by a spring support with a central contactor pin.

4.10.1.1 Construction of vibrotactile devices

A tradeoff is required between the minimum force, and hence displacement, of the pins of a vibrotactile device to ensure that each one registers even after desensitisation, and a force that results in discomfort or even damage to the skin. This tradeoff will vary with the excitation frequency and differ for various parts of the body. For a typical device designed to stimulate Meissner corpuscles in the fingertip the pin spacing need not exceed the spatial discrimination sensitivity of these corpuscles which is about 3 mm (from Table 4.1). Because they are situated about 0.7 mm below the skin surface, a vertical displacement of at least that distance is required. In addition, each actuator pin should be capable of exerting a force greater than the contact force of the fingers during exploration. This is typically between 50 and 100 mN (Velazquez *et al.* 2004).

In theory, to minimise the energy expenditure of the device, the pin and the drive mechanism should be resonant at the frequency of interest. However, as soon as such a device comes into contact with the skin, the skin's elasticity damps the resonance and the advantage is lost. It is therefore practical to develop an actuator that is energised at the required stimulation frequency with sufficient force to obtain the required displacement with each stroke. Such devices, shown in

Figures 4.31 and 4.32, operate using electromagnetic principles, SMA and the piezoelectric effect all of which are described in Chapter 3. However, these are only a few of the many methods to achieve the same result.

The operational frequency of electromagnetic and piezoelectric devices can easily exceed the highest stimulation frequency of 300 Hz required for vibrotactile devices. However, SMA devices, by their nature are limited to a few hertz at most.

In construction, piezoelectric actuators consist of a thin piezoceramic slab with metallised electrodes on the upper and lower surfaces. When a potential difference is applied across the face, the dimensions of the piezoceramic slab are altered. This basic mechanism can be used in isolation if very small displacements are suitable. However, in most cases they are assembled into a stack to increase the displacement for a given voltage, or alternatively, a pair of elements are bonded on either side of a flexible metal beam to produce the bimorph actuator that flexes as shown in Figure 4.32.

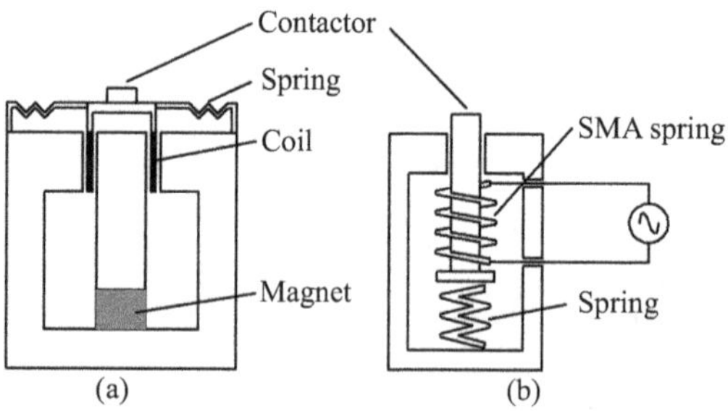

Figure 4.31 Components of individual vibrotactile elements showing (a) conventional electromagnetic actuation, (b) shape memory alloy actuator

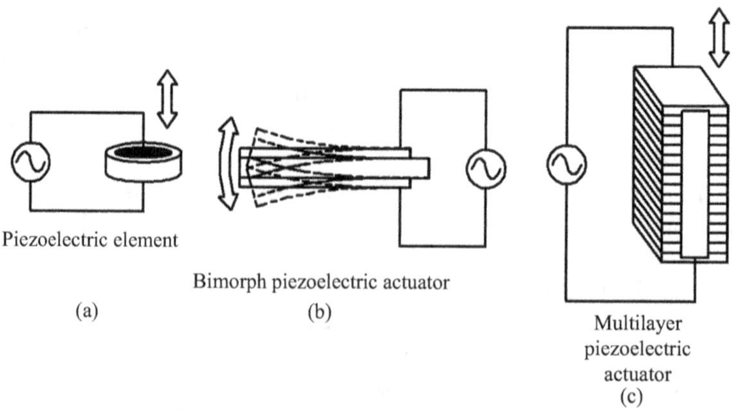

Figure 4.32 Piezoceramic actuator types

One of the more interesting mechanisms uses magneto-rheological fluids (MRF). These fluids respond to a magnetic field with a dramatic change in rheological behaviour. They can reversibly and instantaneously change from a free-flowing liquid to a semi-solid with controllable yield strength that allows them to be used for high-bandwidth vibrotactile devices.

4.10.1.2 Bistatic and shear actuators

Flavin *et al* (2024) discuss a newly developed miniature electromagnetic actuator that when combined with the skin, as an elastic element, supports bistable operation in addition to the conventional actuation shown in Figure 4.33(a). The ability to store and release mechanical energy in skin arises from three components; the core, armature and diaphragm. The core contains a coil embedded in a soft ferromagnetic (iron–cobalt) cylinder, which serves to focus magnetic fields along its central longitudinal axis. This rod interfaces with the skin as a linear shaft that translates through the core. An elastomeric diaphragm, a composite of polydimethylsiloxane (PDMS) and magnetic nanopowder (MNP), encloses the top of the structure. In the compressed state, the magnetic field imposed by the permanent magnet, channelled through the armature, magnetises the soft ferromagnetic core. The strong attraction induced between these elements exceeds the reactive force imposed by the skin, thereby maintaining compression without current applied at the coil. From this

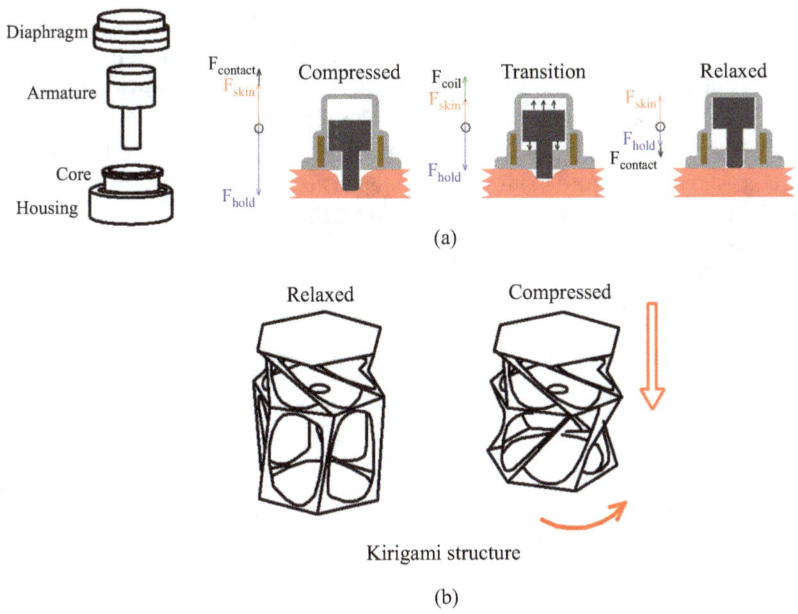

Figure 4.33 Figure showing (a) linear bistatic indentation and (b) shear actuation that uses a Kirigami structure to convert a linear force into a torsional one [adapted from Flavin Ha et al. (2024)]

state, driving the coil with opposite polarity to the permanent magnet substantially weakens its effect on the core, thus allowing the skin to push the armature back into the relaxed state.

This mechanism can provide a static stimulus, but if an alternating current with a DC bias is applied to the electromagnet, the position of the core can be controlled by the DC component, while the AC component will cause the core to oscillate. This targets specific classes of mechanoreceptors in the skin to provide distinct sensory responses.

Generalising the contact surface of the transducer beyond a simple indenting rod, Figure 4.33(b) illustrates a Kirigami structure that translates the normal-directed force of the armature into tangential forces applied parallel to the surface of the skin. This structure is made by patterning, bonding and creasing stiff plastic panels into a thin-walled hexagonal tube.

4.10.2 Electrotactile displays

Most researchers believe that the electric currents passing through the skin from surface electrodes directly stimulate afferent nerve fibres or whole nerve bundles if electrodes are sufficiently large. Depending on the applied voltage, current, waveform electrode size, force and hydration amongst others, subjects describe the sensation as a tingle, itch, vibration, buzz, touch, pressure, pinch, sharp and ultimately as a burning pain (Kaczmarek *et al.* 1991).

As with the electrodes discussed in Chapter 2, the object is to convert electron flow in the lead wire to ionic flow within the tissue. Electrode requirements are therefore similar to those used for monitoring electrical signals. Ideally, the current density should be kept as uniform and as low as possible, but small variations in the contact pressure and sweat concentration make this impractical.

The impedance of electrodes can be modelled as a capacitor and a resistor in parallel as shown in Figure 4.34. For a monophasic 10 mA pulse with a duration of 10 μs the normalised resistance is about 0.1 MΩ/mm^2 and the normalised capacitance about 0.1 nF/mm^2.

Worked Example:

If we increase the area of the electrode by 8 mm^2 from 12 mm^2, what will the new capacitance and resistance be?

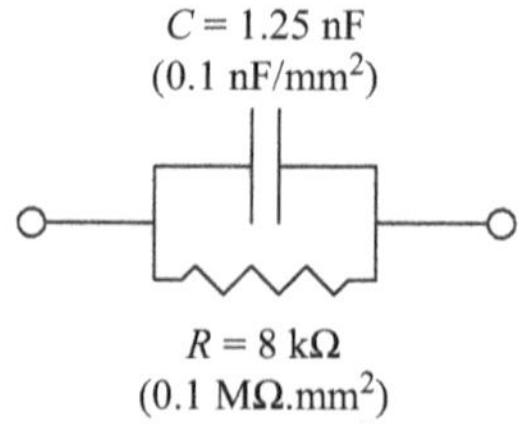

Figure 4.34 Simplified electrical model of a 12 mm^2 electrode [adapted from Kaczmarek et al. (1991)]

The resistance is inversely proportional to the area of the electrode

$$R = \frac{K_r}{A}$$

where R is the resistance (Ω), K_r is the constant of proportionality (Ωmm^2) and A is the area of the electrode (mm^2).

$$\begin{aligned} K_r &= RA \\ &= 8 \times 10^3 \times 12 \\ &= 96 \times 10^3 \ \Omega\text{mm}^2 (\approx 0.1 \ \text{M}\Omega\text{mm}^2) \end{aligned}$$

Therefore, the new resistance will be

$$\begin{aligned} R_{new} &= \frac{K_r}{A_{new}} \\ &= \frac{96 \times 10^3}{12 + 8} \\ &= 4800 \ \Omega \end{aligned}$$

The capacitance is directly proportional to the area of the electrode

$$C = K_c A,$$

where C is the capacitance (nF), K_c is the constant of proportionality (nF/mm^2) and A is the area of the electrode (mm^2).

$$\begin{aligned} K_c &= \frac{C}{A} \\ &= \frac{1.25}{12} \\ &= 0.104 \ \text{nF/mm}^2 \end{aligned}$$

Therefore, the new capacitance will be

$$\begin{aligned} C_{new} &= K_c A_{new} \\ &= 0.104 \times (12 + 8) \\ &= 2.08 \ \text{nF} \end{aligned}$$

The resistive component of the impedance drops sharply with increased current which results in an extremely nonlinear response. For this reason, the electrodes are usually stimulated with constant current rather than constant voltage. Changes in the resistance also alter the effective time constant of the electrode, and this can be a problem for high-frequency stimulation.

The dynamic range of electrotactile stimulation is determined by the ratio of the pain threshold level to the sensation threshold level, P/S (for voltage or current). This can vary from a factor of two (6 dB) to about 10 (20 dB) at best. This range is very limited compared to the other senses. For example, the ear has a dynamic range of 120 dB and the eye, 70 dB.

Dynamic range measurements made by different researchers vary by a large factor because, as are most measurements on human subjects, the thresholds are so subjective. Though the threshold for stimulation is determined with good accuracy, there is no definition for the threshold of "pain". In addition, pain thresholds vary with a whole raft of psychological factors. Furthermore, skin condition and the position of electrodes have a large effect on the dynamic range and comfort of electrical stimulation.

For pulsed stimulation, the sensation threshold increases as the pulse width decreases suggesting that the threshold is determined by the total charge (current × duration) as can be seen in Figure 4.35.

Several different institutions worldwide have researched various aspects of electrotactile excitation. One of these was the Kajimoto laboratory that developed a 32×16 element electrotactile display with an electrode spacing of 3 mm. This display was designed to be mounted on the forehead for ease of installation – it is held in place by a sweatband.

Visual images captured by the camera are converted to tactile information through two processes. The first is spatial outline extraction to enhance edges. The second is a temporal band-pass filtering to enhance time-varying information. In essence, the forehead recognition sensory system (FSRS) imitates functions performed in the retina and visual cortex to facilitate image cognition.

As with many other pieces of new technology, the electrotactile ideas developed within the Kajimoto laboratories have been spun off and developed for production as part of a commercial venture. The technology has been patented and was under development by EyePlusPlus as shown in Figure 4.36. No recent activity from the company has been noted.

For the past two decades, researchers in the Department of Rehabilitation Medicine of the University of Wisconsin have been working on making a device

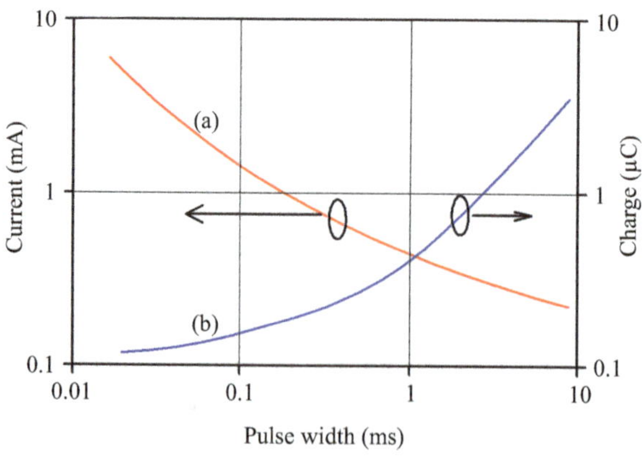

Figure 4.35 Electrotactile sensation threshold as a function of (a) current and (b) charge [adapted from Kaczmarek et al. (1991)]

that delivers electrotactile input to the tongue via a matrix of electrodes worn inside the mouth as shown in the photographs in Figure 4.37. Using a camera, a computer, and the input device, individuals who have been blind their whole lives are now able to use this relatively simple and non-invasive device to see basic images.

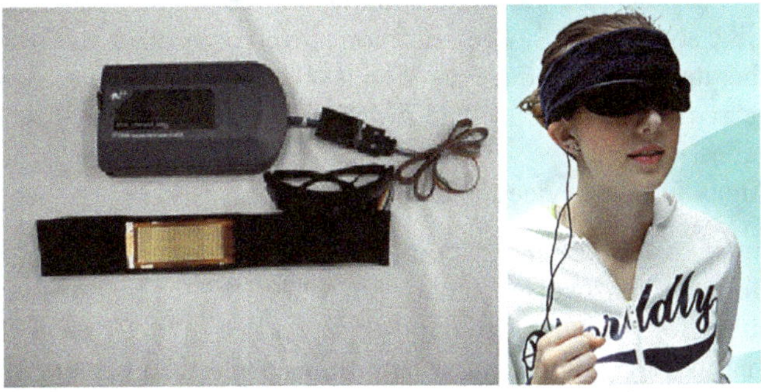

(a) (b)

Figure 4.36 Photograph of a forehead-mounted electrotactile display, reproduced courtesy EyePlusPlus

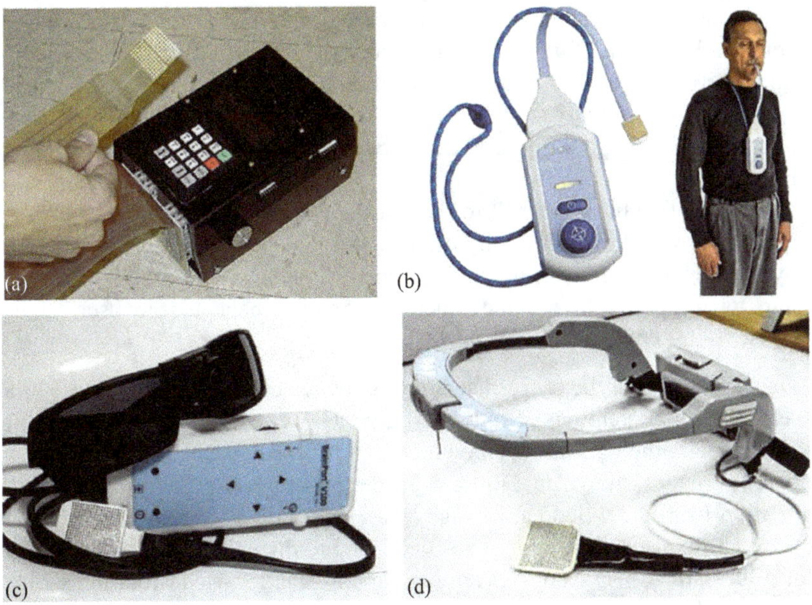

Figure 4.37 Evolution of the BrainPort technology. (a) 2001 – The tongue display unit (TDU) controlled by the keypad. (b) 2006 – The BrainPort balance device used to treat vestibular disorders. (c) 2013 – The BrainPort V100 vision aid with controller to adjust contrast and zoom. (d) 2016 – The latest BrainPort V200 eliminates the handheld controller and features an integrated battery and user interface, reproduced courtesy Wicab, Inc, Middleton, IW, USA with permission.

The research group has developed a generic and flexible way to communicate information to people using these arrays. Initial test results showed that volunteer experimental test subjects could identify very simple geometric patterns such as circles, squares, and triangles, and that they identified these figures as accurately on the tongue as on the fingertips.

Volunteers say that the electrical stimulus on the tongue feels like a tingle or vibration or like soda bubbles. The sensation is well controlled and painless unless the user deliberately turns up the excitation level. Occasionally, it produces a weak metallic taste sensation, but no tissue irritation has been observed with the gold-plated electrodes.

Applications for the device include:

- Provision of vestibular information for people with balance disorders. This is a simple form of sensory substitution, in which the tongue is used to present information from accelerometers or tilt sensors.
- Directional or navigational information for people who operate under central command and control scenarios, such as military and civilian rescue personnel. Providing information via the tongue allows them to fully use their vision and hearing to respond to unforeseen threats or hazards. Experiments have shown that it is possible to navigate a virtual maze using only information received on the tongue.
- Crude visual information is provided through the tongue for persons who are completely blind. Eliana Sampaio at the Louis Pasteur University in Strasbourg, France has used this prosthesis with a small video camera and demonstrated an equivalent visual acuity of about 20-to-830, which is very poor vision, but possibly useful for certain limited activities with enough practice.
- Provision of tactile feedback to the human operators of robots. For example, Nicola Ferrier of the University of Wisconsin is developing a robot controlled by the tongue of persons with quadriplegia that could incorporate touch sensors into its gripper, relaying the touch information back to the user's tongue (Wisc_Edu 2008).

During the prototype stage, users tested arrays containing between 100 and 600 electrotactile elements. Study participants were able to recognise high-contrast objects, their location, movement, and some aspects of perspective and depth. The images shown in Figure 4.38 demonstrate how indoor and outdoor information from the video camera is represented on the tongue. The prototypes had 25×25 pixel resolution on a 3×3 cm tongue display, presented at approximately 30 frames per second, yielding an information-rich stream usable by participants (Wicab 2009).

This technology that was FDA approved in 2015 had already begun to find a much wider market. It is starting to allow divers to "see" using sonar information in

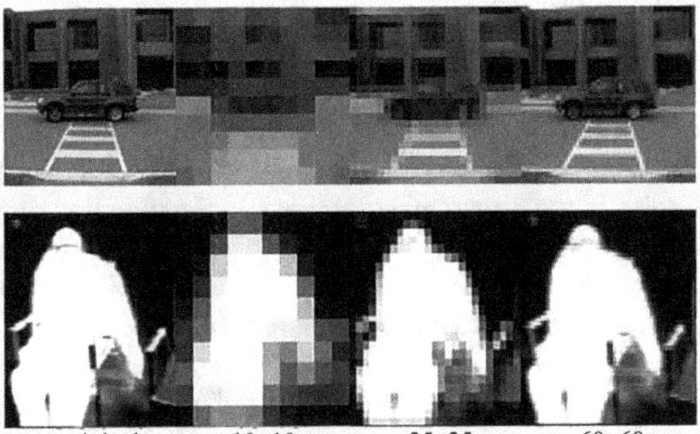

Figure 4.38 Static image acuity as a function of the array size (Wicab 2009), reproduced with permission

murky water and soldiers to obtain 360° night vision using inputs from an infrared camera, amongst many possible applications (Rosenblum 2010).

One of the commercial spin-offs of this technology is the BrainPort V100 vision device developed by Wicab, Inc. This system consists of a postage-stamp-size electrode array for the tongue, a base unit, a digital video camera, and a hand-held controller for zoom and contrast inversion. Visual information is collected from the user-adjustable glasses-mounted camera (FOV range 3°–90°) and sent to the base unit. The base unit translates the visual information into an electrical pattern that is displayed on a 400-element array applied to the dorsal surface of the tongue (Wicab 2024).

The electrode array is arranged in 20 rows of 20 stainless steel electrodes with a diameter of 762 μm that are spaced 558 μm apart measured from edge-to-edge (1320 μm centre-to-centre). The size of the electrodes was chosen based on safety considerations with respect to the maximum charge density. The electrode spacing was chosen on the basis of pilot experiments that revealed the minimal distance where two electrodes could still be spatially separated by test subjects (Stronks *et al.* 2016)

The tactile image is created by presenting white pixels from the camera as strong stimulation, black pixels as no stimulation, and grey levels as medium levels of stimulation, with the ability to invert contrast when appropriate. Users often report the sensation as pictures that are painted on the tongue with Champagne bubbles.

An alternative application for the technology is in the field of balance where it is helping users improve balance, posture and gait. The transducer is divided into four quadrants where impulses to the relevant quadrant, as obtained from an accelerometer, indicate the direction in which the user is leaning (Eagleman and Perrotta 2023).

4.10.3 Thermotactile displays

Peltier elements consist of p-doped and n-doped semiconductor material separated by a metallic interconnect, as shown in Figure 4.39. Current flows through the n-type material before crossing a metallic interconnect and passing into the p-type material. Electrons in the n-type material move against the direction of the current flow while holes in the p-type material move in the direction of the current. At one junction (the cold side), energy is absorbed from the environment as the charge carriers (electrons in the n-type and holes in the p-type) pass from a high to a low energy state. At the opposite junction (the hot side), the process is reversed: charge carriers move from a low to a high energy state, releasing energy in the form of heat. Therefore, this junction heats up. If the polarity is reversed, then the positions of the hot and cold junctions are reversed.

In commercial thermotactile displays, an array of closely spaced Peltier elements can be controlled to generate either hot or cold regions on the upper surface of the array as shown in Figure 4.40. Control electronics monitors the temperature of each to maintain a reasonably constant element temperature. The main advantages of this form of display are that there are no moving parts, and hence nothing to wear out compared to vibrotactile displays. They are also less intrusive than electrotactile displays (Monkman and Taylor 1993).

Because the distribution of temperature-sensitive nerve fibres is lower than that of some of the tactile corpuscles, the practical resolution of thermotactile displays will never be as high.

These devices are still in the experimental phase but can be used to produce Braille or standard letters as well as the simple graphics shown in the thermal images of the display.

Newer technology takes this process much further by making a thermal tactile sensation display that can present various thermal sensations, including cold/cool/warm/hot feelings, by varying the effective thermal conductivity of the display. It

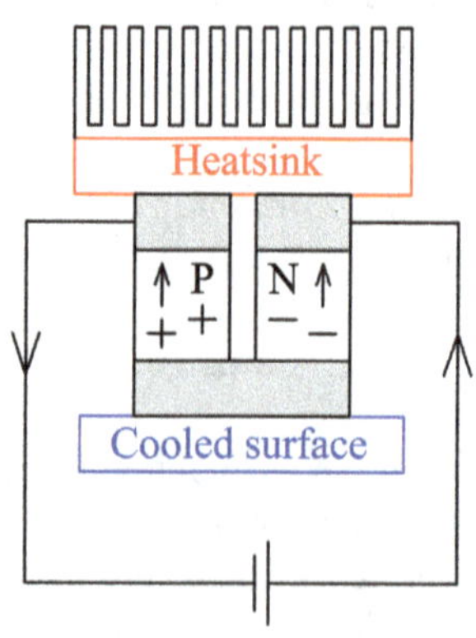

Figure 4.39 Schematic diagram showing the construction of a single Peltier element

(a)

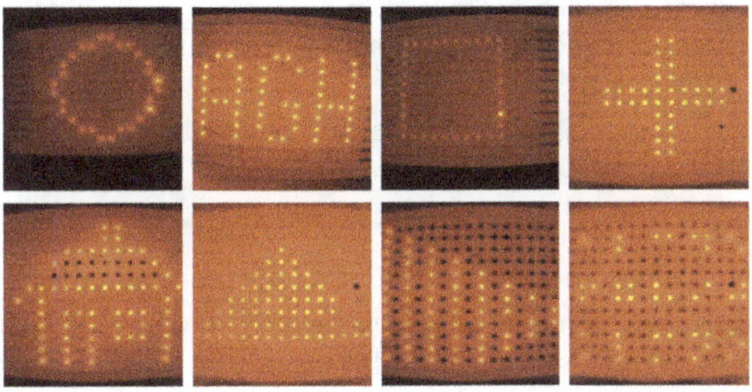

(b)

Figure 4.40 *Photograph showing (a) a prototype thermotactile display and (b) thermal images of the display, reproduced courtesy (Monkman and Taylor 1993)*

functions by using an air-filled cavity that has low thermal conductivity, which then has liquid metal pumped into it to increase the thermal conductivity dramatically. This change leads the user to experience different thermal tactile sensations (Hirai and Miki 2019).

4.11 GPS-based systems

Since the introduction of low-cost Global Positioning System (GPS) units, either stand-alone, or incorporated into a smartphone, blind people have had an effective new navigation method to augment their existing capabilities. The genre is well represented by the System for Wearable Audio Navigation (SWAN) developed at Georgia Tech's School of Psychology and College of Computing. The device was designed to help people get from point A to point B and know what's around them as they move.

SWAN uses GPS to locate the person and an electronic compass to detect the direction they are facing. Both feed into a computer that runs custom software to merge the information and to generate a series of sounds. Using recent developments in audio processing and conventional stereo headphones, the sounds generated by SWAN can appear to come from any direction around the user's head.

A series of beeps (beacons) leads users on the path to their destination. Other beacons convey useful situational awareness information (Walker and Lindsay 2006):

- *Navigation Beacon* sounds guide the user along a predetermined path, from a start point, through several waypoints, to arrive at the user's destination,
- *Object Sounds* indicate the location and type of objects around the user, such as furniture, fountains, doorways, etc.,
- *Surface Transition* sounds signify a change in the walking surface, such as sidewalk to grass, carpet to tile, level corridor to descending stairway, curb cuts, etc.,
- *Locations*, such as offices, classrooms, shops, buildings, bus stops, are also indicated with sounds,
- *Annotations* are brief speech messages recorded by users that provide additional details about the environment. For example, "Deep puddle here when it rains".

Because the system relies on a priori detailed information about the area, any short-term changes, such as road works, are not registered and the user needs to use one of the other prostheses discussed in this chapter. Another limitation is the fact that GPS does not work indoors, so the researchers are developing a camera-based system to help in these situations.

Other related products started to appear on the market including the European Tormes and the rather clunky PERsonal NAvigation System for VIsually disabled People (PERNASVIP) pursued the use of enhanced GPS positioning. Hard to achieve specifications such as location in an urban environment to within 4 m 95% of the time were specified but could often not be met in practice. Additionally, because of limited GPS coverage (particularly indoors) and INS drift, complementary systems were needed to keep track of the user's position. These were generally networks comprising at least one of the following:

- *Ultrasound transmitters*: The University of Florida's Drishti project (2004) applied this technology to people guidance, combining differential GPS outdoors and ultrasound beacons indoors. Indoor localisation accuracies of less than 22 cm were observed in open areas but could be degraded due to signal obstruction, multipath, etc.
- *Optical transmitters*: By 2001, researchers from Tokyo University had developed a guidance system for hospital use based on optical beacons. The transmitters were mounted on the ceiling with each sending a unique ID code associated with its position. Equipment carried by the users read the code then reproduced the appropriate recorded message. Another rather innovative

system developed at the National University of Singapore (2004) used fluorescent lights each modulated by a unique code that was associated with a specific area.

- *RFID tags*: Many of the technical solutions for positioning services were based on arrays of unique but expensive beacons, be they radio frequency or infrared, etc. Unfortunately, the cost of installation and maintenance, or their inability to adapt to changes in the environment (e.g., moved furniture), mitigated against their success. As an alternative, low-cost RFID tag networks costing only a few cents were proposed. Even though their range is limited to a few meters, range measuring techniques based on receive signal strength (RSS), received signal phase (RSP) or time of arrival (TOA) can be used to enhance position accuracy.
- *UWB technology*: This stands out mainly because of its sub-meter accuracy (e.g., 15–20 cm) and robustness to multipath interference, an issue using other sensor modalities inherent to both indoor and outdoor positioning.

Most recently, the application of computer vision for positioning has entered the mainstream. SLAM capability driven by the rapid innovation of autonomous vacuum cleaners now allows cm accuracy for positioning within buildings (Real and Araujo 2019).

4.12 Smartphones and wearables

4.12.1 Smartphones

The inclusion of a wide range of sensors and a powerful processor in modern smartphones has made them particularly useful platforms for sensory substitution development. These sensors include excellent cameras, GPS modules, accelerometers and gyroscopes. In addition, close-range Bluetooth or Wi-Fi communications expand their connectivity to a wide range of external sensors such as 3D cameras, radar and lidar.

Currently, the most popular apps for the blind or visually impaired (BVI) are the following:

- *Moovit*: an easy-to-use free app that offers guidance on the public transport network, managing schedules, notifications, and warnings in real time,
- *BlindSquare*: specifically designed for the BVI community, this application conveys the relative location of previously recorded places of interest (POIs) through speech. It makes use of Foursquare's and OpenStreetMap's databases,
- *Lazzus*: a paid application, again designed for BVI users, which coordinates GPS and built in motion capture and orientation sensors to provide users with intuitive cues about the locations of a range of POIs including zebra crossings in the surrounding area,. Its main sources of data are Google Places and OpenStreetMap (Real and Araujo 2019).

4.12.2 Wearables

Early wearables required that the users manually scan their environments to build up an "image". This was best achieved by mounting sensors to parts of the user that facilitated this scanning process. The head, wrist or attached to a cane were the obvious choices. Camera-based sensors attached to smart glasses provide an intrinsically wide field of view, but without depth information and require significant processing capability to be useful. In addition to their basic localisation and navigation function (in conjunction with sensors in the smartphone), they can also provide more situational awareness such as the reliable detection of obstacles and stairs in the user's path.

Bone conduction headphones and smart glasses with built-in cameras are two of the primary wearable sensors used in BVI mobility support. Some of the advantages of these wearables are their generally wide fields of view, immersive interfaces as well as discrete hands-free capability (Real and Araujo 2019).

4.13 The future of sensory substitution

4.13.1 Technological advancements

With the ongoing miniaturisation of both sensors and actuators, sensory substitution devices will become more compact allowing for integration seamlessly into everyday items. As an example, tongue-based systems like BrainPort should evolve into more discrete and possibly implantable solutions.

One of the current problems with sensory substitution relates to poor-quality feedback. Whereas camera technology has improved exponentially, suitable actuators have not. For this reason, it is imperative that haptic technology improves sufficiently to create more immersive and relevant sensory experiences. Ideally, these experiences should be adaptive to both the individual user and the circumstances.

4.13.2 Expanded applications

Haptic devices are expected to play a more significant role in medical training, allowing professionals to practice procedures in safe, realistic environments. Along with virtual and augmented reality, these technologies will also be increasingly used in physical rehabilitation.

It is almost certain that the technology will move beyond merely compensating for sensory loss to expanding normal senses, making them "better", "wider" and "faster" as well as creating entirely new sensory experiences (Eagleman and Perrotta 2023).

References

Becker, H., T. Jang, T. Debus and P. Dupont. (2009). "Multi-Channel Vibrotactile Display." Retrieved September 2009 from http://robotics.bu.edu/dupont/research_projects/vibrotactile/index.htm.

Benjamin, J., N. Ali and A. Schepis (1973). "A laser cane for the blind." *Proceedings of the San Diego Biomedical Symposium* 12: 53–57.

Bissitt, D. and A. Heyes (1980). "An application of biofeedback in the rehabilitation of the blind." *Applied Ergonomics* 11(1): 31–33.

Borenstein, J. and I. Ulrich (1997). *The GuideCane- A Computerized Travel Aid for Active Guidance of Blind Pedestrians*. IEEE International Conference on Robotics and Automation, Albuquerque, NM.

Brabyn, J. (1982). "New developments in mobility and orientation aids for the blind." *IEEE Transactions on Biomedical Engineering BME* 29(4): 285–289.

Brodal, A. (1969). *Neurological Anatomy in Relation to Clinical Medicine*. New York: Oxford University Press.

Brooker, G. (2021). *Sensors for Ranging and Imaging, 2nd ed*. Stevenage: The Institution of Engineering and Technology (IET).

Burnside, N. (2004). "A Function that Returns the Atmospheric Attenuation of Sound, *MATLAB Central*." http://www.mathworks.com/matlabcentral/fileexchange/loadFile.do?objectId=6000&objectType=FILE.

Busaeed, S., I. Katib, A. Albeshri, J. M. Corchado, T. Yigitcanlar and R. Mehmood (2022). "LidSonic V2.0: a LiDAR and deep-learning-based green assistive edge device to enhance mobility for the visually impaired." *Sensors* 22(19): 7435.

Capelle, C., C. Trullemans, P. Arno and C. Veraart (1998). "A real-time experimental prototype for enhancement of vision rehabilitation using auditory substitution." *IEEE Transactions on Biomedical Engineering* 45(10): 1279.

Caraiman, S., O. Zvoristeanu, A. Burlacu and P. Herghelegiu (2019). "Stereo vision based sensory substitution for the visually impaired." *Sensors* 19(12): 2771.

Cincotti, F., L. Kauhanen, F. Aloise, *et al.* (2007). "Vibrotactile Feedback for Brain-Computer Interface Operation." *PubMed* Retrieved September 2008 from http://www.pubmedcentral.nih.gov/articlerender.fcgi?artid=2267023.

Eagleman, D. M. and M. V. Perrotta (2023). "The future of sensory substitution, addition, and expansion via haptic devices." *Frontiers of Human Neuroscience - Brain-Computer Interfaces* 16: 1055546.

Efron, N. (1977). "Optacon – a replacement for Braille?" *Australian Journal of Optometry* 60(118): 118–129.

Flavin, M. T., K.-H. Ha, Z. Guo, *et al.* (2024). "Bioelastic state recovery for haptic sensory substitution." *Nature* 635(8038): 345–352.

Geldard, F. (1957). "Adventures in tactile literacy." *The American Psychologist* 12: 115–124.

Hirai, S. and N. Miki (2019). "Thermal tactile sensation display with controllable thermal conductivity." *Micromachines (Basel)* 29(10): 359–368.

K-Sonar, B. (2008). "Bay Advanced Technologies." Retrieved August 2008 from http://www.batforblind.co.nz/.

Kaczmarek, K. A., J. G. Webster, P. Bach-y-Rita and W. J. Tompkins (1991). "Electrotactile and vibrotactile displays for sensory substitution systems." *IEEE Transactions on Biomedical Engineering* 38(1): 1.

Kay, L. (1974). "A sonar aid to enhance spatial perception of the blind: engineering design and evaluation." *Radio and Electronic Engineer* 44: 605–627.

Knudsen, V. and C. Harris (1950). *Acoustical Designing in Architecture*. New York: John Wiley & Sons.

Lampton, M. (1981). "The microchannel image intensifier." *Scientific American* 245(5): 62.

Lebedev, V. and V. Sheiman (1980). "Assessment of the possibilities of building an echo locator for the blind." *Telecommunications and Radio Engineering* 34–35(3): 97–100.

Leonard, J. and H. Durrant-Whyte (1991). "Simultaneous Map Building and Localisation for an Autonomous Mobile Robot." *IEEE International Workshop on Intelligent Robots and Systems*.

Levesque, V., J. Pasquero and V. Haward (2005). "Display of virtual Braille dots by lateral skin deformation." *ACM Transactions on Applied Perception* 2(2): 132–139.

Malone, P., S. Eberhardt, E. Auer, R. Klein, L. Bernstein and M. Riesenhuber (2021). "Neural Basis of Learning to Perceive Speech through Touch Using an Acoustic-to-Vibrotactile Speech Sensory Substitution." *bioRxiv*.

Meijer, P. (2006). "Seeing with Sound." Retrieved August 2008 from http://www.seeingwithsound.com.

Monkman, G. and P. Taylor (1993). "Thermal tactile sensing." *IEEE Transactions on Robotics and Automation 8*(3): 313–318.

Nagel, S., C. Carl, T. Kringe, R. Märtin and P. König (2005). "Beyond sensory substitution – learning the sixth sense." *Journal of Neural Engineering* 2(4): R13–26.

Novich, S. and D. Eagleman (2015). "Using space and time to encode vibrotactile information: toward an estimate of the skin's achievable throughput." *Experimental Brain Research* 23: 2777–2788.

Pasquero, J. (2003). *STRESS: A Tactile Display using Lateral Skin Stretch*. M.Eng, McGill University.

Pasquero, J. (2006). "*Survey of Communication through Touch: TR-CIM 06.04.*" Retrieved August 2008 from http://www.cim.mcgill.ca/~jay/index_files/new-research.htm.

Player, P. (2021). *Extrasensory Belt. BE Honours Thesis*, Sydney.

Rasouli Kahaki, Z., M. Karimi, M. Taherian and R. Simi (2023). "Development and validation of a white cane use perceived advantages and disadvantages (WCPAD) questionnaire." *BMC Psychology* 11(1): 253.

Real, S. and A. Araujo (2019). "Navigation systems for the blind and visually impaired: past work, challenges, and open problems." *Sensors (Basel)* 19: 3404.

Rosenblum, L. (2010). *See What I'm Saying: The Extraordinary Powers of Our Five Senses*. New York: W.W. Norton and Company, Inc.

Senscomp. "40LT16." Retrieved October 2007 from http://www.senscomp.com/specs/40LT16%20%20spec.pdf.

Senscomp. "600 Series Transducer." Retrieved October 2007 from http://www.senscomp.com/specs/600%20instrument%20spec.pdf.

Spence, C. (2014). "The skin as a medium for sensory substitution." *Multisense Res.* 27(5–6): 293–312.

Storer, A. (2006). "Visual perception through Sensory Substitution." Retrieved August 2008 from http://cns.bu.edu/~storer/sensorysubstitution/Alex%20Storer%20-%20Final%20Paper%20-%20Updated.pdf.

Stronks, H., E. Mitchell, A. Nau and N. Barnes (2016). "Visual task performance in the blind with the BrainPort V100 Vision Aid." *Expert Review of Medical Devices* 13(10): 919–931.

Stronks, H., A. Nau, M. Ibbotson and N. Barnes (2015). "The role of visual deprivation and experience on the performance of sensory substitution devices." *Brain Research* 1624: 140–152.

Thaler, L., G. M. Reich, X. Zhang, *et al.* (2017). "Mouth-clicks used by blind expert human echolocators – signal description and model based signal synthesis." *PLOS Computational Biology* 13(8): e1005670. doi: 10.1371/journal.pcbi.1005670.

Trivedi, B. (2010). Ear today, eye tomorrow. *New Scientist* 43–45.

Velazquez, R., J. Pissaloux, J. Szewczyk and M. Hafez (2004). "Design and Characterisation of a Shape Memory Alloy Based Micro-Actuator for Tactile Stimulation." *IEEE International Symposium on Industrial Electronics*. Ajaccio, France. 1: 3–8.

Walker, B. and J. Lindsay (2006). "Navigation performance with a virtual auditory display: effects of beacon sound, capture radius, and practice." *Human Factors* 48(2): 265–278.

Warren, D. and E. Strelow (1984). "Learning spatial dimensions with a visual sensory aid: molyneux revisited." *Perception* 13: 331–350.

Weinstein, S., Ed. (1968). *Intensive and Extensive Aspects of Tactile Sensitivity as a Function of Body Part, Sex and Laterality. The Skin Senses*. Springfield, IL, Charles Thomas.

Wicab. (2009). "Brainport Technologies – Vision Device." Retrieved September 2009 from http://wicab.us/technology/vision.php.

Wicab. (2024). "BrainPort – Vision Pro." Retrieved April 21 2024 from https://www.wicab.com/brainport-vision-pro.

Wilson, C. (2017). This is how some blind people are able to echolocate like bats. *New Scientist*.

Wisc_Edu. (2008). "*Tongue Display Unit*." Retrieved July 2008 from http://kaz.med.wisc.edu/projects_tdu.php.

Chapter 5
Signal processing

5.1 Introduction

This chapter deals with two separate aspects of biomechatronic signal acquisition and processing. The first is concerned with signals obtained directly from the organism including electrical, chemical, pressure, etc. The second is concerned with the remainder of the signals that are generated as part of a biomechatronic process. These would include tactile signals from a prosthetic hand, outputs of potentiometers, rate gyros, etc., that are used for control or feedback.

These signals are manipulated in various ways using analogue circuitry or digital algorithms to produce outputs that can be used to actuate or control a mechatronic device such as a prosthetic limb. Alternatively, the output can be used as an indicator of the state of the organism, of which an example would be an electrocardiograph (ECG) trace that is indicative of the condition of the heart.

5.2 Biomedical signals

Biomedical signals originate from a number of sources including the following (Bronzino 2006):

- **Bioelectric Signals:** This is a generic term for all of the electrical signals generated by nerve and muscle cells. The source is the membrane potential which, under certain circumstances, may generate an action potential. In single-cell measurements, the action potential is the bioelectric signal. However, in most cases, an embedded or surface electrode measures the sum of the action potentials of a large number of cells. The latter include myoelectric and ECG signals amongst others.
- **Bioimpedance Signals:** The electrical impedance of tissue contains important information concerning its makeup, blood volume, endocrine activity etc. These signals are normally obtained by injecting low-current (<20 mA) sinusoidal electrical signals into the body at frequencies between 50 kHz and 1 MHz and monitoring the relationship between the current and the voltage.
- **Bioacoustic Signals:** Many biological phenomena generate acoustic outputs, and these are indicative of the function being performed. A good example of this is the "lub-dub" sound produced by the pumping heart. Other sounds are generated by blood flow, airflow and the transit of solids, liquids and gas through the digestive system.

- **Biomagnetic Signals:** All of the organs in which electrical activity occurs generate magnetic fields as a result of this activity. These organs include the brain, heart and lungs as well as the skeletal muscles. Unfortunately, the amplitude of these signals is very small, and they are difficult to measure.
- **Biomechanical Signals:** These signals include motion and displacement as well as pressure, tension and flow within the organism. Measurement of these requires the use of the sensors and transducers discussed in Volume 1, Chapter 2. Unlike electrical and magnetic signals, these signals generally do not propagate (with the exception of pressure) and so are mostly measured at source.
- **Biochemical Signals:** These result from chemical activity within the organism which can alter its chemical composition in both subtle and gross ways. Chemical composition is measured using the sensors discussed in Volume 1, Chapter 2 and can include gases such as CO_2 and O_2 as well as dissolved solids like glucose and various salts.
- **Bio-optical Signals:** These are signals related to the optical reflectance or transmission of the tissue. For example, blood oxygen levels may be determined from the relationship between the reflectance of IR and visible wavelengths. Other information may be gathered using fluorescence characteristics based on injected dye materials.

The signals can be classified with regard to their source, application or in terms of the signal characteristics. Biological signals can be considered to be continuous or discrete. Continuous signals include temperature, pressure and chemical concentration, while discrete signals include electrical impulses generated by individual nerve cells.

These signals can be classified into broad classes dependent on the rate and nature of the variations that take place (Carr 1997).

5.2.1 Bioelectric signals

The idea that electricity is generated by the body was first introduced by Luigi Galvani in 1786, but it was not until 1903 when William Einthoven was able to measure these potentials using an improved string galvanometer that their true potential as a diagnostic tool became apparent. This potential was further enhanced by the invention of the vacuum tube amplifier a few years later.

Bioelectric potentials are ionic voltages produced as a result of electrochemical activity in some types of specialist cells. Conductive solutions consisting of dissolved salts surround the cells in the body. The principal ions in solution are sodium (Na+), potassium (K+) and chloride (Cl-). The semi-permeable membranes of nerve and muscle cells allow the entry of potassium and chloride ions but block sodium ions. Because the ions seek to balance both potential and concentration across the membrane, the restriction on the diffusion of sodium ions results in an imbalance in the concentration of sodium ions, with fewer within the cell and more in the intercellular fluid. In an attempt to balance the charge, additional potassium ions enter the cell. At equilibrium, a potential difference of -70 mV exists across the cell membrane, with the interior being negative with respect to the exterior. This potential difference, which has been measured at between -60 and -100 mV is called the resting potential of the cell, and cells at this potential are referred to as polarised.

When a section of the cell membrane is excited by the flow of ionic current or by some other form of excitation energy, the membrane becomes permeable to sodium ions, and they begin to flow across the boundary as illustrated in Figure 5.1. This ionic current increases the permeability further with the result that the current flow increases exponentially (the avalanche effect) and sodium ions rush into the cell to try to reach equilibrium. At the same time, some potassium ions begin to move out of the cell for the same reasons. Potassium ions are slower than sodium ions, and as a result the cell is left with a slight potassium imbalance, which results in a positive potential difference of about +20 mV across the cell membrane. This is known as the action potential of the cell, and one that exhibits this potential is said to have been depolarised.

At this time, the cell reverts to its semi-permeable state, and an active process called a sodium pump quickly transports sodium ions back out of the cell during the process known as repolarisation.

The time periods involved in this process vary with different cell types. For example, in nerve and muscle cells, the repolarisation process is so quick that the action potential appears as a sharp spike as little as 1 ms wide. Heart muscle repolarises much more slowly, with the action potential lasting from 150 to 300 ms. However, as shown in Figure 5.2, regardless of the duration, the resting and action potentials are always the same.

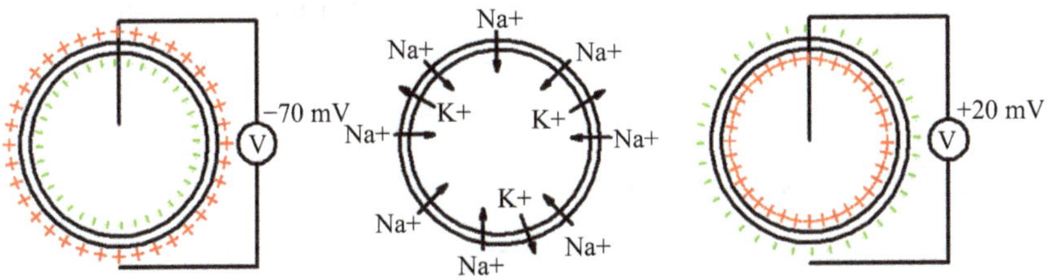

Figure 5.1 Process of triggering a cell action potential

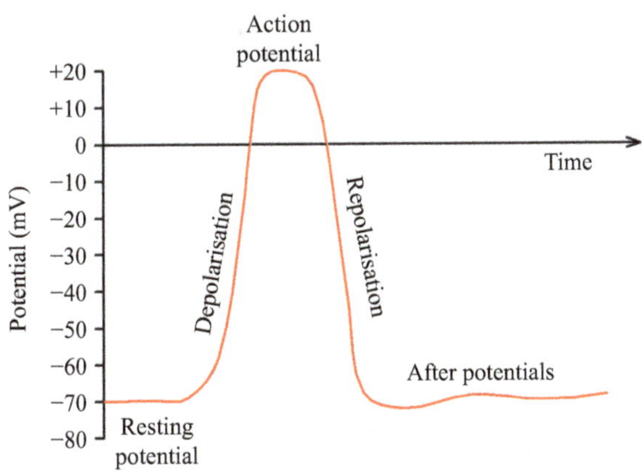

Figure 5.2 Waveform showing the potential difference across a cell membrane as a function of time

When a cell is excited and generates an action potential, ionic currents flow in the intercellular fluid or in adjacent areas of the same cell, and these can excite neighbouring cells. In the case of nerve cells with a long axon fibre, the action potential is generated in a very small segment of the fibre but propagates rapidly in both directions from the trigger point. Under normal conditions, nerve fibres are only excited near their input end and so propagate in one direction only.

As an action potential travels down the fibre, it cannot re-excite the portion of the fibre immediately prior because of a refractory period that follows the action potential. The rate at which the potential propagates down the fibre is called the propagation rate, the nerve conduction rate, or conduction velocity. This rate varies widely depending on the type and diameter of the nerve fibre, but it is usually something between 20 and 140 m/s in nerves. Propagation through heart muscle is much slower, with an average rate of only 0.2 to 0.4 m/s, and some time-delay fibres between the atria and ventricles propagate even more slowly at between 0.03 and 0.05 m/s (Cromwell *et al.* 1973).

5.2.2 *Signals characterised by source*

As can be seen from Table 5.1, a range of potentials from less than 1 μV right up to 100 mV and frequencies right up to 10 kHz must be accommodated. One of the major difficulties arises where a small amplitude signal must be examined in the presence of noise or another much larger signal.

Table 5.1 Characteristics of some bioelectric signals

Classification	Acquisition	Frequency range (Hz)	Dynamic range	Description
Action potential	Micro electrode	100–2000	10 μV–100 mV	Cell membrane potential
Electroneurogram (ENG)	Needle electrode	100–1000	5 μV–10 mV	Potential of a nerve bundle
Electroretinogram (ERG)	Micro electrode	0.2–200	0.5 μV–1 mV	Evoked flash potential
Electroencephalogram (EEG)				
Surface	Surface electrode	0.5–100	2–100 μV	Multichannel scalp potential
Delta range		0.5–4		Children, deep sleep
Theta range		4–8		Temporal & central – alert state
Alpha range		8–13		Awake, relaxed, closed eyes
Beta range		13–22		
Sleep spindles		6–15	50–100 μV	Bursts of 0.2–0.6 s
K-complexes		12–14	100–200 μV	Bursts during deep sleep
Evoked potentials (EP)	Surface electrode		0.1–20 μV	Brain response to stimulus
Visual (VEP)		1–300	1–20 μV	Occipital lobe 200 ms duration
Somatosensory (SEP)		2–3000		Sensory cortex
Auditory (AEP)		100–3000	0.5–10 μV	Vertex
Electrocorticogram	Needle electrode	100–5000		Exposed brain surface potentials
Electromyogram (EMG)				
Single fibre	Needle electrode	500–10,000	1–10 μV	Action potential – single fibre
Motor unit action pot.	Needle electrode	5–10,000	100 μV–2 mV	
Surface EMG (SEMG)	Surface electrode			
Skeletal muscle		2–500	50 μV–5 mV	
Smooth muscle		0.01–1		
Electrocardiogram (ECG)	Surface electrode	0.05–100	1–10 mV	

5.2.3 Signals characterised by type

5.2.3.1 Static and quasi-static signals

Static signals are by definition unchanging over a long period of time. Such signals are essentially DC levels and in isolation convey very little information. Quasi-static signals are those that change very slowly with time; examples of such signals include the long-term drift on a sensor, or the decreasing voltage on a slowly discharging battery.

5.2.3.2 Periodic and repetitive signals

Periodic signals are those that repeat themselves on a regular basis, though the timescale for repetition can be from femtoseconds up to days, or even years. These include sine, square and sawtooth waves amongst others, their defining nature being that each waveform is identical. Repetitive signals are periodic in nature, but the exact shape may change slightly with time. Ultimately, very few biological signals are truly periodic, so ECG signals and oestrogen levels in the blood are good examples of repetitive signals.

5.2.3.3 Transient and quasi-transient signals

Transient signals are, by definition, one-time only signals, while quasi-transient signals are those which are periodic but with a duration which is very short compared to the period of the waveform. Once again, this classification is rather arbitrary, with the differences between quasi-transient and periodic signal types being rather vague.

5.2.3.4 Stochastic signals

Stochastic signals are generated by a stochastic biological process that produces sample functions each of which differs from the others from a temporal perspective but shares the same distribution characteristics. These signals cannot be classified exactly and are described by the statistics of the whole collection of sample functions (referred to as an ensemble). These classifications can include the probability density function (PDF) which describes the amplitude characteristics of the signal, and the autocorrelation function, or its associated power spectral density (PSD). They are often also characterised in terms of their mean and variance.

5.3 Signal acquisition

The acquisition of electrical signals generated by the organism can be achieved using one of the myriad electrode types discussed in Chapter 2. Any of the other signal types are converted to electrical signals by one of the sensors discussed in that chapter. These signals are generally acquired in a continuous manner after which they are processed using some analogue circuitry before sampling and digitisation. This is followed by storage or display as for EEGs, or as inputs into some form of actuation, such as an insulin dispenser if the blood glycogen levels are incorrect.

5.3.1 Noise

In all the examples, the ultimate electrical signal will be accompanied by some form of noise. This noise could come from other physiological activities, from external interference or, for the smaller signals, from thermal noise generated by the acquisition electronics. It may be random, or it may be repetitive or even periodic, depending on its source. There is therefore no single signal processing technique that can be used to minimise the noise and hence maximise the signal-to-noise ratio.

5.3.1.1 Thermal noise

Thermal noise is electrical noise generated by random fluctuations of the voltage or current due to the thermal agitation of electrons within a conductor. It is therefore common to any electrical circuit associated with biomechatronic sensors.

More formally, if $v(t)$ is the thermal noise voltage across the terminals of a resistor, R, then if this voltage is measured at regular intervals over a long period the mean value, $\bar{v}$, is

$$\bar{v} = \frac{1}{m}\sum_{j=1}^{m} v_j \tag{5.1}$$

Measurements show that in the limit as $m \to \infty$ the mean value approaches zero. This result can be justified by considering the random motion of large numbers of electrons which produce fluctuations in the potential. These must average out to zero in the long term otherwise they would result in the flow of a current.

The time-averaged squared signal $\bar{v}^2$ is determined in a similar way

$$\bar{v}^2 = \frac{1}{m}\sum_{j=1}^{m} v_j^2 \tag{5.2}$$

In the limit as $m \to \infty$ this mean squared value, $\bar{v}^2$ (V^2), can be shown to approach

$$\bar{v}^2 = 4kTR\Delta f \tag{5.3}$$

where k is Boltzmann's constant (1.38×10^{-23} J/K), T (Kelvin) is the absolute temperature, R (ohms) is the resistance value and Δf (Hz) is the bandwidth (Young 1990).

If samples of the noise voltage are taken over a long period, and the results plotted as a histogram with bin widths dV, a distribution of the form shown in Figure 5.3 is produced (Brooker 2021).

The probability $p\{V\}dV$ that any future measurements will fall in the range $V \to V + dV$ is given by this plot that is known as the PDF. This function approximates the Normal or Gaussian distribution (Walpole and Myers 1978) which can be described by

$$p\{V\}dV = \frac{1}{\sigma\sqrt{2\pi}}e^{-V^2/2\sigma^2} \tag{5.4}$$

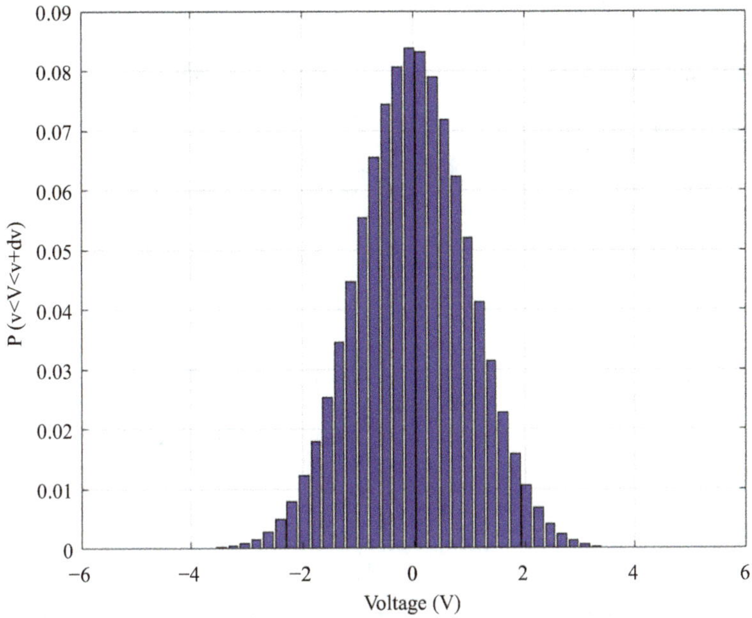

Figure 5.3 Histogram showing the PDF for thermal noise with unity variance

The time averaged squared value, $\bar{v}^2$, equates to the variance, σ^2, because the distribution is Gaussian. Its value is a measure of how wide the distribution is, hence, it is a useful indicator of the amount of noise present. However, in practice, it is more common to specify the noise level in terms of the root mean square (RMS) quantity where v_{rms} is

$$v_{rms} = \sqrt{\bar{v}^2} \tag{5.5}$$

Noise Power Spectrum for Thermal Noise: In theory, this is completely flat spectrally and is known as "white noise" as an analogy to white light which comprises a uniform mix of all the colours. Strictly speaking, however, it is not possible to produce a power spectrum that is truly white over an infinite frequency range, as the total power integrated over this bandwidth would be infinite. In reality, all noise-generating processes are subject to some band-limiting mechanism which produces a finite noise bandwidth. In addition, the measurement process is also band limited which limits the measured value for the total noise power still further.

It is often convenient to remove the bandwidth dependence on the power spectrum, this is referred to as the PSD or voltage variance per hertz, V^2/Hz. Electronic noise levels are often quoted as volts per root hertz ($\mathrm{V}/\sqrt{\mathrm{Hz}}$) as this serves as a reminder that the RMS voltage increases with the square root of the noise bandwidth.

Example

Determine the noise PSD of a 100 kΩ resistor at a temperature of 25 °C

$$\begin{aligned}\bar{v}^2 &= 4kTR\ (\mathrm{V}^2/\mathrm{Hz})\\ &= 4 \times 1.38 \times 10^{-23} \times (273 + 25) \times 100 \times 10^3\\ &= 1.65 \times 10^{-15}\mathrm{V}^2/\mathrm{Hz}\end{aligned}$$

By taking the square root, it is possible to obtain the PSD in the more common form

$$v_{rms} = 40.56\ \mathrm{nV}/\sqrt{\mathrm{Hz}}$$

A data acquisition system has a 20 kHz bandwidth, and a very high input impedance compared to the resistor value, so it would measure the RMS voltage across the resistor to be

$$\begin{aligned}v_{rms} &= 40.56 \times 10^{-9} \times \sqrt{20 \times 10^3}\\ &= 5.7\ \mu V\end{aligned}$$

5.3.1.2 Shot noise

Shot or Schottky noise is typically generated by the current flowing across a barrier such as the action potential developed across a cell membrane. The noise is generated by the migration of individual charge elements across the barrier at random intervals, so even though, on average, the current flow may be constant, fluctuations around the average take the form of a Poisson distribution (Walpole and Myers 1978)

$$p(n, \gamma) = \frac{e^{-\gamma}\gamma^n}{n!} \tag{5.6}$$

where e is the base of the natural log (2.71828), n is the actual number of occurrences and γ is the expected number of occurrences during a given interval.

Figure 5.4 shows several examples of the distribution for different values of γ. Note that the occurrences must be discrete integers and so the lines joining these points are for illustration only.

The Poisson distribution has several interesting characteristics, one of which is that the mean and the variance are both equal to γ. In addition, the figure shows that, as the expected number of occurrences in a given time interval increases, the distribution becomes more Normal. For $\gamma > 1000$, a Gaussian distribution with both mean and variance equal to γ is an excellent approximation of the Poisson distribution.

One of the results of this relationship is that as the mean current, I_{dc} (A), along a cluster of nerve fibres increases, the RMS noise current, i_{rms} (A), increases proportionally

$$i_{rms} = \sqrt{2qI_{dc}\Delta f} \tag{5.7}$$

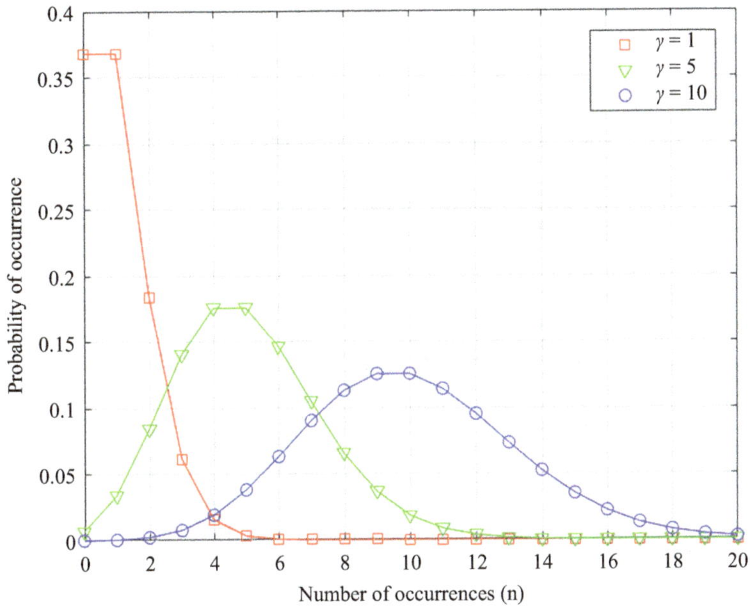

Figure 5.4 Poisson distributions for differing occurrence expectations (Brooker 2008)

where q is the electron charge (1.6×10^{-19} C) and Δf (Hz) is the bandwidth. As with the thermal noise case, the magnitude of shot noise is also proportional to the measurement bandwidth (Young 1990).

If this current is measured using an electrode, buffered, and flows through a load resistor, R_{load} (Ω), then the RMS noise voltage, v_{rms} (V) will be

$$v_{rms} = i_{rms} R_{load} \tag{5.8}$$

Noise Power Spectrum for Shot Noise: This can be determined by examining the noise generation process. As each charge element flows across a cell membrane, it generates a current pulse. Because the duration of each pulse is relatively short, its effective bandwidth is reasonably wide (by biometric standards). It can therefore be concluded that shot noise is "white". A good analogy of the process is the sound of rain on a corrugated iron roof.

Example:

Determine the noise PSD, i_{rms} ($\mathrm{A}\sqrt{\mathrm{Hz}}$)of the shot noise for the spinal cord carrying a current of 0.5 mA.

$$\begin{aligned} i_{rms} &= \sqrt{2qI_{dc}} \\ &= \sqrt{2 \times 1.6 \times 10^{-19} \times 0.5 \times 10^{-3}} \\ &= 12.6\ \mathrm{pA}/\sqrt{\mathrm{Hz}} \end{aligned}$$

If this current flows through a load resistor with $R_{load} = 1\ \mathrm{k\Omega}$ the RMS voltage spectral density will be $v_{rms} = 12.6\ \mathrm{nV}/\sqrt{\mathrm{Hz}}$

The RMS noise voltage measured by the data acquisition system with a 20 kHz bandwidth will be

$$v_{rms} = 12.6 \times 10^{-9} \times \sqrt{20 \times 10^3} = 1.79\ \mu V$$

5.3.1.3 Common mode and differential mode noise

Measurement of a single fibre electromyogram with a magnitude of only 10 μV in the presence of an ECG signal one thousand times as large would seem to be a lost cause. However, if the electrodes are placed with care, the vast majority of the ECG signal will be common-mode noise which can be eliminated using a combination of common-mode filtering and an amplifier with a differential input. This process is illustrated in Figure 5.5.

Differential mode noise generated by electrode movement or any other mechanism is indistinguishable from the EMG signal and cannot be removed by this technique. It is therefore output along with the EMG signal and would have to be removed by the signal processor.

5.3.2 Amplifiers

Most analogue signal processing is performed using combinations of operational amplifiers, as shown in Figure 5.6. Known as op-amps, these versatile integrated

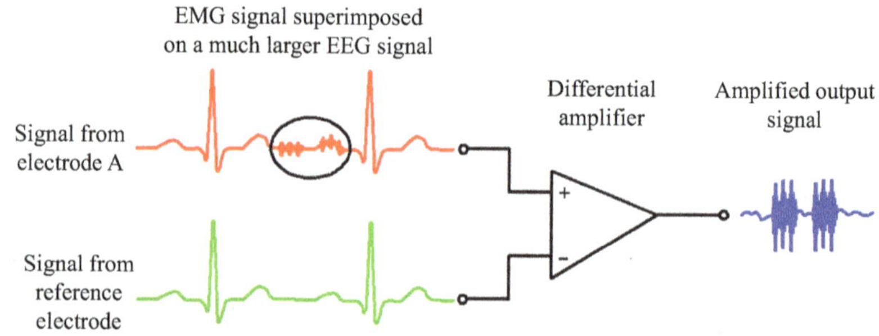

Figure 5.5 Common mode noise rejection of an EEG signal using a differential amplifier

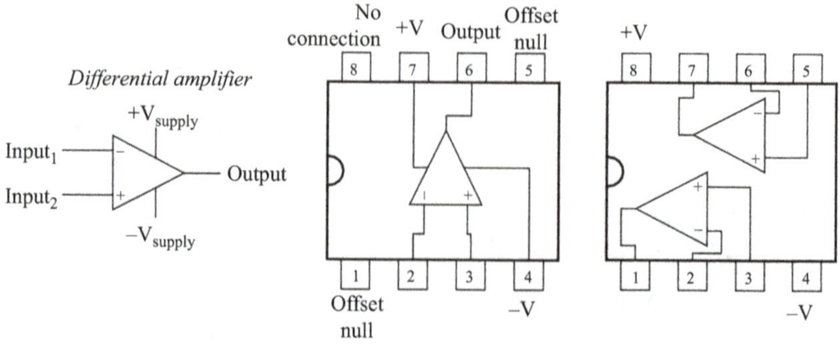

Figure 5.6 Op-amp symbol and common pin configurations

circuits consist of hundreds of transistors, resistors and capacitors to produce an extremely high-gain, wide bandwidth amplifier with a differential input.

The ideal op-amp is considered to have infinite gain at DC, infinite input impedance and zero output impedance. As a consequence of the infinite input impedance, no current is drawn by the inputs, and as a result of the infinite gain, any difference between the two inputs results in an infinite output voltage. The zero output impedance results in an output voltage that is independent of the output current.

These characteristics may seem to be illogical, but they provide a good approximation of the actual performance of an op-amp in the case where negative feedback is used to reduce the actual gain.

5.3.2.1 Negative feedback

If the output of an op-amp is connected to its inverting input and a voltage signal is applied to the non-inverting input, as shown in Figure 5.7, the output voltage of the op-amp closely follows that input voltage.

As V_{in} increases, V_{out} will increase in accordance with the differential gain. However, as V_{out} increases, that output voltage is fed back to the inverting input, thereby acting to decrease the voltage difference between inputs. This acts to reduce the output. What happens for any given voltage input is that the op-amp will output a voltage very nearly equal to V_{in}, but just low enough for the voltage difference left between V_{in} and the negative (−) input to be amplified to generate the output voltage.

The circuit will quickly reach equilibrium at a point where the output voltage is exactly the right value to maintain the correct amount of differential, which in turn produces the right amount of output voltage. This technique is known as negative feedback, and it is the key to having a self-stabilising system[1]. This stability gives the op-amp the capacity to work in its linear region.

For an op-amp with a gain A_v, the output voltage, V_{out}, is the product of the voltage gain and the differential input voltage

$$V_{out} = (V_{in} - V_{out})A_V \tag{5.9}$$

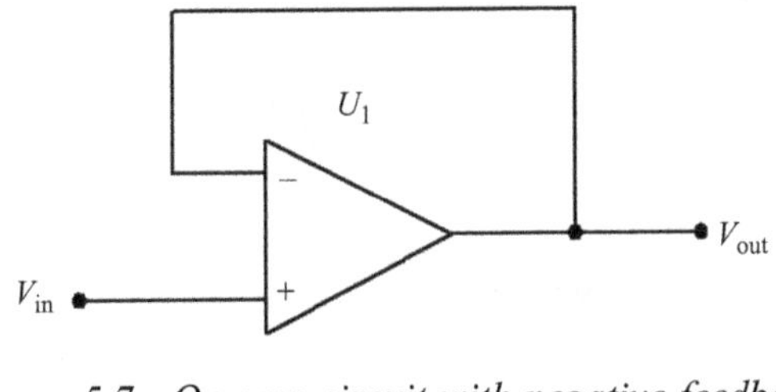

Figure 5.7 Op-amp circuit with negative feedback

[1]This is true not only of op-amps, but of any dynamic system in general.

This can be rewritten as

$$V_{out} = V_{in}\frac{A_V}{1 + A_V} \tag{5.10}$$

Typical op-amps have DC gains of $A_v = 10^6$ or more; therefore to all intents, the output voltage will be equal to the input voltage, and the differential voltage will be zero (Kuphaldt 2003).

5.3.2.2 Inverting amplifier

The relationship between the input and output voltages can easily be determined if it is remembered that the input impedance to the amplifier is extremely high, and therefore the current into node A of the circuit shown in Figure 5.8 must equal the current exiting it (Kirchhoff's current law). Therefore

$$-\frac{V_{in}}{R_1} = \frac{V_{out}}{R_2} \tag{5.11}$$

from which it is simple to determine that the output voltage will be

$$V_{out} = -V_{in}\frac{R_2}{R_1} \tag{5.12}$$

In the example shown in the figure, the gain $A_V = -R_2/R_1 = -10$, therefore the output voltage will be ten times as large as the input voltage, just inverted.

In most modern op-amps it is possible to produce a gain of less than one and still maintain the stability of the circuit. However, in most cases op-amp circuits are used to increase the level of the signal and $R_2 \gg R_1$.

One further consideration is that because the –ve input to the op-amp is a "virtual" earth, the input impedance of the inverting amplifier is equal to R_1 which is usually less than 100 kΩ or so (depending on the actual input impedance of the op-amp and the bias current required). This can result in this amplifier circuit loading the input, with unforeseen consequences (Kuphaldt 2003).

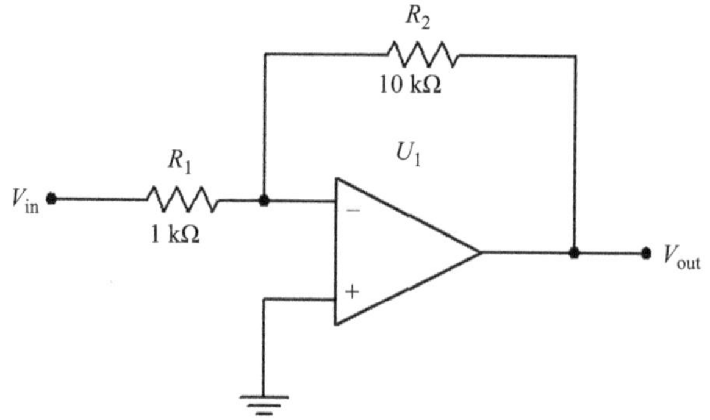

Figure 5.8 Op-amp circuit for an inverting amplifier

5.3.2.3 Non-inverting amplifier

The relationship between the input and output voltage can easily be derived from Figure 5.9. Consider that the input impedance of the op-amp is extremely high, therefore, the current flowing into node A must equal the current flowing out.

$$\frac{V_{in}}{R_1} = \frac{V_{out} - V_{in}}{R_2} \tag{5.13}$$

from which it is easy to determine that

$$V_{out} = V_{in}\frac{R_1 + R_2}{R_1} \tag{5.14}$$

For the component values shown in the figure, the gain $A_V = 11$, making the output 1.1 V for an input of 100 mV.

If the resistance of R_2 is reduced to zero, the circuit reduces to that in Figure 5.7, and the gain reduces to unity.

In this configuration, the input impedance of the amplifier is equal to the input impedance of the op-amp, which is extremely high and hence no loading to any previous stages will occur.

5.3.2.4 Differential amplifier

An op-amp with no feedback is already a differential amplifier, amplifying the voltage difference between the two inputs. However, its gain cannot be controlled, and it is generally too high to be of any use, except as a comparator. In the previous examples, the application of negative feedback to op-amps has resulted in the practical loss of one of the inputs, with the resulting amplifier only good for amplifying a single voltage signal input. However, an op-amp circuit maintaining both voltage inputs with a controlled gain set by external resistors can be constructed as shown in Figure 5.10.

If all the resistor values are equal, this amplifier will have a differential voltage gain of 1. The analysis of this circuit is similar to that of an inverting amplifier, except that the non-inverting input (+) of the op-amp is at a voltage equal to a fraction of V_2, rather than being connected directly to ground. In this configuration, V_2 functions as the non-inverting input and V_1 functions as the inverting input of

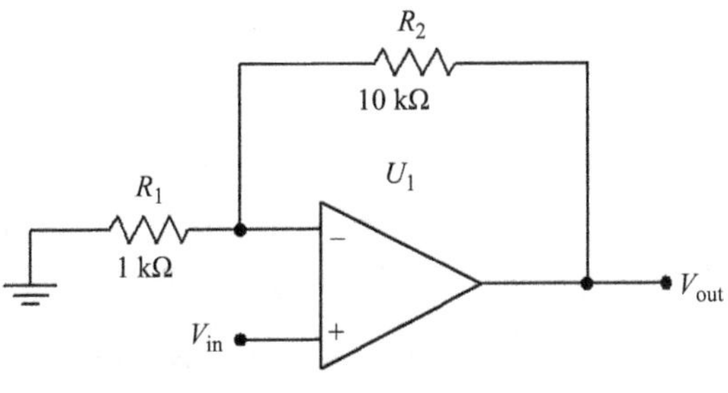

Figure 5.9 Op-amp circuit for a non-inverting amplifier

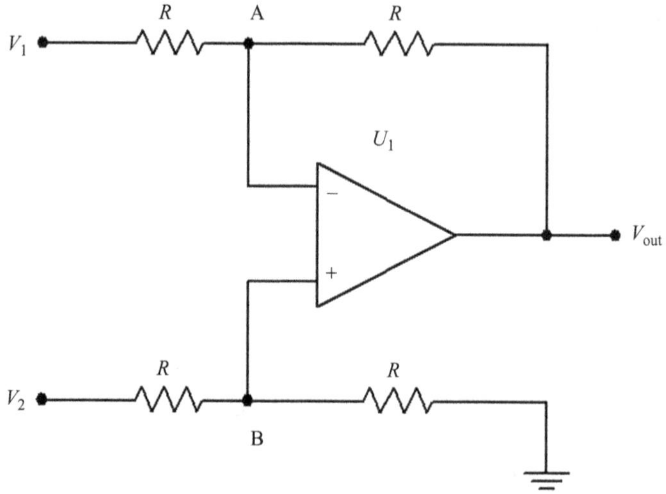

Figure 5.10 Op-amp circuit for a differential amplifier

the final amplifier circuit. The voltage, V_B at node B is

$$V_B = V_2 \frac{R}{2R} = \frac{V_2}{2}$$

Using Kirchoff's law for currents into node A

$$\frac{V_A - V_1}{R} = \frac{V_{out} - V_A}{R}$$

Because the gain is large the voltage at node A is equal to that at node B, $V_A = V_B$.

Substituting for V_A and simplifying gives

$$V_{out} = V_2 - V_1 \tag{5.15}$$

If a differential gain of anything other than unity is required, then the resistances of both the upper and lower voltage dividers would have to be adjusted simultaneously for balanced operation, and that is not practical. As with the inverting amplifier, the gain would be determined by the ratio of the resistors, $A_v = R_2/R_1$ and the output voltage would be

$$V_{out} = (V_2 - V_1)\frac{R_2}{R_1} \tag{5.16}$$

Another limitation of this amplifier design is the fact that its input impedances are rather low compared to that of some other op-amp configurations. Each input voltage source has to drive current through a resistance to a virtual ground, which constitutes a far lower impedance than the extremely high input impedance of the op-amp alone. This problem is solved by buffering the inputs as shown in Figure 5.11.

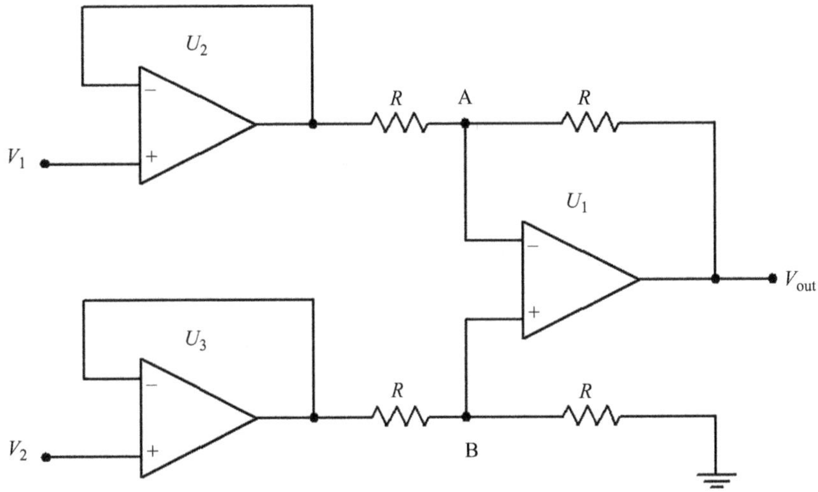

Figure 5.11 Op-amp circuit for a buffered input differential amplifier

Now the V_1 and V_2 input lines are connected straight to the inputs of two voltage-follower op-amps, giving very high impedance. The two op-amps on the left now handle the driving of current through the resistors instead of letting the input voltage sources do it (Kuphaldt 2003).

5.3.2.5 Instrumentation amplifier

To construct a practical instrumentation amplifier, it is necessary to modify the previous circuit to simplify the gain control. This is achieved using three new resistors linking the two buffer circuits together as shown in Figure 5.12. Consider all resistors to be of equal value except for R_{gain}. The negative feedback of the upper-left op-amp causes the voltage at node 1 to be equal to V_1. Likewise, the voltage at node 2 is held to a value equal to V_2. This establishes a voltage drop across R_{gain} equal to the voltage difference between V_1 and V_2. That voltage drop causes a current through R_{gain}, and since the feedback loops of the two input op-amps draw no current, an identical current to that flowing through R_{gain} must pass through the two "R" resistors above and below it. This produces a voltage drop between nodes 3 and 4 equal to

$$V_{3-4} = (V_2 - V_1)\left(1 + \frac{2R}{R_{gain}}\right) \tag{5.17}$$

The differential amplifier on the right-hand side of the circuit then takes this voltage drop between nodes 3 and 4 and amplifies it by a gain of 1 (assuming again that all "R" resistors are of equal value).

Though this may appear to be a cumbersome way to build a differential amplifier, it has the distinct advantages of possessing extremely high input

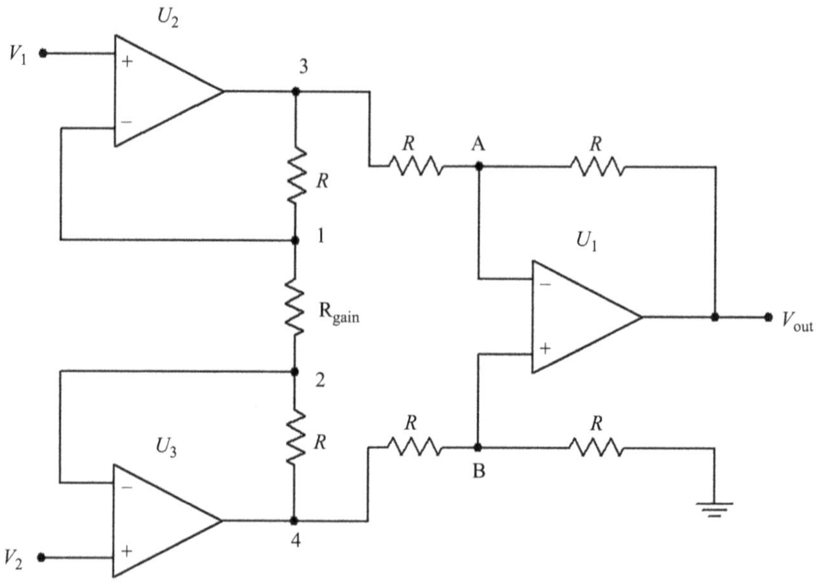

Figure 5.12 Practical differential input instrumentation amplifier

impedances on the V_1 and V_2 inputs, and adjustable gain that can be set by a single resistor. The voltage gain of the instrumentation op-amp is

$$A_V = \left(1 + \frac{2R}{R_{gain}}\right) \tag{5.18}$$

The overall gain can still be adjusted by changing the values of the other resistors as well as R_{gain}, but this would necessitate balanced resistor value changes for the circuit to remain symmetrical.

Note that the lowest gain possible with the above circuit is obtained with R_{gain} completely open (infinite resistance), and that gain value is 1 (Kuphaldt 2003).

One of the major problems with this circuit implementation using discrete components is that, even with precision resistors, imbalances remain and the common mode rejection ratio (CMRR) remains poor. Fortunately, instrumentation amplifiers are available as commercial integrated circuits with laser-trimmed internal resistors to maximise the CMRR. These include the AD524 and AD624 from Analogue Devices as well as the LM623 from National Semiconductor. The CMRR for the AD524 increases from 90 dB at unity gain up to 120 dB at a gain of 1000.

5.3.2.6 Charge amplifier

A charge amplifier has as its input a capacitance that provides an extremely high input impedance at low frequencies. Contrary to what the name suggests, charge amplifiers do not amplify electric charge, but convert the input charge into a voltage and present it as a low impedance output. It should therefore be called a charge-to-voltage converter. Common applications include capacitive accelerometers and piezoelectric sensors.

In effect a charge amplifier consists of a high-gain inverting voltage amplifier with a MOSFET or JFET at its input to achieve a sufficiently high input impedance (see Figure 5.13).

In this schematic C_t is the sensor capacitance; C_c is the cable capacitance; C_r is the feedback capacitor R_t is the time constant resistor (or insulation resistance of range capacitor); R_i is the insulation resistance of input circuit (cable and sensor); q is the charge generated by sensor and A_V is the open loop gain of the op-amp.

Neglecting the effects of R_t and R_i, the resulting output voltage, V_{out}, becomes

$$V_{out} = \frac{-q}{C_r} \frac{1}{1 + \frac{1}{A_V C_r}(C_t + C_r + C_c)} \tag{5.19}$$

For sufficiently high open-loop gain, the cable and sensor capacitance can be neglected, leaving the output voltage dependent only on the input charge and the range capacitance

$$V_{out} = -\frac{q}{C_r} \tag{5.20}$$

In short, the amplifier acts as a charge integrator that balances the sensor's electrical charge with a charge of equal magnitude and opposite polarity. This ultimately produces a voltage across the range capacitor. In effect, the purpose of the charge amplifier is to convert the high impedance charge input, q, into a usable output voltage, V_{out}.

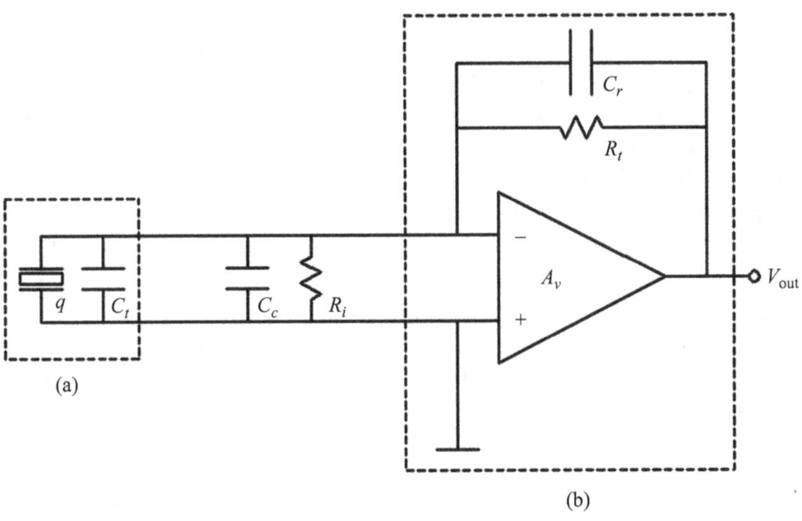

Figure 5.13 Schematic showing (a) piezoelectric accelerometer connected to (b) a charge amplifier

Two of the more important considerations in the practical use of charge amplifiers are time constant and drift. The time constant, τ_c, is defined as the discharge time of an AC-coupled circuit. In a period of time equivalent to one time constant, a step input will decay to 37% of its original value. The time constant of a charge amplifier is determined by the product of the range capacitor, C_r, and the time constant resistor, R_t:

$$\tau_c = R_t C_r \tag{5.21}$$

Drift is defined as an undesirable change in output signal over time and is independent of the input signal. In a charge amplifier, it can be caused by low insulation resistance at the input, R_i, or by leakage current of the input MOSFET or JFET in the op-amp.

Drift and the time constant simultaneously affect a charge amplifier's output. One or the other will be dominant. Either the charge amplifier output will drift toward saturation (power supply rail) at the drift rate, or it will decay toward zero at the time constant rate.

5.3.3 Practical considerations

It must be remembered that op-amps do not really have infinite gain or infinite input impedances, amongst other problems.

In addition, real op-amps exhibit an imbalance caused by a small mismatch between the input transistors, which results in unequal bias currents flowing through the input terminals, and results in an offset in the output voltage. This offset can be balanced out by introducing a compensating bias voltage.

As mentioned earlier, one of the main uses of differential amplifiers in biomechatronic applications is to eliminate common-mode voltages. In reality, this rejection is not perfect because of circuit imbalances, and some of the common-mode signal leaks through to the output. Op-amp specifications include their performance in this regard. It is referred to as the CMRR and is defined as the difference between the differential mode gain and the common mode gain of the amplifier.

Because of internal compensation capacitors within the op-amp that keep it stable, gain rolls off with frequency as shown in Figure 5.14. This limits the highest frequency that can be amplified by an op-amp, and the maximum effective closed-loop gain.

5.3.4 Op-amp specifications

There are literally hundreds of op-amp models to choose from. Many sell for less than $1 apiece, while special-purpose instrumentation and radiofrequency (RF) op-amps may be quite a bit more expensive. Table 5.2 lists the performance of a range of general purpose and specialist op-amps.

There is substantial variation in performance between some of these units. Take for instance the parameter of input bias current: the CA3130 wins the prize for lowest, at 0.05 nA (or 50 pA), and the LM833 has the highest at slightly over 1 μA. The model CA3130 achieves its incredibly low bias current through the use of MOSFET transistors in its input stage. One manufacturer advertises the CA3130's input impedance as 1.5 terra-ohms, or 1.5×10^{12} Ω! Other op-amps shown here, with low bias current figures, use JFET input transistors, while the high bias-current models use bipolar input transistors.

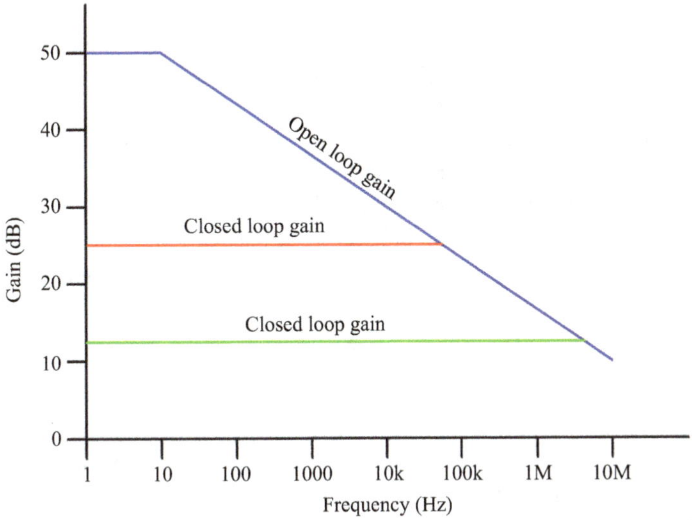

Figure 5.14 Bandwidth and closed-loop gain of an op-amp

Table 5.2 Some common op-amp and their specifications [adapted from Kuphaldt (2003)]

Model Number	Amps in package	Supply voltage (min/max)	Bandwidth (MHz)	Bias current (nA)	Slew rate (V/μs)	Output current (mA)
TL082	2	12/36	4	8	13	17
LM301A	1	10/36	1	250	0.5	25
LM318	1	10/40	15	500	70	20
LM324	4	3/32	1	45	0.25	20
LF353	2	12/36	4	8	12	20
LF356	1	10/36	5	8	12	25
LF411	1	10/36	4	20	15	25
LM741C	1	10/36	1	500	0.5	25
LM833	2	10/36	15	1050	7	40
LM1458	2	6/36	1	800	10	45
CA3130	1	5/16	15	0.05	10	20
CLC404	1	10/14	232	44000	2600	70
CLC425	1	5/14	1900	40000	350	90
LM12CL	1	15/80	0.7	1000	9	13000
LM7171	1	5.5/36	200	12000	4100	100

While the 741 is specified in many electronic circuit schematics, its performance has long been surpassed by other devices. Even some designs originally based on the 741 have been improved over the years to beat the original design specifications. Op-amps with JFET and MOSFET input transistors far exceed the 741's performance in terms of bias current and generally manage to beat the 741 in terms of bandwidth and slew rate as well.

When low bias current is a priority (such as in low-speed integrator circuits), choose the CA3130. For general-purpose DC amplifier work, the LM1458 offers good performance. For an upgrade in performance, choose the LF353, as it is a pin-compatible replacement for the LM1458. The LF353 is designed with JFET input circuitry for very low bias current and has a bandwidth four times greater than the LM1458, although its output current limit is lower (but still short-circuit protected).

If a low power supply voltage is a requirement, the model LM324 is suitable, as it functions on as low as 3 volts DC. Its input bias current requirements are also low, and it provides four op-amps in a single 14-pin package. Its major weakness is speed, limited to 1 MHz bandwidth and an output slew rate of only 0.25 volts per μs. For high-frequency AC amplifier circuits, the LM318 is a good "general purpose" device.

Special-purpose op-amps are available for modest cost, which provide better performance specifications. Many of these are tailored for a specific type of performance advantage, such as maximum bandwidth or minimum bias current. The CLC402 and CLC425 have bandwidths of 250 MHz and 1.9 GHz, respectively. In both cases high speed is achieved at the expense of high bias currents and restrictive power supply voltage ranges.

The last two op-amps in the table provide high output current capabilities for driving low impedance loads.

5.4 Analogue signal processing

5.4.1 Frequency content of a signal

In the frequency domain, a continuous sinusoidal signal of infinite duration can be represented in terms of its position on the frequency continuum and its amplitude only. However, most practical signals are not of infinite duration and so there is some uncertainty in the measured frequency, and this is represented in the frequency domain by a finite spectral width.

From a mathematical perspective, this is equivalent to windowing the continuous sinusoidal signal using a rectangular pulse of duration τ (s). Because windowing, or multiplication, in the time domain becomes convolution in the frequency domain, the continuous signal spectrum must be convolved by the spectrum of a rectangular pulse to obtain the spectrum of the windowed signal.

The spectrum of a rectangular pulse is the Sync function (Carlson 1998)

$$F(\omega) = \tau \frac{\sin(\omega\tau/2)}{\omega\tau/2} \tag{5.22}$$

and the spectrum of a continuous sinusoidal signal is an impulse $\delta(\omega)$, so the resultant convolution is just the Sync function.

It can be seen from equation (5.22) that as the duration of the signal decreases, $\tau \to 0$, its spectral width increases until, in the limit, when the signal can be represented by an impulse $\delta(t)$, the spectral width is infinite. This relationship is shown graphically in Figure 5.15.

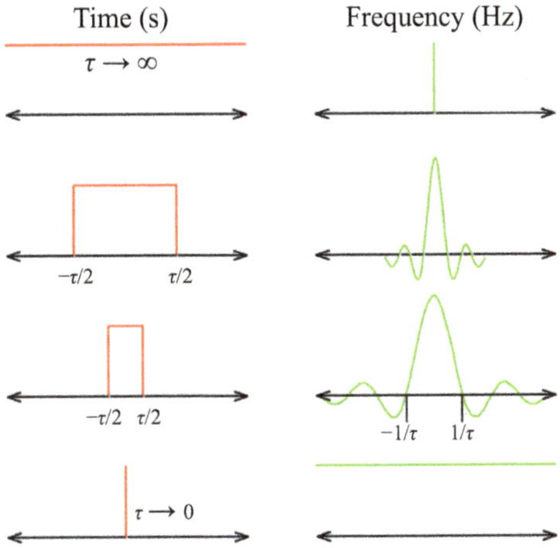

Figure 5.15 Mapping the relationship between the duration of a pulse and its spectrum

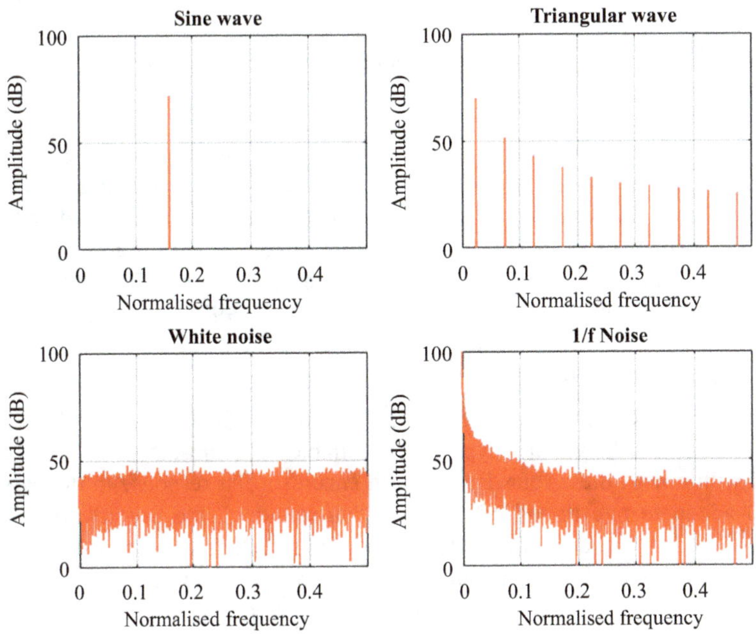

Figure 5.16 Spectra of various types of signal and noise

More complex signals can usually be made up of several sinusoidal components of varying amplitudes. These can be calculated using the Fourier series, so it is often easier to identify the spectrum of a time domain signal by processing it through a Fourier Transform and then examining the amplitudes of the resultant components. Some examples of this process are shown in Figure 5.16.

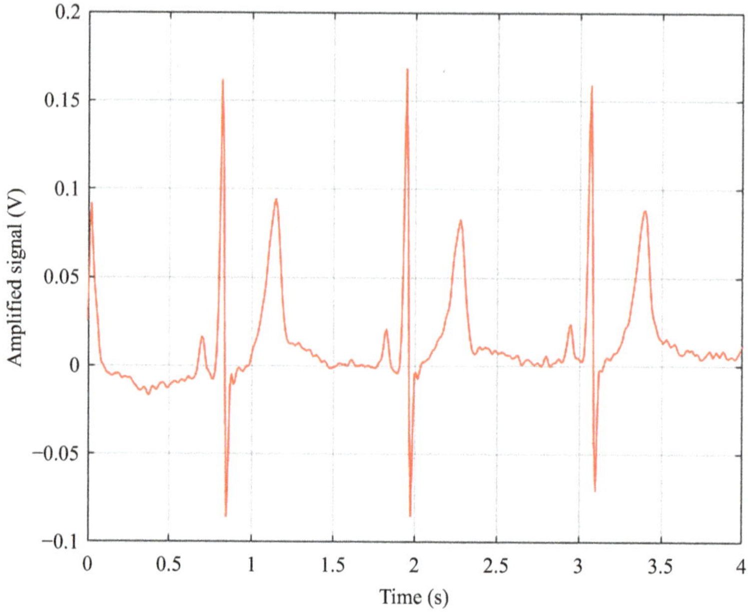

Figure 5.17 Measured electrocardiogram trace of the author

As can be seen from the harmonics for the triangular wave signal shown in Figure 5.16, even though the series is infinite, the coefficients decrease in amplitude and eventually become so small that their contribution is considered to be negligible. For example, the electrocardiogram trace shown in Figure 5.17, with a fundamental frequency of about 1.2 Hz, can be reproduced with 70 to 80 harmonics which equates to a bandwidth of about 100 Hz (Carr 1997).

5.4.2 Analogue filters

The most common filter responses are the Butterworth, Chebyshev, Bessel and Elliptical types. Many other types are available, but most applications can be solved with one of these four. Butterworth ensures a flat response in the passband and an adequate rate of roll-off. A good "all-rounder," the Butterworth filter is simple to understand and suitable for applications such as audio processing.

The Chebyshev gives a much steeper roll-off, but passband ripple makes it unsuitable for audio systems. It is superior for applications in which the passband includes only one frequency of interest (e.g., the derivation of a sine wave from a square wave, by filtering out the harmonics).

The Bessel filter gives a constant propagation delay across the input frequency spectrum. Therefore, applying a square wave (consisting of a fundamental and many harmonics) to the input of a Bessel filter yields an output square wave with no overshoot (all the frequencies are delayed by the same amount). Other filters delay the harmonics by different amounts, resulting in an overshoot on the output waveform.

The elliptical filter is more complicated than the other types but because it has the steepest roll-off, it is often used in anti-aliasing filters.

5.4.2.1 Low-pass filter

A low-pass filter is a filter that passes low frequencies and attenuates high frequencies as shown in Figure 5.18. The amplitude response of a low-pass filter is flat from DC or near DC to a point where it begins to roll off. A standard reference point for this roll-off is the point where the amplitude has decreased by 3 dB, to 70.7% of its original amplitude (volts).

The region from around DC to the point where the amplitude is down 3 dB is defined as the passband of the filter. The range of frequencies from the 3 dB point to infinity is defined as the stopband of the filter.

The amplitude of the filter response at ten times the 3 dB frequency is attenuated a total of 20 dB for a single pole filter, and a total of 40 dB for a two-pole Butterworth filter. At higher frequencies, the amplitude continues to roll-off in a linear fashion, where the slope of the line is −20 dB per decade (10 times frequency) for a single pole filter and −40 dB per decade for a two-pole filter.

The term Butterworth refers to a type of filter response, not a type of filter. It is sometimes called the Maximally Flat approximation, because for a response of order n, the first $(2n-1)$ derivatives of the gain with respect to frequency are zero at frequency = 0. There is no ripple in the passband, and DC gain is maximally flat.

The term Chebyshev also refers to a type of filter response, not a type of filter. It is sometimes referred to as an equal-ripple approximation. It features superior

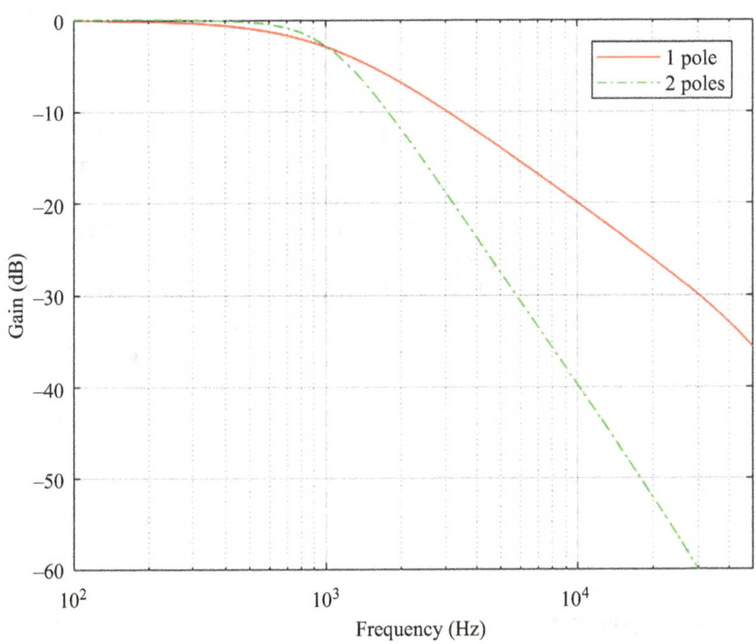

Figure 5.18 Frequency response of Butterworth low-pass filters

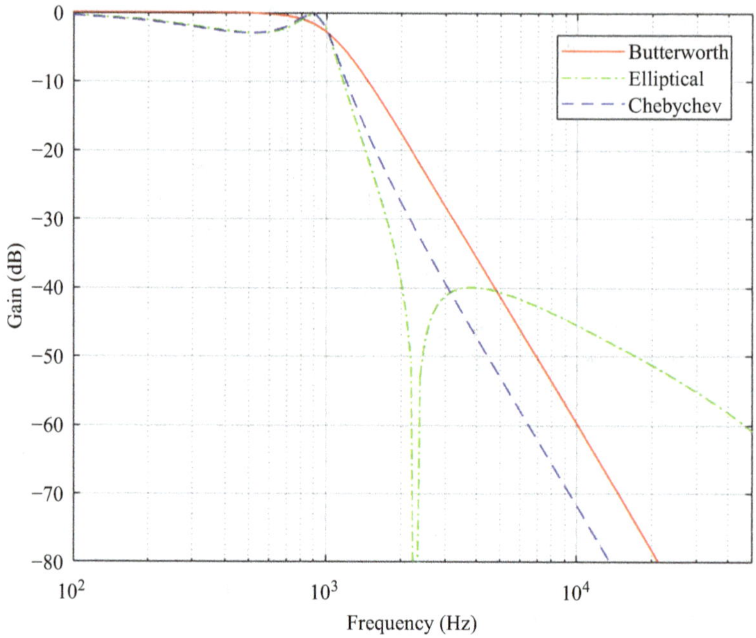

Figure 5.19 Comparison between the frequency response of Butterworth, Chebyshev and Elliptical low-pass filters

attenuation in the stop band, at the expense of ripple in the passband as shown in Figure 5.19. Generally, the designer will choose a ripple depth of between 0.1 dB and 3 dB. Chebyshev filter response, therefore, is not limited to a single value of response.

Elliptical filters have the best roll-off characteristics even for low order filters as shown in Figure 5.19. This characteristic makes them very useful for anti-aliasing filters prior to digitisation, or to remove clock spurs in direct digital synthesis (DDS) systems. In these applications the filter zeros are generally placed at multiples of the clock frequency.

Unfortunately, all these filter types suffer from group delay[2] problems as shown in Figure 5.20. However, very few filters are designed with square waves in mind because most of the time the signals filtered are sine waves, or close enough that the effect of harmonics can be ignored. If a waveform with high harmonic content is filtered, such as a square wave, the harmonics can be delayed with respect to the fundamental frequency and distortion will result.

To counteract this problem filters with a Bessel response are used. This response features flat group delay in the passband, and this is the characteristic of Bessel filters that makes them valuable to digital designers.

[2]Group delay measures the average time delay of a filter as a function of frequency, essentially the time delay experienced by the envelope of a signal.

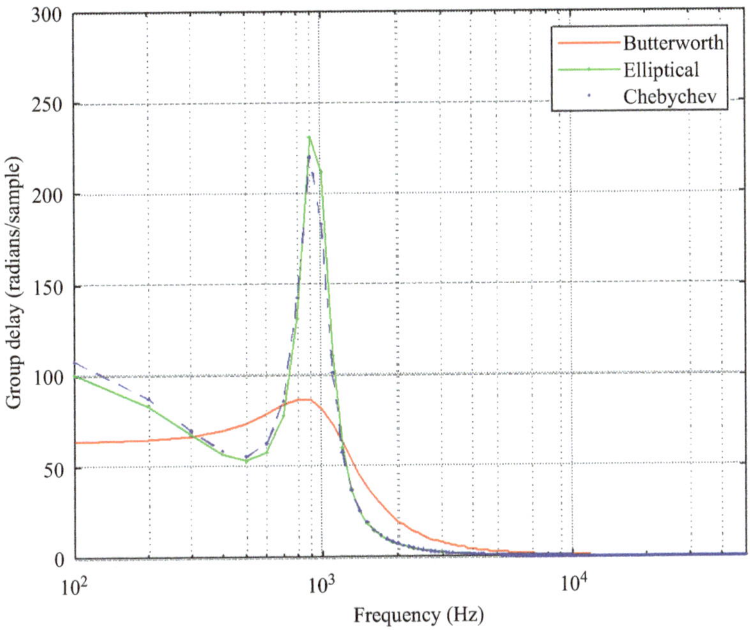

Figure 5.20 Comparison of the group delay of Butterworth, Chebyshev and Bessel low-pass filters

5.4.2.2 High-pass filters

A high-pass filter is a filter that passes high frequencies and attenuates low frequencies as illustrated in Figure 5.21. The amplitude response of a high-pass filter is flat from infinity down to a point where it begins to roll off. A standard reference point for this roll-off is the point where the amplitude has decreased by 3 dB, to 70.7% of its original amplitude. The region from infinity to the point where the amplitude is down 3 dB is defined as the passband of the filter. The range of frequencies from the 3 dB point down to zero (or near zero) is defined as the stopband of the filter.

The amplitude of the filter at one-tenth the 3 dB frequency is attenuated a total of 20 dB for a one-pole filter, and a total of 40 dB for a two-pole Butterworth filter. At lower frequencies, the amplitude continues to roll off in a linear fashion, where the slope of the line is −20 dB per decade (10 times frequency) for a single pole filter and −40 dB per decade for a two-pole filter.

5.4.2.3 Band-pass filters

Band-pass filters are those that pass a range of frequencies above a minimum and below a maximum as shown in Figure 5.22. These filters can be made using cascaded band-pass stages, or by cascading high and low-pass sections.

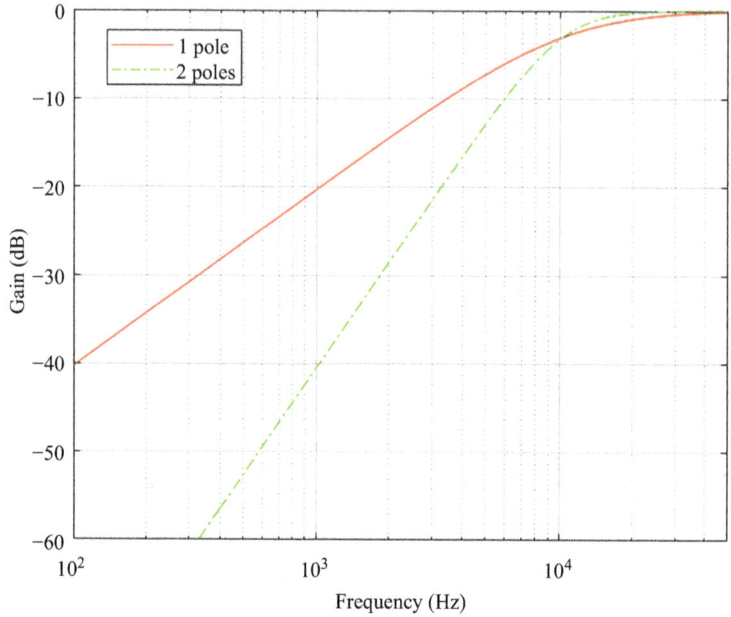

Figure 5.21 Frequency response of Butterworth high-pass filters

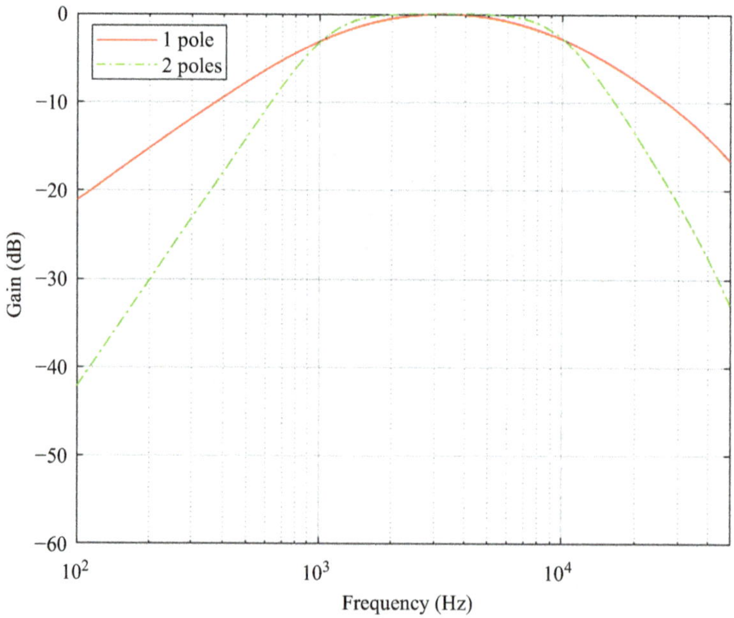

Figure 5.22 Band-pass filter frequency response – 2nd order band-pass compared to cascaded 2nd order high and low pass stages

- The figure shows the effect of cascading double-pole Butterworth filters.
- Both the high-pass and low-pass filters must have the same gain.
- The high-pass stage must come first, followed by the low-pass. In this way, high-frequency noise from the high-pass filter will be attenuated by the low-pass filter.
- Cascaded high-pass and low-pass filters probably take the same number of op-amps as cascading band-pass filters, yet the response is clearly sharper, giving a double-pole characteristic at the low- and high-frequency 3 dB points, instead of 6 dB of roll off caused by cascading two band-pass filters each with single-pole response.

5.4.2.4 Notch and band reject filters

A Notch filter is a filter that passes all frequencies except those in a stop band centred on a specific frequency. High Q notch filters eliminate a single frequency or narrow band of frequencies, while a band reject filter eliminates a wider range of frequencies

The amplitude response of a notch filter is flat at all frequencies except at the stop band on either side of the centre frequency. The standard reference points for the roll-offs on each side of the stop band are the points where the amplitude has decreased by 3 dB, to 70.7% of its original amplitude.

The −3 dB points and −20 dB points are determined by the size of the stop band in relation to the centre frequency, in other words the Q of the filter. The Q is the centre frequency divided by the bandwidth.

In the case of Figure 5.23 the following points should be noted:

- The −3 dB points are at about 300 Hz and 30 kHz for a Q of 0.1 and a centre frequency of 3.1 kHz. At one-tenth the bandwidth the amplitude is down 20 dB.
- The response of the band-pass filter with a $Q = 1$ is also shown. The −3 dB points are at about the same frequencies as the −20 dB levels for a Q of 0.1.
- At very high Q values, the response of the circuit will begin to have overshoot and undershoot, and that will destroy the integrity of the notch. The frequency that was supposed to be rejected may actually be amplified!

A band reject filter is constructed by summing the outputs of two parallel low-pass and high-pass filters as illustrated in Figure 5.24.

Some points are worth noting with regard to the relative merits of notch and band reject filters as shown in Figure 5.25:

- The performance increase that comes with summing low-pass and high-pass filter outputs comes at the expense of an additional op-amp, the op-amp that performs the summing function.
- Higher-order low-pass and high-pass filters will improve the performance of the band-reject filter.
- The farther apart the passbands are, the better the performance of the band reject filter.

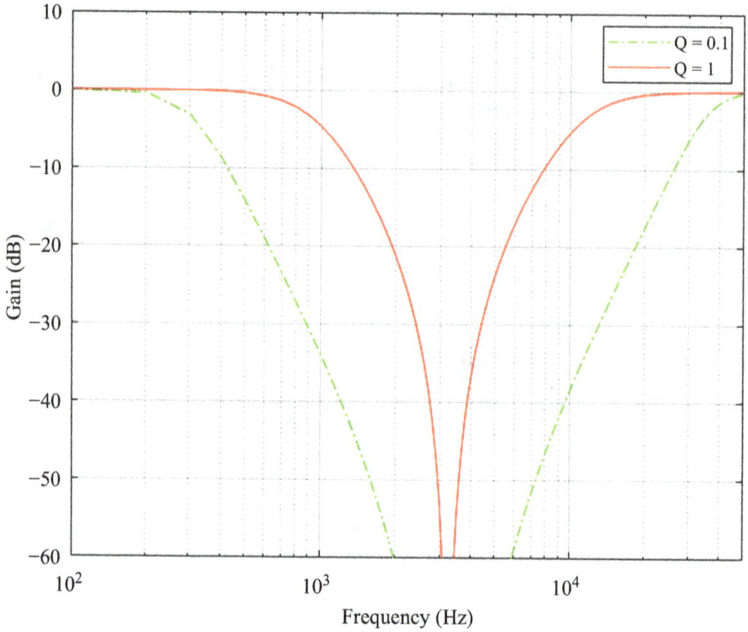

Figure 5.23 Notch filter frequency responses

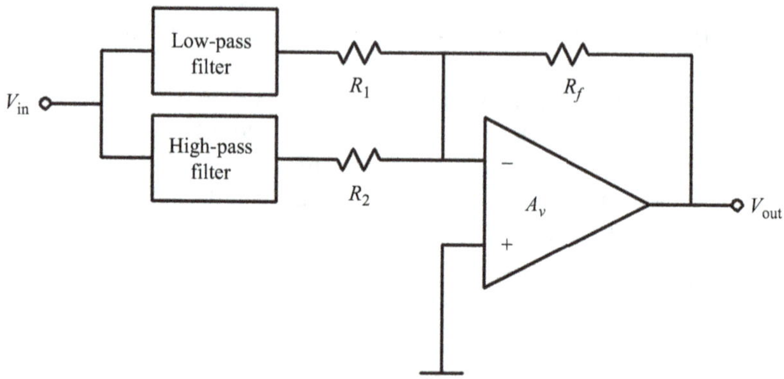

Figure 5.24 Construction of a band-reject filter

5.4.2.5 Active filter implementation

There are several different topologies that can be used to implement the filters discussed in the previous section. Some of these use the standard op-amp feedback mechanisms. Practical realisations of analogue filters are usually based on

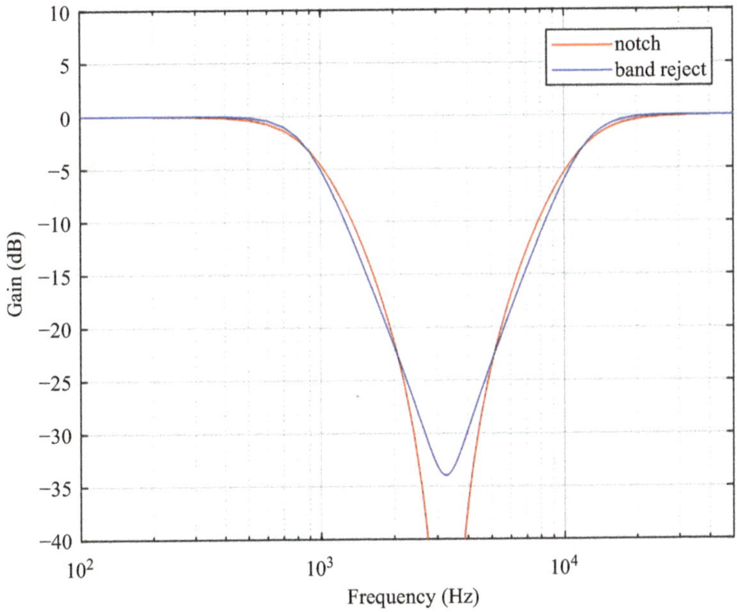

Figure 5.25 Comparison between a band-reject and notch filter

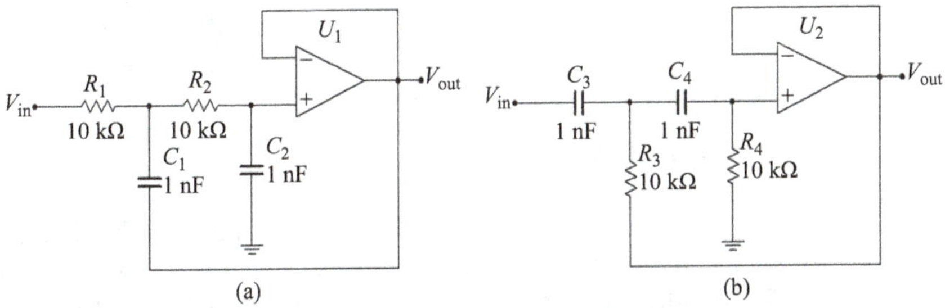

Figure 5.26 Sallen Key filter topology (a) low-pass and (b) high-pass

cascading second-order sections based on a complex conjugate pole-pair or a pair of real poles. A first-order section is then added if an odd-order filter is required.

One is the Sallen Key filter topology which is particularly valued for its simplicity. The circuit produces a second-order (12 dB/octave) low-pass or high-pass response using only two capacitors and two resistors and a unity gain buffer as shown in Figure 5.26.

The band-pass configuration is slightly more complex, and it is shown in Figure 5.27 for a non-unity gain configuration.

Higher order filters can be obtained by cascading these building blocks.

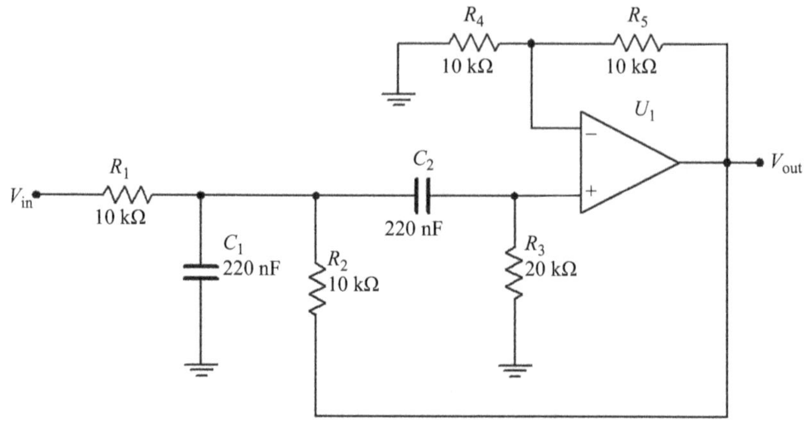

Figure 5.27 Sallen Key band-pass filter

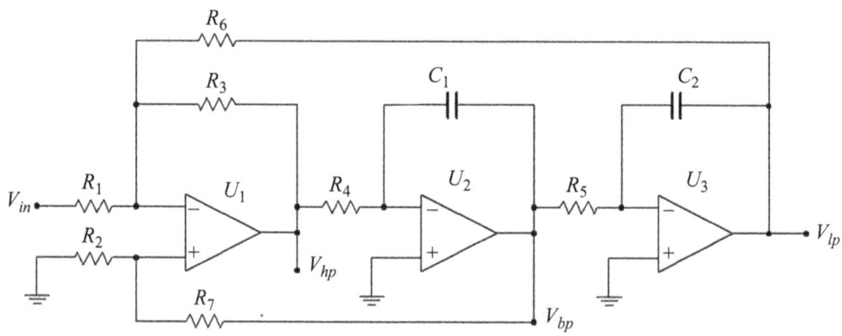

Figure 5.28 State variable filter topology

One of the more interesting topologies is that of the state variable filter. As shown in Figure 5.28, it consists of two cascaded integrators fed back into a summing amplifier. This configuration emulates the state-space model of a linear time-invariant (LTI) system exactly. The outputs of the various integrators therefore correspond to the state-space model's state variables and become the low-pass, high-pass and band-pass filter outputs.

Design of these filters is mostly done using any of the many web-based packages, or applets that are supplied with EDA software.

5.4.3 Other analogue circuits

5.4.3.1 Current to voltage converter

Many sensor types output a current rather than a voltage. These include photomultiplier and photodiode-based devices amongst others. However, it is inconvenient to process currents using the analogue techniques discussed in this section, or to digitise them for further processing. The circuit shown in Figure 5.29, based

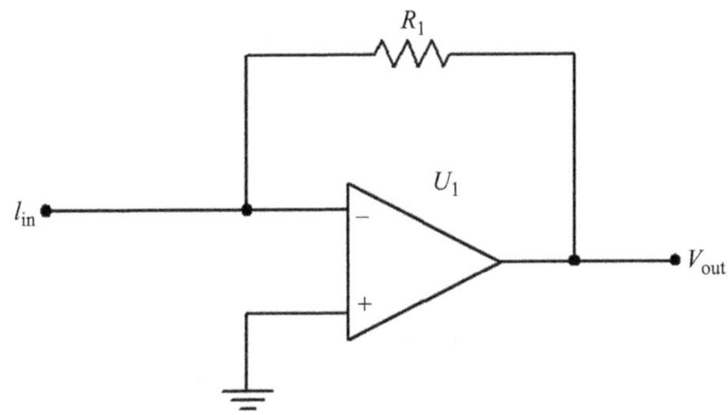

Figure 5.29 Current to voltage converter

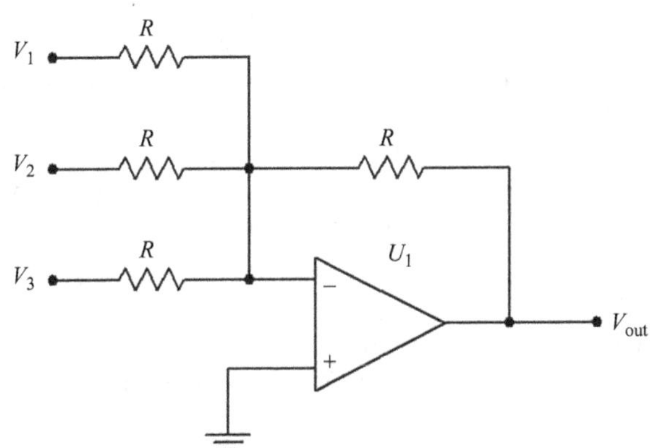

Figure 5.30 Summing amplifier

on the inverting amplifier is a simple method to convert a current to a voltage

$$V_{out} = I_{in}R \tag{5.23}$$

The advantage of using this circuit rather than a simple resistor is that the current flows into a virtual ground, eliminating the effects of the internal resistance of the current source.

5.4.3.2 Summing amplifier

The circuit shown in Figure 5.30 operates in a similar manner to the standard inverting amplifier except that the current at node A comprises the currents from the three inputs and that fed back from the output. With all resistor values equal, the currents through each of the three resistors will be proportional to their respective input voltages. Since those three currents input added at the virtual ground node, the algebraic sum of those currents through the feedback resistor will produce a voltage at V_{out} equal to

$$V_{out} = -(V_1 + V_2 + V_3) \tag{5.24}$$

5.4.3.3 Integrator and differentiator

By introducing electrical reactance into the feedback loops of op-amp amplifier circuits, it is possible to cause the output to respond to changes in the input voltage over time. Drawing their names from their respective calculus functions, an integrator produces a voltage output proportional to the product of the input voltage and time, and a differentiator produces a voltage output proportional to the input voltage's rate of change.

An op-amp circuit that determines the change in voltage by measuring current through a capacitor and outputs a voltage proportional to that current can be constructed, as shown in Figure 5.31.

The right-hand side of the capacitor is held to a voltage of 0 volts, due to the "virtual" ground. Therefore, the current flowing through the capacitor is solely due to *change* in the input voltage. A steady input voltage will not cause a current through C, but a changing input voltage will.

From a more rigorous perspective, consider first the current that flows through the feedback resistor to the virtual ground to produce a voltage drop across it. This will be identical to the output voltage. Therefore

$$V_{out} = iR$$

From the charge relationship and the definition of capacitance, it can be shown that the voltage developed across a capacitor is equal to the integral of the current flowing into it

$$V_{cap} = \frac{1}{C}\int idt$$

From Kirchhoff's current law, the two currents that flow into the negative input node must balance. Therefore, the current through the capacitor must equal

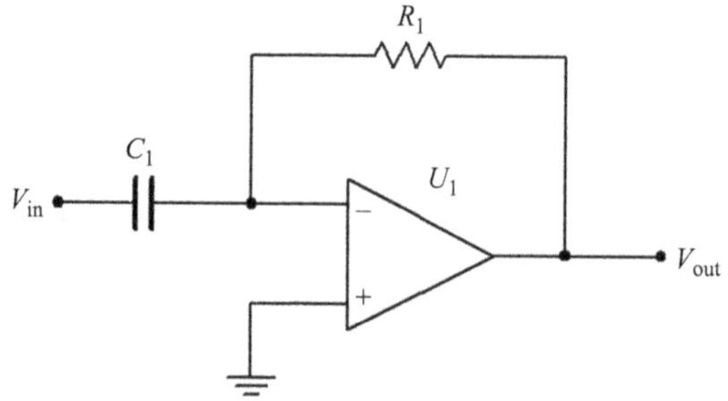

Figure 5.31 Differentiator

the current through the resistor, $i_{cap} = -i_{res}$, which makes it

$$i_{cap} = -\frac{V_{out}}{R}$$

Substituting this current into the integral gives

$$V_{cap} = -\frac{1}{RC}\int V_{out}dt$$

Taking the derivative of both the LHS and the RHS gives

$$\frac{dV_{cap}}{dt} = -\frac{1}{RC}V_{out}$$

Finally, because one terminal of the capacitor is the virtual earth $V_{in} = V_{cap}$, so

$$V_{out} = -RC\frac{dV_{in}}{dt}. \tag{5.25}$$

A positive rate of change of input voltage will result in a steady negative voltage at the output of the op-amp if the rate of change is linear. Conversely, a linear, negative rate of change of input voltage will result in a steady positive voltage at the output of the op-amp. This polarity inversion from input to output is due to the fact that the input signal is applied to the inverting input of the op-amp, so it acts like the inverting amplifier analysed previously. The faster the rate of change of input voltage, the greater the voltage at the output.

Applications for this, besides representing the derivative function in an analogue computer, include rate-of-change indicators for process instrumentation. One such rate-of-change signal application might be for monitoring (or controlling) the rate of temperature change in an incubator. In a prosthetic arm, it could be used to limit the slew rate of the elbow joint.

To produce an integrator, the op-amp circuit must generate an output voltage proportional to the magnitude and duration of the input voltage signal. Stated differently, a constant input signal will generate a certain rate of change in the output voltage. To achieve this, all that has to be done is to swap the capacitor and resistor in the previous circuit as shown in Figure 5.32.

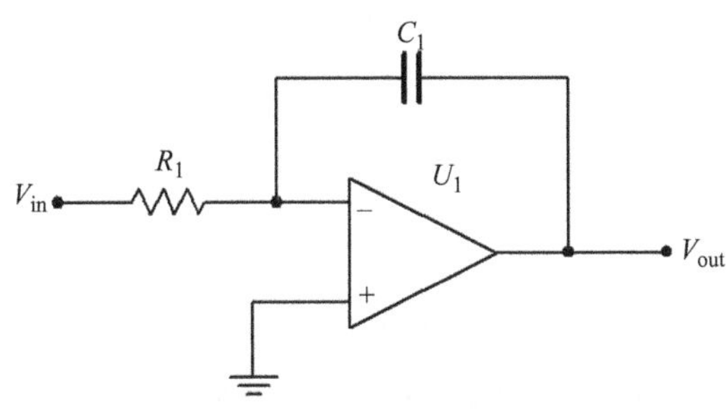

Figure 5.32 Integrator

The negative feedback of the op-amp ensures that the inverting input will be held at 0 volts (the virtual ground). If the input voltage is exactly 0 volts, there will be no current through the resistor; therefore, no charging of the capacitor, and the output voltage will not change. However, the output voltage may not be zero if the capacitor has been previously charged.

If a constant positive voltage is applied to the input, the op-amp output will fall negative at a linear rate, in an attempt to produce the changing voltage across the capacitor necessary to maintain the current established by the voltage difference across the resistor. Conversely, a constant, negative voltage at the input results in a linear, rising (positive) voltage at the output. The output voltage rate-of-change will be proportional to the value of the input voltage.

The equation describing this relationship is

$$V_{out} = -\frac{1}{RC}\int V_{in}dt + V_c \tag{5.26}$$

where V_c is the initial charge on the capacitor at $t = 0$.

Real integrator circuits generally have a switch to short out the capacitor at the start of an integration cycle, and sometimes also a large value resistor across it to drain away any residual voltage that may have accumulated due to leakage.

5.4.3.4 Envelope detection

For an amplitude-modulated signal, the envelope is a construct that joins the positive peaks of the signal to recreate the original modulation waveform. The process of envelope detection starts with rectification of the bipolar signal. If the signal amplitude is sufficiently large, a conventional half or full-wave rectifier can be used – notwithstanding the volt-drop across the diodes. A conventional envelope detector consisting of a diode and a low-pass filter is shown in Figure 5.33.

However, if the signal is small, then a precision full-wave rectifier such as the one shown in Figure 5.34 must be used.

This circuit is very common and has been around for many years. The tolerance of R2, 3, 4 and 5 is critical for good performance, and all four resistors should be 1% or better. Note that the diodes have been reversed to obtain a positive

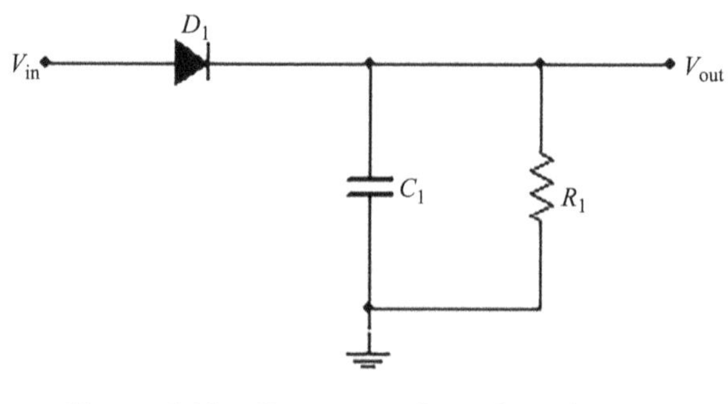

Figure 5.33 Conventional envelope detector

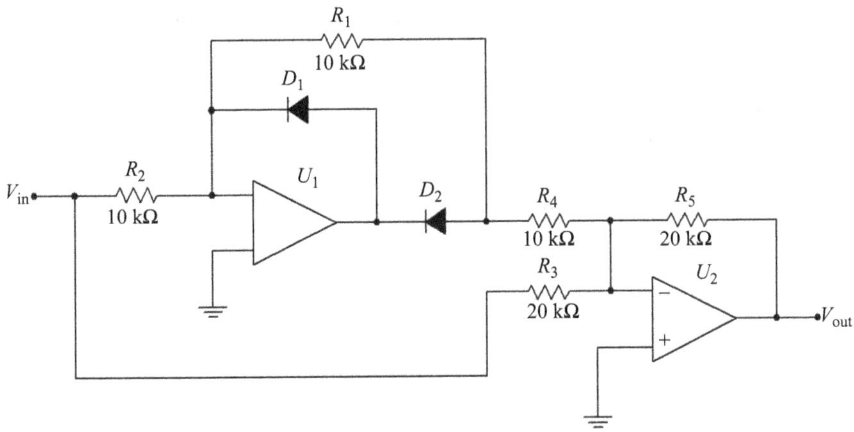

Figure 5.34 Precision full-wave rectifier

rectified signal. The second stage inverts the signal polarity. To obtain an improved high-frequency response, the resistor values should be reduced.

This circuit is sensitive to source impedance, so it is important to ensure that it is driven from a low impedance, such as an op-amp buffer stage. The circuit has good linearity down to a couple of mV at low frequencies but has a limited high-frequency response. Use of high-speed diodes, lower resistance values and faster op-amps is recommended when greater sensitivity and/or higher frequency response is required.

The output of the rectifier is passed through a low-pass filter to produce the envelope.

5.4.3.5 Worked example: myoelectric signal processing

Developing a differential amplifier for electromyography (EMG) poses few real problems if the appropriate precautions are taken. When a muscle is caused to contract, the distribution of electrolytes within the tissue changes, and this induces small voltages on the surface of the skin that can be picked up using the appropriate electrodes. From the specifications in Table 5.1, the voltage levels vary from 50 μV up to a maximum of 5 mV. However, this signal will probably be contaminated by other much larger biopotential signals and certainly by induced "mains hum".

The problem is to sense and isolate this signal so that it can be used to control the movement of a prosthetic device by driving a pneumatic artificial muscle (PAM) discussed in Chapter 3.

The first requirement is to select an appropriate set of electrodes to detect this small voltage with the addition of as few measurement artefacts as possible. Electrode types were discussed in Volume 1, Chapter 2, and from this discussion, it is obvious that a silver–silver chloride surface electrode and a conductive gel are best. These electrodes can be applied directly over the muscle complex of interest.

The electrode interface is reasonably high impedance, and they also pick up 50 Hz mains, ECG signals and a myriad of other electrical noise generated by the body and from extraneous sources.

An instrumentation amplifier is required to provide a high input impedance, a good CMRR and the appropriate gain to amplify the small signal to a useable level for further processing. This amplifier must be provided with a pair of differential inputs to sense the myoelectric signal and a ground reference as shown in Figure 5.35.

This instrumentation configuration can easily achieve a CMRR of more than 60 dB to eliminate most of the common mode 50 Hz, while still providing a voltage gain of 100. This will provide an output of between 0.5 mV and 500 mV.

The next problem to be solved is that of electrode movement. As discussed in Volume 1, Chapter 2, electrode movement alters the cell potential of the conductive gel, which results in the generation of large electrical signals called motion artefacts. These can saturate any further stages of amplification and need to be removed.

Fortunately, motion artefacts are found at the lower end of the spectral response of the EMG signal and can therefore be removed by high-pass filtering without removing too much of the real signal. A first-order high-pass filter with a

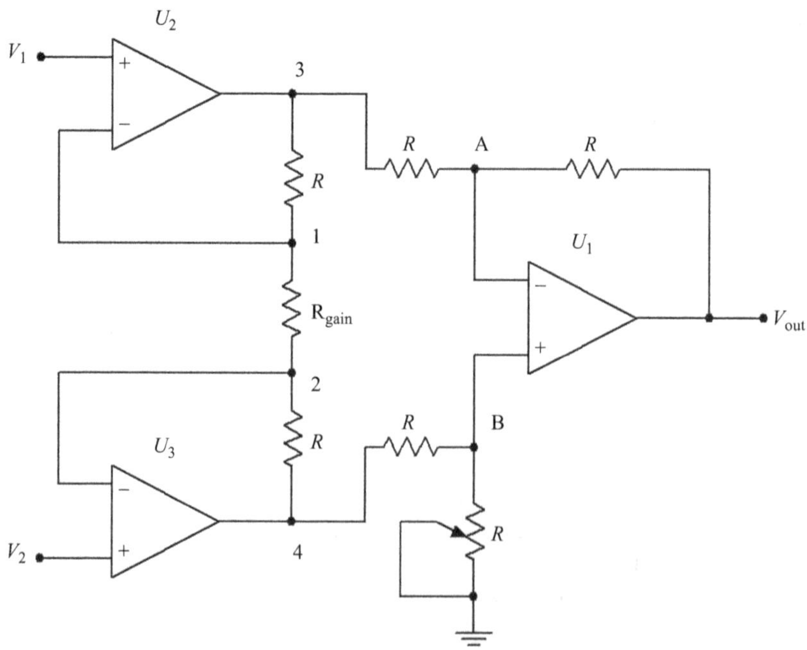

Figure 5.35 Instrumentation op-amp with ground reference electrode

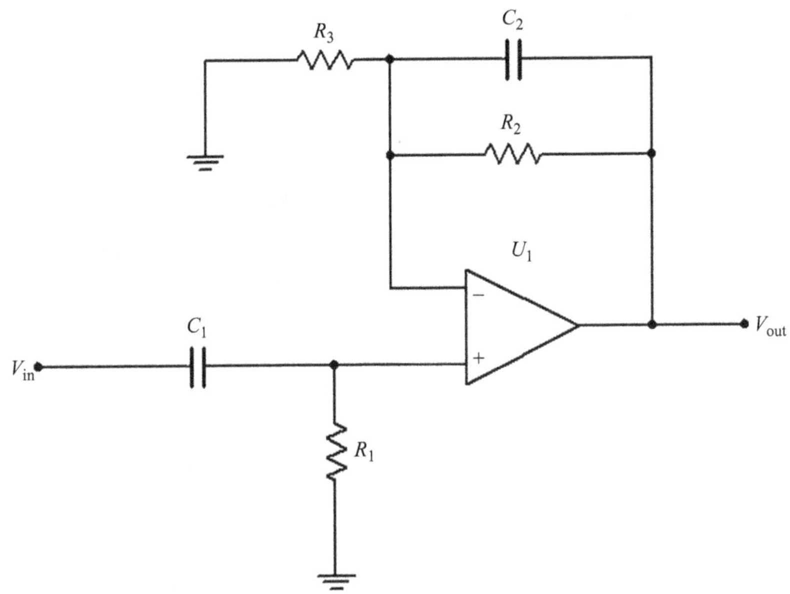

Figure 5.36 First-order high-pass filter followed by further amplification

5 Hz cutoff frequency realised by C_1 and R_1 followed by an amplifier with a voltage gain of 10, can easily be designed to implement this function, as shown in Figure 5.36. Note that a small capacitor, C_2, is included in the feedback path to reduce the high-frequency gain of the amplifier. This eliminates high-frequency noise above the maximum bandwidth of the EMG signal.

If mains hum is a problem, it is expedient to include a notch or band reject filter to remove components of the signal around 50 Hz, after which three bursts of an increasing EMG signal should have the features shown in Figure 5.37 Its frequency content should vary between 5 Hz and 300 Hz with a peak amplitude that varies between 5 mV and a maximum of about 2.5 V.

The next step in the process is to extract the envelope of the EMG signal. A simple rectifier diode requires a turn-on voltage of 0.7 V, which is larger than the maximum amplitude of the signal shown. Therefore, a precision full-wave rectifier shown in the previous section should be used. This is followed by a low-pass filter to produce only the signal envelope as shown in Figure 5.38.

Various techniques are available for using this envelope to actuate prosthetic limbs and these are addressed in detail in Volume 2, Chapter 5. The simplest method is to open the pneumatic solenoid valve to the actuator if the envelope exceeds a threshold voltage, and to keep it closed if the level is below this. A comparator, discussed in the following section, followed by a power transistor, can be used to control the solenoid as shown in Figure 5.39.

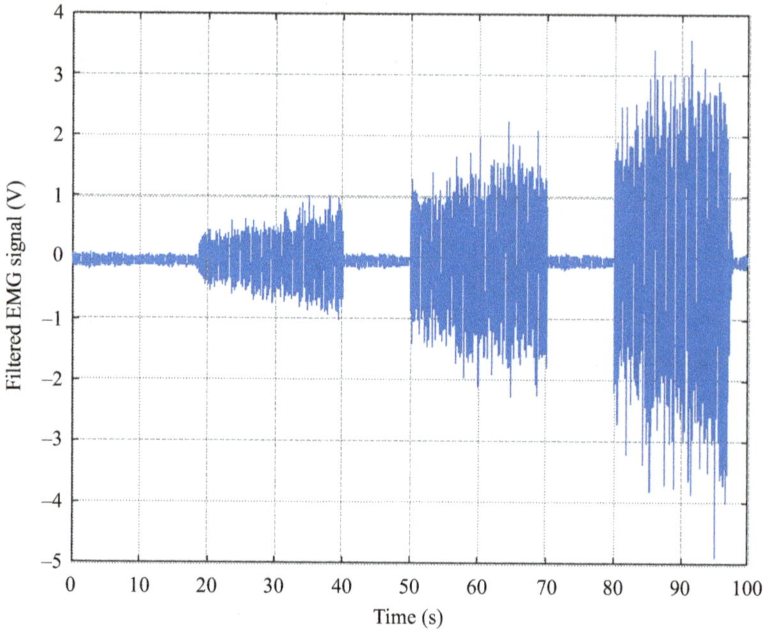

Figure 5.37 EMG Signal after amplification and filtering

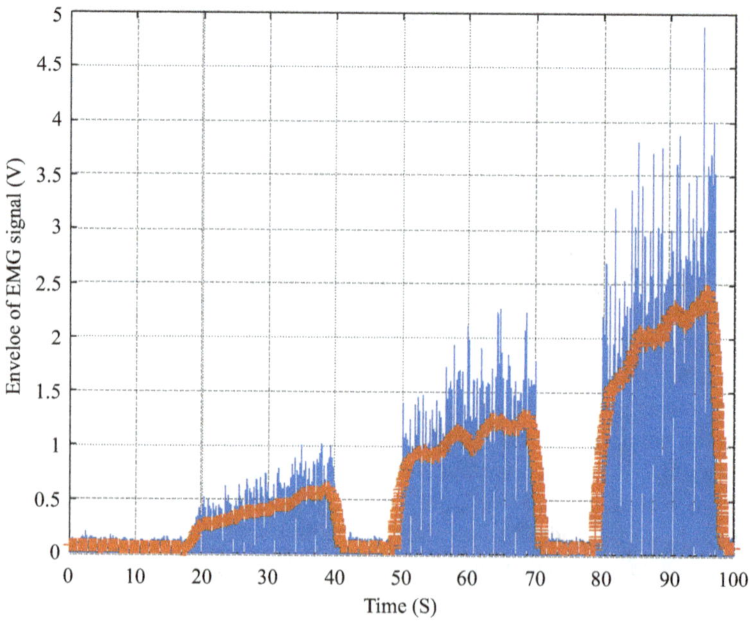

Figure 5.38 The red trace shows the output from the rectifier and the envelope detector

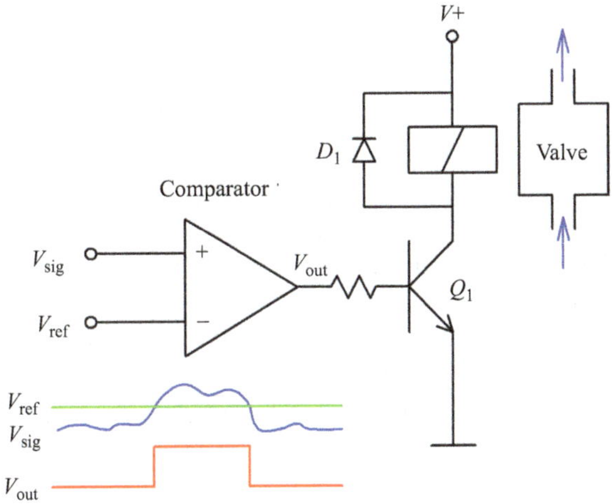

Figure 5.39 Threshold voltage-based control of a pneumatic solenoid valve

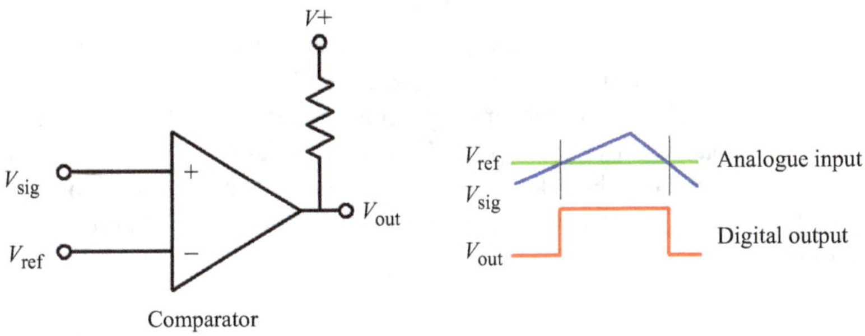

Figure 5.40 Comparator circuit

5.5 Digital signal processing

5.5.1 The comparator

The comparator can be considered to be the simplest analogue to digital converter (ADC) as it converts an analogue signal with an infinite number of amplitudes to a single-bit digital output. If the analogue signal is above a given threshold, the output is high, and if it is lower, the output is low as shown in Figure 5.40.

In reality, the analogue signal almost always includes a small amount of noise, so as it nears the threshold level, multiple triggering may occur. This can be eliminated by including some positive feedback from the comparator output to add hysteresis to the threshold level. This circuit configuration is referred to as a Schmitt trigger.

5.5.2 Signal acquisition and processing overview

Processing signals within a computer (digital signal processing) requires that they be sampled periodically and then converted to a digital representation using an ADC. The result is that a continuous signal is reduced to one that is only defined at discrete points, as shown in Figure 5.41. The sampling theorem (also known as Shannon's sampling theorem) states that to ensure accurate representation, the signal must be sampled at the Nyquist rate which is defined as at least double the highest significant frequency component of the signal.

$$f_s \leq 2f_{\text{max}} \tag{5.27}$$

In addition, the number of discrete levels to which the signal is quantised must also be sufficient to represent variations in the amplitude to the required accuracy. Most ADCs quantise to 12 or 16 bits that represent $2^{12} = 4096$ or $2^{16} = 65536$ discrete levels, respectively.

The signal can then be manipulated within a computer in various ways. This process often produces intermediate results that are much larger than the input values requiring that a larger word size be used if fixed-point values are used or, as is more common, a floating-point representation is used. Once the signal has been processed, the final result is often passed through a digital to analogue converter (DAC) to convert it from the internal binary representation back to a voltage. It is therefore possible to represent a complete DSP, as shown in Figure 5.42, with ADCs a processor and DACs.

The DAC only outputs new values at discrete times and so a continuous signal needs to be reconstructed. This generally involves holding the signal constant (zero-order hold) during the period between samples as shown in Figure 5.43. This signal is then cleaned up by being passed through a low-pass filter to remove high-frequency components generated by the sampling process.

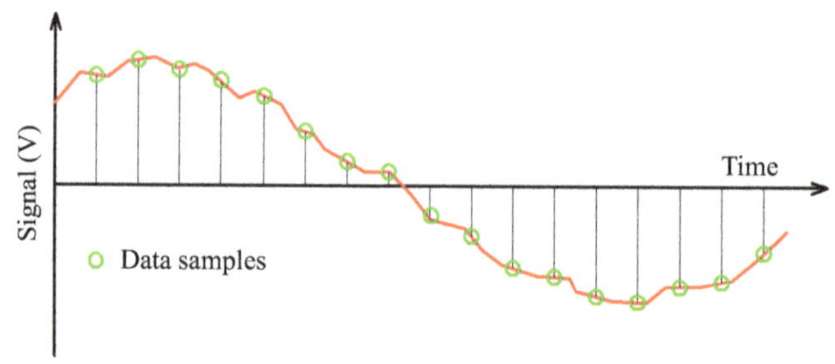

Figure 5.41 Digitising a signal

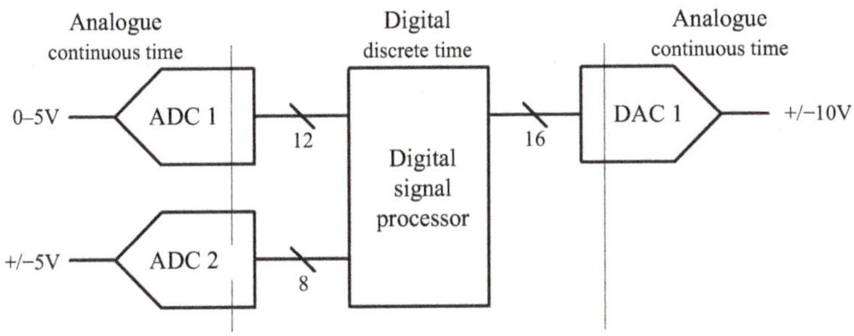

Figure 5.42 Typical configuration for a digital signal processing application (Brooker 2008)

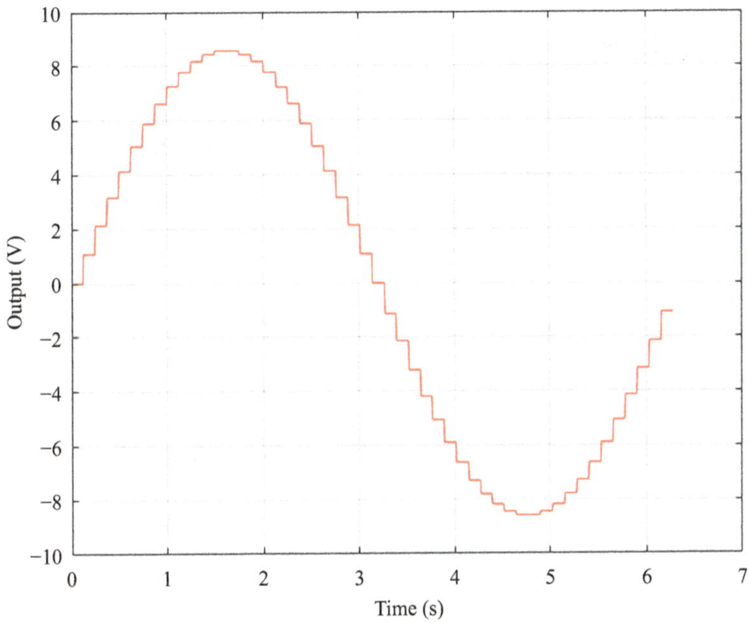

Figure 5.43 Analogue reconstruction of a sampled signal using a zero-order hold

Other common methods of interfacing to the outside world include manipulating a single bit or a complete digital word to drive some peripheral device like a video display unit. Alternatively, it is possible to output clock signal at a specific frequency, or with a particular duty cycle, to represent a number.

5.5.3 *ADCs and DACs*

There are many ways of converting an analogue signal to its digital equivalent. These are selected depending on the number of digital bits required and the rate of conversion. Two of the most common, used in commercial designs, are the successive approximation and flash conversion, though others including single- and dual-slope integration, switched capacitor and delta-sigma conversion, are implemented for specialist applications.

The successive approximation method is used in bioelectrical applications, for relatively low conversion rates, because it is low cost. As can be seen in the block diagram shown in Figure 5.44, it consists of several modules including a DAC in the feedback path.

The process of conversion is as follows:

- At the start of the conversion, the analogue signal is latched (captured and held) using a sample and hold (S&H),
- The successive approximation register (SAR) sets its most significant bit (MSB) which causes the DAC to output an analogue value voltage equal to $V_{max}/2$,
- The DAC value is compared to the analogue input value using a comparator,
- If the analogue signal is still larger than this value, then the output of the comparator remains high and the bit remains set. If it is smaller, then the output of the comparator goes low and the bit is reset,
- The process is then repeated with the next MSB, and then the one after, until at last the least significant bit (LSB) has been output,
- The SAR then contains a digital equivalent of the analogue input signal that is clocked into the output register and latched so that the whole process can be repeated.

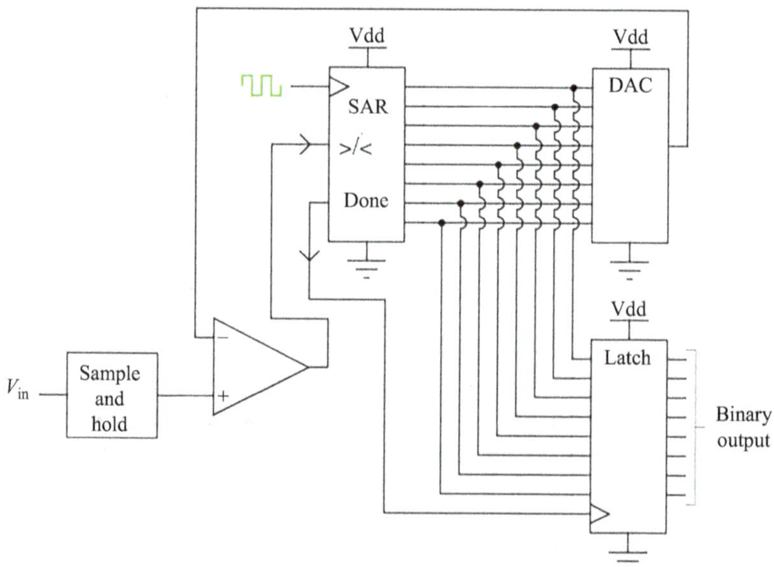

Figure 5.44 Successive approximation ADC

An n-bit ADC converter requires $n\Delta T$ seconds to perform one conversion, where ΔT is the cycle time for the DAC and control unit. Typical conversion times for 8-bit to 12-bit successive approximation ADCs range from 1 to 100 μs (Alciatore and Histand 2003).

The fastest ADCs are known as flash converters. As shown in Figure 5.45, these consist of a bank of comparators acting in parallel to identify the signal level. A resistive potential divider chain forms the references for each of the comparators, so that the moment an analogue input is presented, the respective comparator outputs will be encoded for it. A priority encoder then converts the comparator output format to a binary word.

As the name implies, DACs convert a digital word into an analogue output. The simplest kind uses an R-$2R$ resistor ladder network connected to an inverting summer op-amp circuit as shown in Figure 5.46.

This configuration is convenient and accurate because it uses only two precision resistor values, R and $2R$ to produce a series of reference voltages, each half the size of the previous one, with the MSB value being V_{REF}, the next being $V_{REF}/2$ etc, down to $V_{REF}/8$ in this example.

The analogue output is generated by switching the $2R$ resistor, either to the negative input of the op-amp (virtual ground) if the bit is high, or to the true ground if the bit is low.

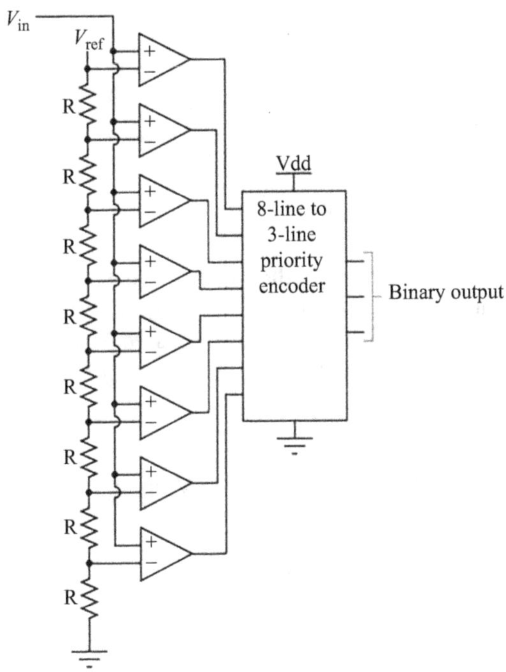

Figure 5.45 Flash ADC

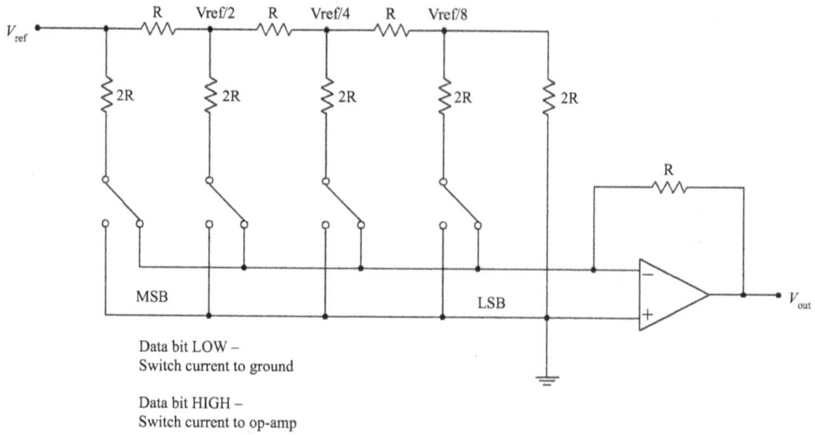

Figure 5.46 Ladder resistor DAC

Consider the case where only the LSB is high, in which case D_o switches the voltage $V_{REF}/8$ to the negative input of the op-amp through $2R$ and the remaining resistors are grounded.

If the gain of the op-amp is A_V

$$A_V = -\frac{R}{2R} = -\frac{1}{2} \tag{5.28}$$

then the output will be

$$\begin{aligned} V_{out_0} &= \frac{V_{REF}}{8} A_V \\ &= -\frac{V_{REF}}{16} \end{aligned} \tag{5.29}$$

In a similar manner, if D_3 is high, the MSB switches V_{REF} to the negative input and the output of the op-amp goes to $V_{out3} = -V_{REF}/2$. If both are switched, the output will be the sum of the two values, as discussed earlier in this chapter.

In the general case for a 4-bit DAC with a binary input word $b_3b_2b_1b_0$

$$V_{out} = b_3 V_{out3} + b_2 V_{out2} + b_1 V_{out1} + b_0 V_{out0} \tag{5.30}$$

If $V_{REF} = 10$ V, then V_{out} can vary from 0 V up to a maximum of

$$\begin{aligned} V_{out} &= -V_{REF}\left(\frac{1}{2} + \frac{1}{4} + \frac{1}{8} + \frac{1}{16}\right) \\ &= -\frac{15}{16} V_{REF} \end{aligned}$$

5.5.4 Signal aliasing

If the Nyquist criterion is not satisfied and the analogue signal is not sampled at least twice the frequency of the highest frequency component, then these

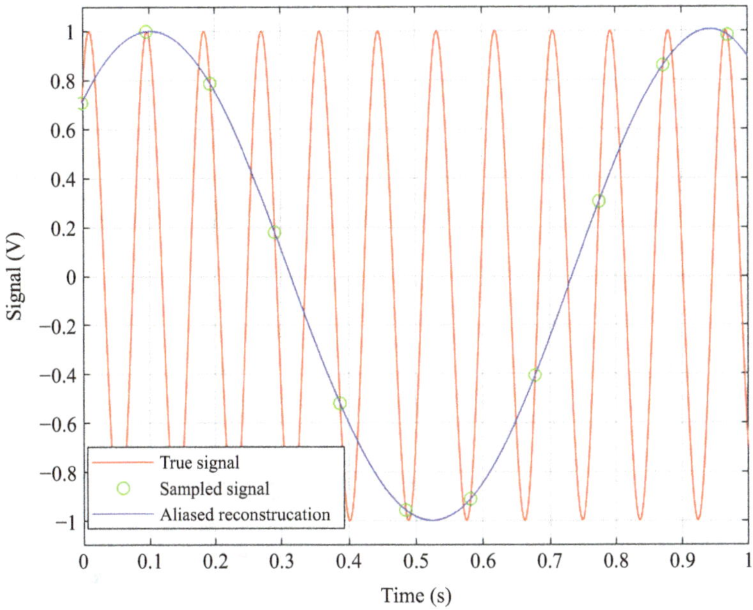

Figure 5.47 Interpretation of aliasing effects in the time domain

high-frequency signals are folded or *aliased* down to a lower frequency as illustrated in the time domain representation shown in Figure 5.47.

In the frequency domain, a generic analogue signal may be represented in terms of its amplitude and total bandwidth as shown in Figure 5.48(a). A sampled version of the same signal can be represented by a repeated sequence spaced at the sample frequency (generally denoted f_s) and shown in Figure 5.48(b) and (c). If the sample rate is not sufficiently high, then the sequences will overlap, and high-frequency components will appear at a lower frequency (albeit with reduced amplitude).

In reality, the finite roll-off of the filter response requires that a guard band be maintained between the spectra. This is achieved by selecting an anti-aliasing filter with a cutoff frequency that is less than 0.4 times the sample frequency. Using this ratio as a rule of thumb A typical third-order low-pass filter will attenuate these unwanted signals by between 40 and 60 dB (1/100 to 1/1000) in voltage.

Worked example: electrocardiogram

Consider the filtered electrocardiogram signal shown in Figure 5.49. At a glance, it would appear to be reasonably noise free.

If the signal is sampled at about 143 samples per second and the spectrum examined, it can be seen in Figure 5.50 that there is a significant frequency component at 50 Hz. This is caused by the residual mains hum that was not removed by the CMRR of the differential amplifier and any subsequent filtering. Because most

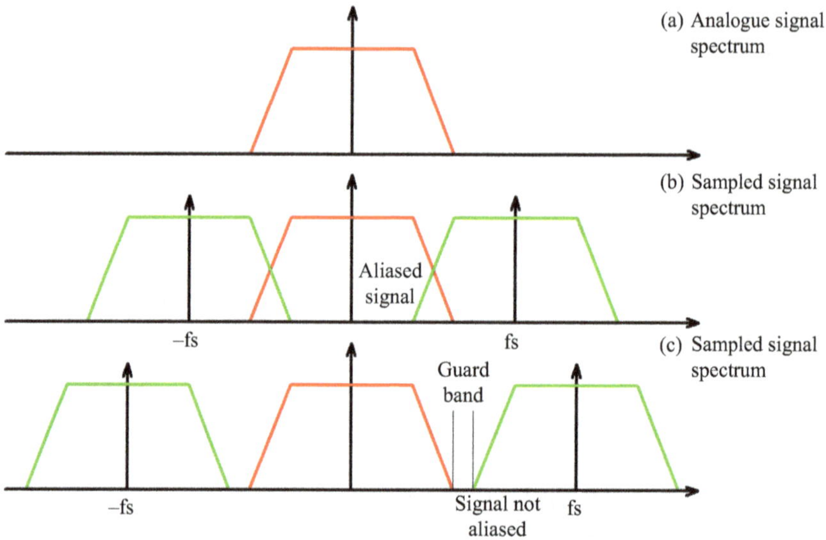

Figure 5.48 Interpretation of aliasing effects in the frequency domain (Brooker 2008)

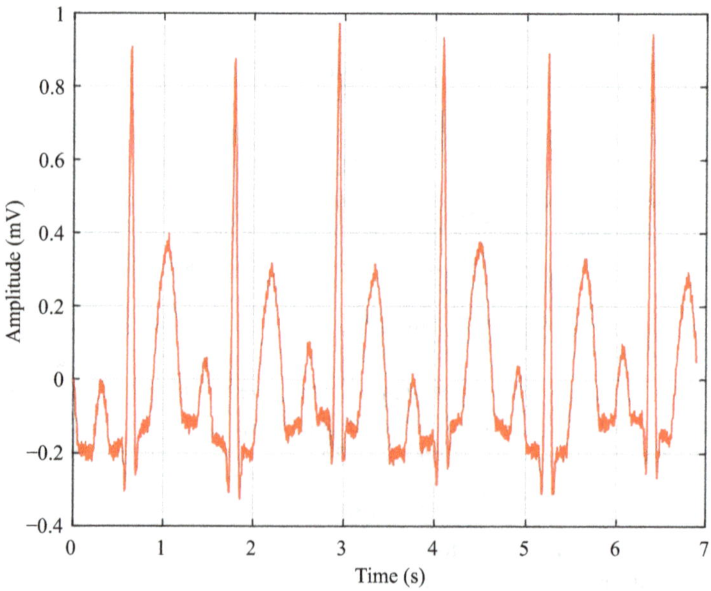

Figure 5.49 Measured electrocardiogram

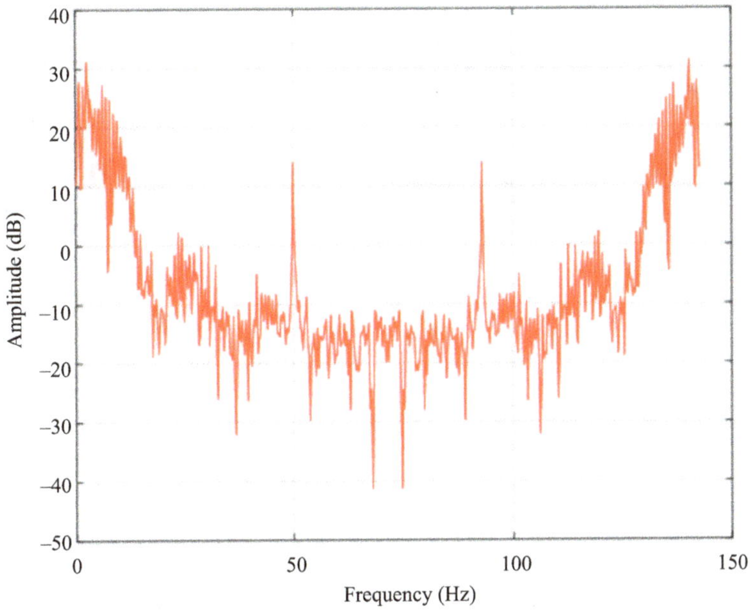

Figure 5.50 Double-sided ECG spectrum for a 143 Hz sample rate (folding frequency $f_s/2$ = 71.5 Hz)

of the signal content of the actual ECG resides at frequencies below 30 Hz, it would seem reasonable to sample this signal at 75 to 85 Hz.

If this same ECG signal is sampled at 83 samples per second, and the spectrum is generated, the 50 Hz component has been aliased down to about 33 Hz (83–50 Hz) as shown in Figure 5.51. In this instance, there is no way of knowing that the 33 Hz artefact is not a genuine component of the ECG signal.

The only way to eliminate this aliasing is to ensure that the analogue signal is filtered to less than $f_s/2$ prior to sampling and digitisation.

5.5.5 *Digital filters*

Digital filters perform the same functions as their analogue counterparts; they can be configured as low or high-pass filters or any of the other functions discussed earlier, as well as a whole lot more. However, instead of manipulating actual voltages, digital filters manipulate numbers – the digital representations of the sampled analogue signals.

5.5.5.1 Difference equations and transfer functions

Any filter can be written in terms of a difference equation, which is the weighted sum of a sequence of input values and output values that together generate the next value of the output sequence. This is shown in general terms (Stanley 1975)

$$y(n) = \sum_{k=0}^{M} (b_k x(n-k)) - \sum_{k=1}^{N} (a_k y(n-k)) \tag{5.31}$$

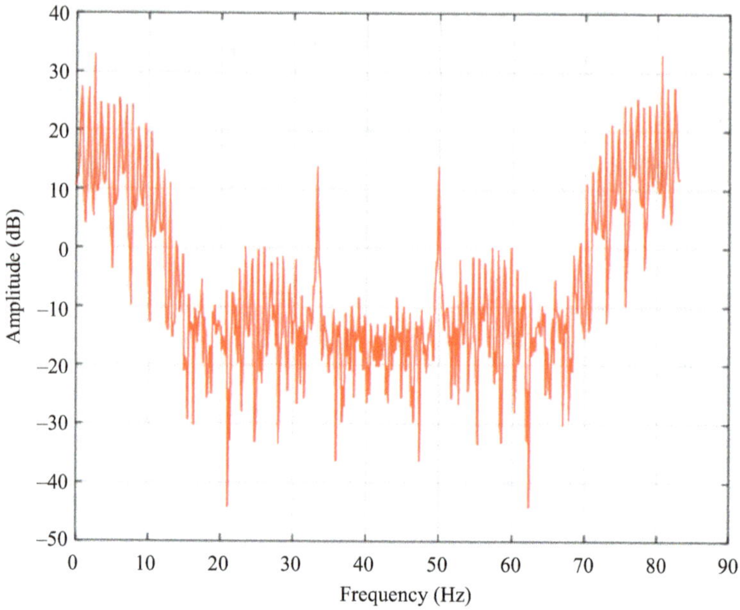

Figure 5.51 Double-sided ECG spectrum for an 83.3 Hz sample rate

For this equation, $y(n-k)$ represents the outputs and $x(n-k)$ represents the inputs. The value of N gives the order of the difference equation and corresponds to the depth of the memory required to implement it. Because this equation relies on past values of the output, it is recursive, and these initial values (initial conditions) must be known. It is in this form that the actual numbers are manipulated by the DSP to perform the required function, but in general, the filters are described by their transfer functions.

The z-transform is useful for the manipulation of discrete data sequences and has acquired a new significance in the formulation and analysis of discrete-time systems since the advent of digital computers. It is used extensively in the areas of applied mathematics and digital signal processing amongst others. In these cases, discrete models are solved with difference equations in a manner that is analogous to solving continuous models with differential equations. The role played by the z-transform in the solution of difference equations corresponds to that played by the Laplace transforms in the solution of differential equations.

The transfer function $H(z)$ is easily obtained from the difference equation. The first step involves taking the Fourier Transform before using the linearity property to move the transform to within the summation. The time-shifting property of the z-transform is used to change the time-shifting terms to exponentials.

For $a_o = 1$, this process results in the following

$$Y(z) = \sum_{k=0}^{M} \left(b_k X(z) z^{-k}\right) - \sum_{k=1}^{N} \left(a_k Y(z) z^{-k}\right) \tag{5.32}$$

The transfer function $H(z)$ is then

$$H(z) = \frac{Y(z)}{X(z)} = \frac{\sum_{k=0}^{M} (b_k z^{-k})}{1 + \sum_{k=1}^{N} (a_k z^{-k})} \tag{5.33}$$

The frequency response is then obtained by replacing every instance of z with $e^{j\omega}$.

$$H(\omega) = \frac{\sum_{k=0}^{M} (b_k e^{-j\omega k})}{\sum_{k=0}^{N} (a_k e^{-j\omega k})} \tag{5.34}$$

Fortunately, this process of obtaining the transfer function is easily achieved in MATLAB®.

Example:

Obtaining a Filter Transfer Function using MATLAB

The difference equation for a Butterworth low-pass filter can be written in terms of the following difference equation

$$\begin{aligned} y(n) &= 0.0029x(n) + 0.0087x(n-1) + 0.0087x(n-2) + 0.0029x(n-3) \\ &\quad -[2.3741y(n-1) - 1.9294y(n-2) + 0.5321y(n-3)] \end{aligned}$$

The coefficients of the difference equation can be written as arrays, where A includes the coefficients of y and B includes the coefficients of x

$$A = (a_0, a_1, a_2, a_3)$$
$$B = (b_0, b_1, b_2, b_3)$$

Remember that $a_0 = 1$, so

$$A = (1, \ -2.3741, 1.9294, \ -0.5321)$$
$$B = (0.0029, 0.0087, 0.0087, 0.0029)$$

The frequency response can then be obtained and plotted, as shown in Figure 5.52, using the following MATLAB code.

```
[h,w] = freqz(B,A,512);     % extract the transfer function from DC to fs/2
freq = (0:511)/512;
plot(freq,abs(h));
xlabel('Normalised Frequency')
ylabel('|H(w)|')
```

5.5.5.2 Synthesising infinite impulse response (IIR) filters

There is a large body of literature that deals with the synthesis of the common and less common IIR filter algorithms. However, once again MATLAB will synthesise all of the common filter types with a minimum of effort. For example, the filter for the previous example was generated with a single command

```
[B,A]=butter(3,0.1)
```

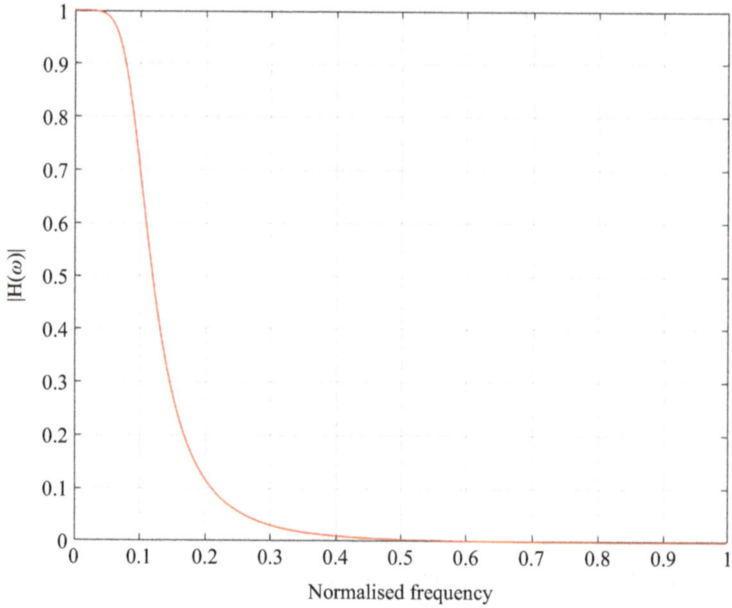

Figure 5.52 Frequency response of a low-pass Butterworth filter

This generates the coefficients of a 3rd order Butterworth filter with a cutoff frequency of 0.1 (normalised to $f_s/2$).

In addition to a range of Butterworth filters, MATLAB will synthesise Bessel, Chebyshev and Elliptical low-pass, high-pass and band-pass filters of whatever order is required. However, some care needs to be exercised as numerical issues tend to make high-order filters unstable.

5.5.5.3 Synthesising finite impulse response (FIR) filters

In FIR filters, only the input values are used to produce an output; they are therefore not recursive.

In this implementation, the shape of the filter is described roughly in terms of the required amplitude response. So, if the frequency response of the Butterworth filter described in the previous section was to be synthesised using a FIR filter, then the following MATLAB code could be used to produce the frequency response shown in Figure 5.53.

```
f=[0,0.1,0.2,0.3,0.4,0.5,0.6,0.7,0.8,1.0];   % define frequency response
a=[1,0.707,0.1,0.02,0.01,0,0,0,0,0];          % from DC to fs/2

B=firls(10,f,a);                % generate coefficients for 10th order FIR filter
A=[1];
[h,w] = freqz(B,A,512);         % extract the transfer function from DC to fs/2
freq = (0:511)/512;
plot(freq,abs(h));
xlabel('Normalised Frequency')
ylabel('|H(w)|')
```

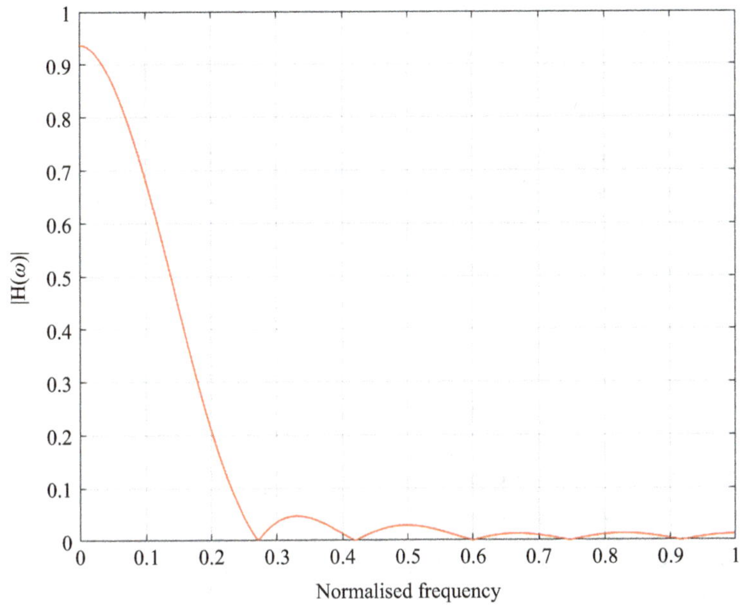

Figure 5.53 Frequency response of a tenth order FIR filter with a normalised cutoff frequency of 0.1

5.5.5.4 Tracking filters

Consider a prosthetic arm in which only a noisy measurement of the angle of the elbow joint is measured, but effective control requires that the angular rate be estimated. This can be achieved in isolation by using a differentiator, a state variable filter or using a tracking filter. The latter can be used to produce an estimate of the angular rate (and higher order derivatives) as well as a filtered estimate of the actual angle.

The α-β tracker is a fixed-gain formulation of the Kalman filter and is still widely used because it is easy to implement and performs well under most circumstances (Blackman 1986). The implementation of most tracking filters follows a two-stage process, first of smoothing and then of prediction. The equations defining these two stages are shown below for the α-β tracker implementation:

Smoothing

$$\widehat{R}_n = \widehat{R}_{pn} + \alpha\left(R_n - \widehat{R}_{pn}\right) \tag{5.35}$$

$$\widehat{V}_n = \widehat{V}_{pn} + \frac{\beta}{T_s}\left(R_n - \widehat{R}_{pn}\right) \tag{5.36}$$

Prediction

$$\widehat{R}_{p(n+1)} = \widehat{R}_n + \widehat{V}_n \cdot T_s \tag{5.37}$$

$$\widehat{V}_{p(n+1)} = \widehat{V}_n \tag{5.38}$$

where:

$\widehat{R}_n$ = Smoothed estimate of position,
$\widehat{V}_n$ = Smoothed estimate of rate,
R_n = Measured position,
$\widehat{R}_{p(n+1)}$ = Predicted position after T_s seconds (one sample ahead),
$\widehat{V}_{p(n+1)}$ = Predicted position rate after T_s seconds (one sample ahead),
$\widehat{R}_{pn}$ = Predicted position at the measurement time,
$\widehat{V}_{pn}$ = Predicted rate at the measurement time,
T_s = Sample time,
α, β = Smoothing constants.

The equations shown above are only one of the ways that the tracking filter can be described. Other ways include the block diagram representation (Mahafza 2000), state-space representations or rewriting the equations as a position transfer function from the difference equations.

$$R_p(z) = R(z)z^{-1} + T_s\dot{R}(z)z^{-1} \tag{5.39}$$

$$R(z) = R_p(z) + \alpha\left[R_m(z) - R_p(z)\right] \tag{5.40}$$

$$\dot{R}(z) = \dot{R}(z)z^{-1} + \frac{\beta}{T_s}\left[R_m(z) - R_p(z)\right] \tag{5.41}$$

where $R_p(z)$ is the predicted position, $R(z)$ is the estimated position, $\dot{R}(z)$ is the rate estimate and $R_m(z)$ is the measured position.

The position transfer function will therefore be

$$H_1(z) = \frac{R(z)}{R_p(z)} \tag{5.42}$$

and it can be obtained by manipulation of the equations (5.38) to (5.40)

$$H_1(z) = \frac{\alpha + (\beta - \alpha)z^{-1}}{1 + (\beta + \alpha - 2)z^{-1} + (1 - \alpha)z^{-2}} \tag{5.43}$$

This is a convenient form of the filter to be used in MATLAB where the filter coefficients described earlier are

$$B = [\alpha, (\beta - \alpha)]$$
$$A = [1, (\beta + \alpha - 2), (1 - \alpha)]$$

One possible filter optimisation (Benedict and Bordner 1962) is to minimise the output noise variance at steady state, and the transient response to a changing input is modelled by a ramp function. This results in the following relationship between the two gain coefficients

$$\beta = \frac{\alpha^2}{2 - \alpha} \tag{5.44}$$

Other criteria can also be used. For example, it has been suggested that the filter should have the fastest possible step response (critical damping), in which

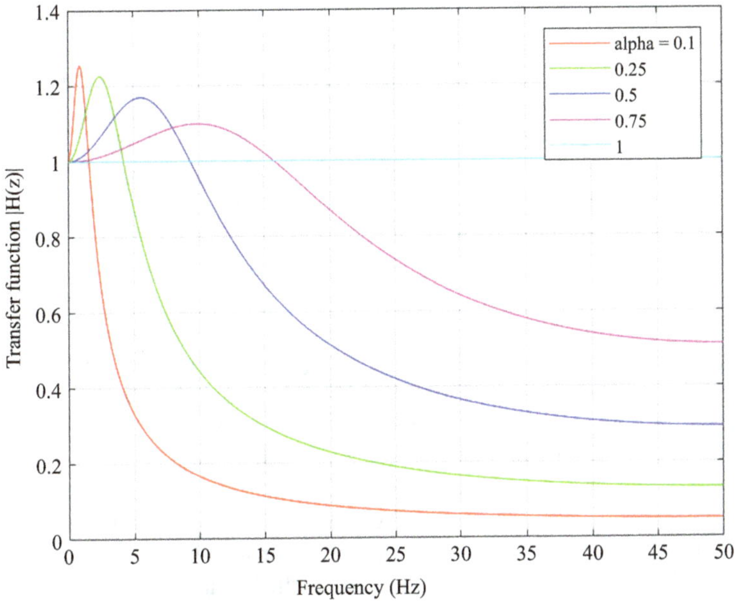

Figure 5.54 Frequency response of a Benedict Bordner α–β filter at f_s = 1 kHz

case the coefficients are related as

$$\alpha = 2\sqrt{\beta} - \beta \tag{5.45}$$

The actual values of the gains depend on the sample period, the predicted dynamics and the required loop bandwidth.

Figure 5.54 shows the filter transfer function for the position estimate of the tracking filter using various values of α and β (Benedict Bordner relationship). This is obtained using equation (5.42) and the Freqz function in MATLAB. Even though the transfer function does not conform to the classical low-pass filter shape discussed earlier, the half-power bandwidth is still measured at |H(z)| = 0.707.

```
% plot the alpha beta filter position transfer function
% ab_trans.m
%
% Variables
fs = 1000;                      % Sample frequency (Hz)
alpha = 0.1                     % Position gain
beta = alpha.^2 /(2-alpha);     % Velocity gain (Benedict Bordner Relationship

% Calculate the filter coefficients for the position transfer function
B=([alpha, beta-alpha])
A=([1,(beta+alpha-2),(1-alpha)]);
[H1,F]=freqz(B,A,1024,fs);

% Plot the magnitude of the transfer function
plot(F,abs(H1),'k')
grid
xlabel('Frequency (Hz)')
ylabel('Transfer Function |H(z)|')
```

One of the main disadvantages of the fixed gain α-β filter is that it estimates the position of an accelerating input with a constant lag. However, because the magnitude of this lag can easily be estimated, the filter can use adaptive gains to improve the RMS tracking accuracy under these circumstances. The cost function that is minimised in this case is $[lag^2 + \sigma_r^2]$.

For estimates that include accelerations, the α-β-γ tracker can be substituted. However, its performance for constant velocity motion is poorer than that for the simple α-β tracker because of the additional noise introduced by the acceleration estimate.

5.5.5.5 Differentiator

A simple differentiator can be obtained by taking the difference between successive input samples and dividing by the interval between them. This algorithm produces a high-pass filter and therefore adds noise to the measurement. Some improvement can be obtained by following the differentiator with a higher-order low-pass filter to attenuate high-frequency noise.

An alternative is to use the same MATLAB process that was used to synthesise the FIR low-pass filter, to synthesise a differentiator with good characteristics. The frequency response of the filter generated by the following code is shown in Figure 5.55.

```
b=firls(30,[0,0.9],[0,0.9], 'differentiator');
```

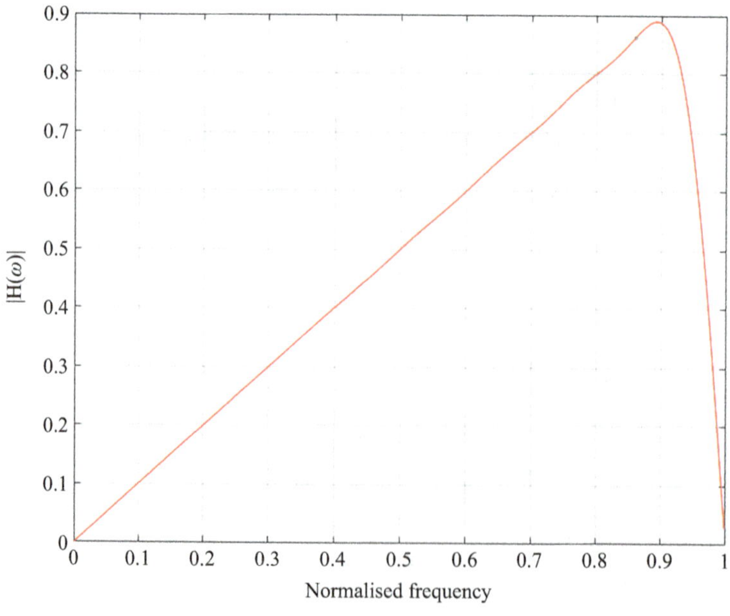

Figure 5.55 Frequency response of a 30th order FIR differentiator

5.5.5.6 Integrator

The common methods of integration use the forward or backward Euler (also known as rectangular) and the Trapezoidal algorithms.

The difference equation for the forward Euler integration is defined in terms of the latest input, $x(n-1)$, and the sample interval, T

$$y(n) = y(n-1) + Tx(n-1) \tag{5.46}$$

This can be written as a transfer function

$$H(z) = \frac{T}{z-1} \tag{5.47}$$

For the backward Euler method, the difference equation is

$$y(n) = y(n-1) + Tx(n) \tag{5.48}$$

and the transfer function is

$$H(z) = \frac{Tz}{z-1} \tag{5.49}$$

In the trapezoidal case, the difference equation is a little more complicated

$$y(n) = y(n-1) + \frac{T[x(n) + x(n+1)]}{2} \tag{5.50}$$

and the transfer function can be written as

$$H(z) = \frac{T(z+1)}{2(z-1)} \tag{5.51}$$

The frequency response of the trapezoidal integrator is shown in Figure 5.56.

In theory the DC gain will be infinite because the output increases indefinitely for a DC input. However, as shown in the figure, this gain decreases exponentially with increasing frequency.

5.5.5.7 Averager

The process of averaging is one way of generating a simple low-pass filter. This process can be implemented as a block algorithm in which N input samples are processed to generate a single output

$$y(k) = \frac{1}{N} \sum_{n=(k-1)N}^{kN} x(n) \tag{5.52}$$

This reduces both the bandwidth and the sample rate of the output. However, if the sample rate is to be maintained, the moving-average algorithm is used. In this case, the average is formed by adding the most recent input and subtracting the oldest.

$$y(n) = y(n-1) + \frac{1}{N}[x(n-1) - x(n-N-1)] \tag{5.53}$$

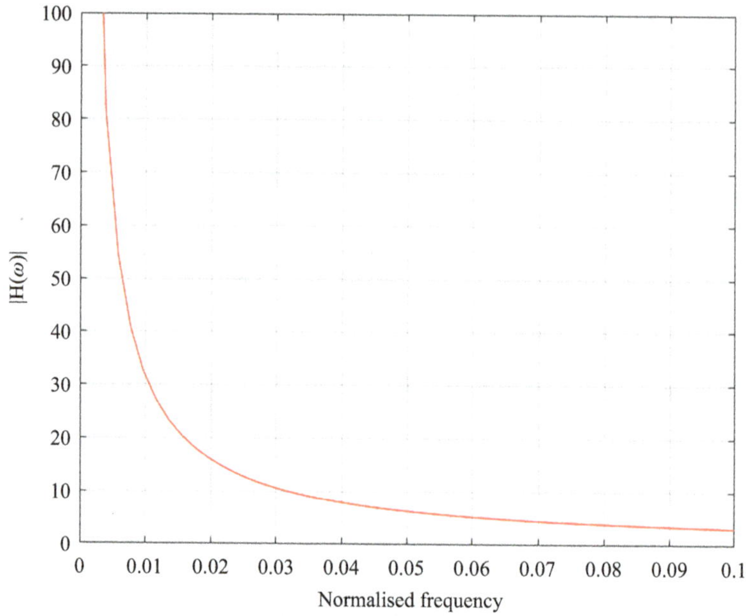

Figure 5.56 Integrator frequency response for $T = 1$

It can be seen that this algorithm is very similar to the integrator discussed in the previous section. The frequency response of the moving averager is shown in Figure 5.57.

If the signal spectral content and that of the noise overlap significantly, there is often no alternative but to use an averaging process. For example, brain potentials generated by external visual or acoustic stimulation, called evoked potentials (EPs), have amplitudes that are much lower than the background EEG activity, but contain information in the same bands, so they cannot be separated using conventional frequency-based filters (Bronzino 2006).

When the desired signal repeats almost identically at each iteration, a form of block averaging can solve the problem. In this case, a sequence of points is sampled, covering the duration and synchronised to the triggered event. This process is repeated many times as illustrated in Figure 5.58 and the sequences are averaged each time. As N increases, the signal-to-noise ratio is improved by a factor $\sqrt{N}$ (in rms value) until the EP characteristics emerge from the random EEG background.

This averaging process can be described by

$$y(n) = \frac{1}{N}\{x(n) + x(n-h) + x(n-2h) + x(n-4h) + \ldots + x(n-[N-1]h)\} \tag{5.54}$$

where N is the number of elements in the averaging process and h is the number of points in the sequence. The results of a visually evoked potential (VEP) experiment are shown in Figure 5.59. A two second burst of the data is shown first. It is clear

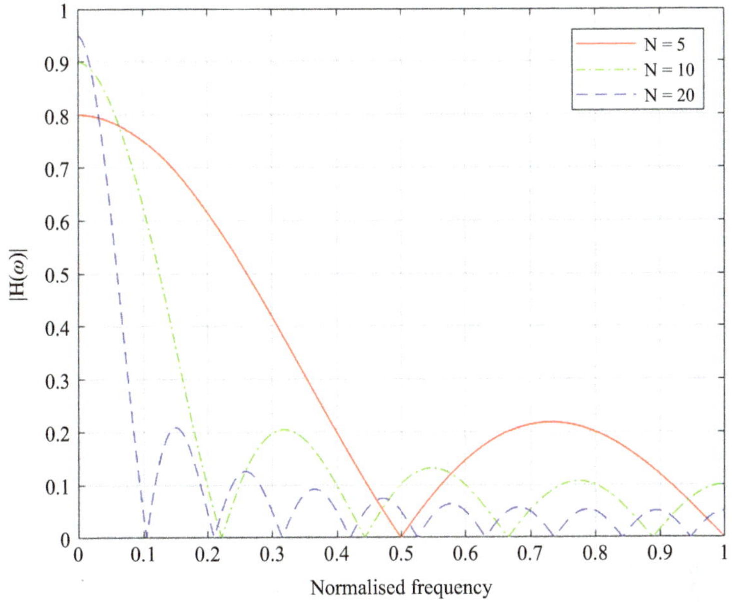

Figure 5.57 Frequency response of a moving averager for different N

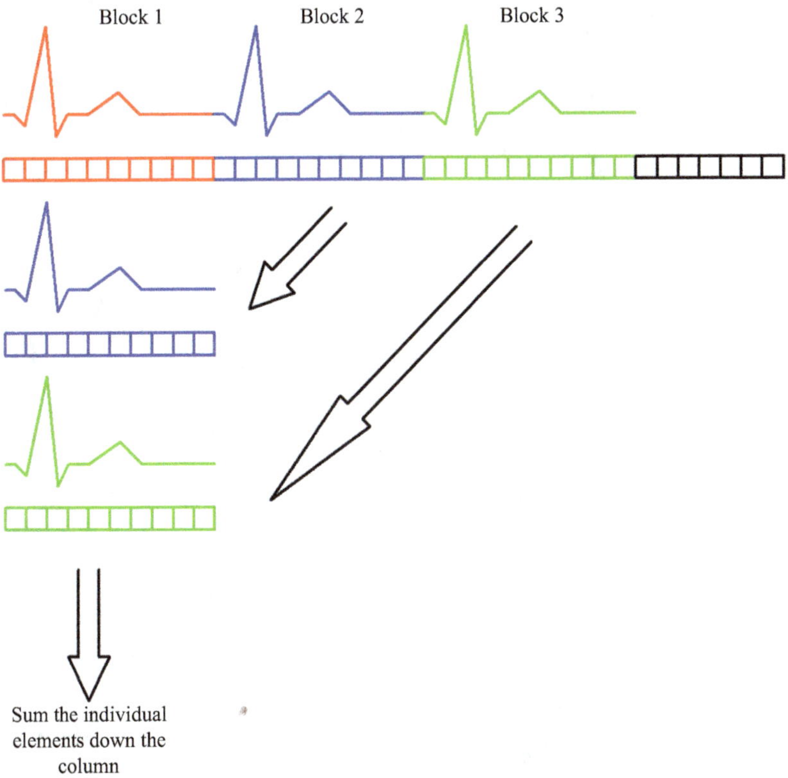

Figure 5.58 Graphical illustration of the block averaging procedure

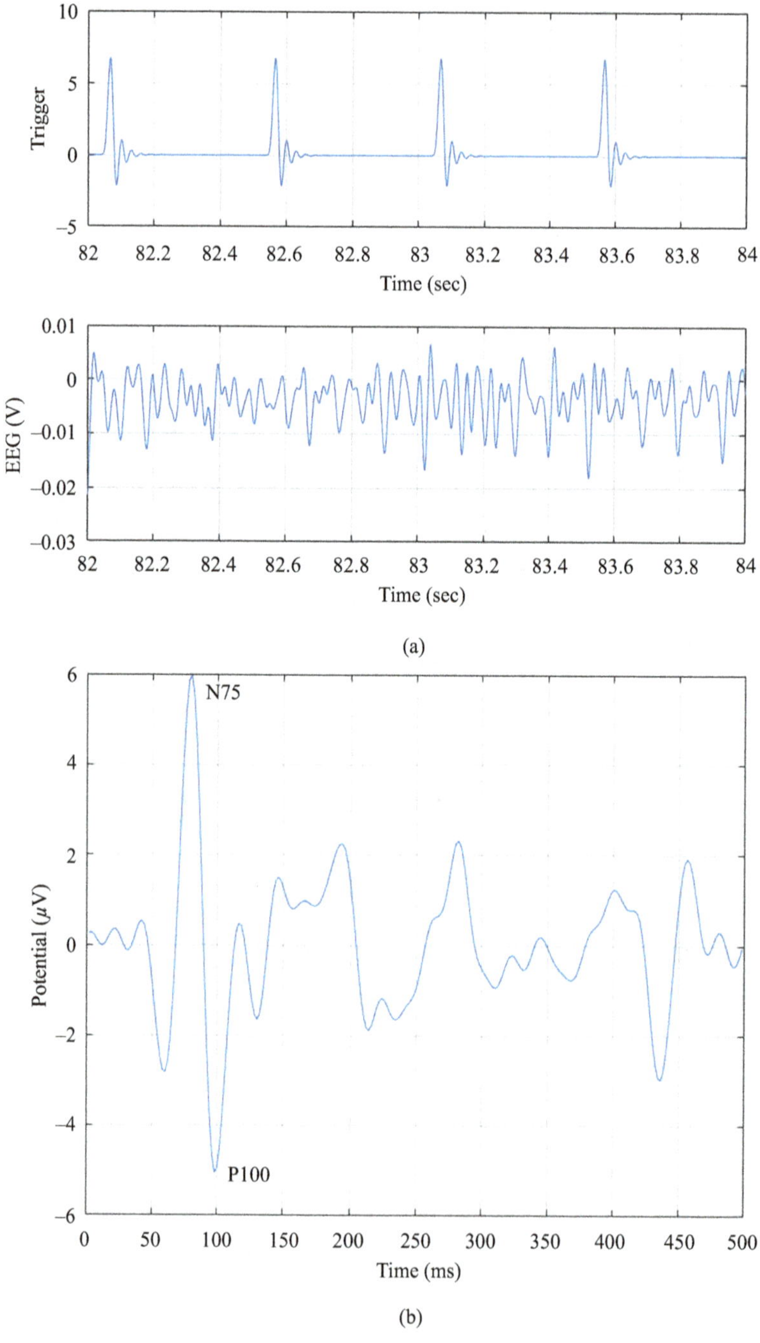

Figure 5.59 Block averaging process used to extract visual evoked potentials from background EEG activity

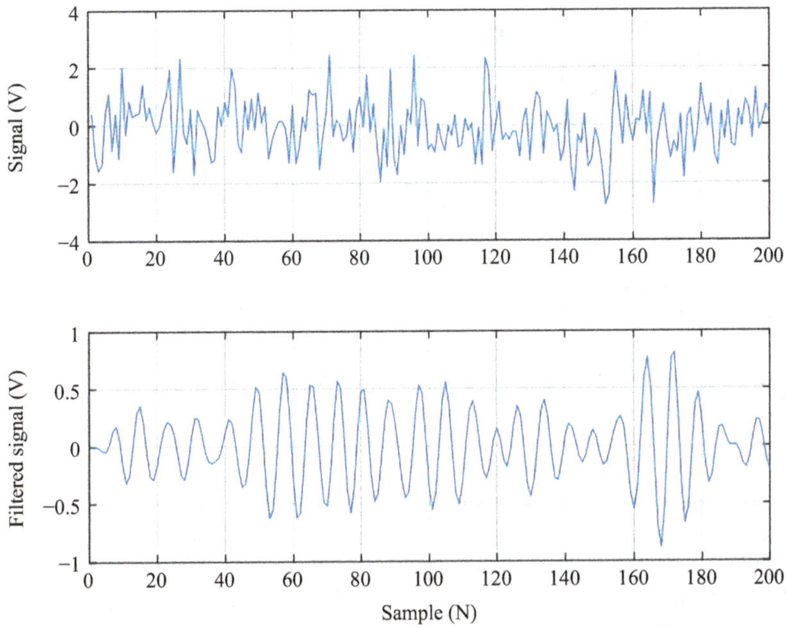

Figure 5.60 Band-pass filter response to white noise

that the raw EEG signal is too noisy to see the VEP, but after a few hundred blocks have been averaged, the N75 and P100 signals begin to appear.

5.5.6 *Filter time domain response*

The time-domain response of both analogue and digital filters is obtained by exciting the filter with the signal of choice and monitoring the output waveform. Common methods of stimulating filters include white noise, impulses, steps and ramps. All these inputs are easily generated in MATLAB as is the filter. Figure 5.60 shows the response of a Butterworth band-pass filter to white noise, as generated by the code.

```
% Band-pass filter
% bandpass2.m
fs = 200e+03;                % sample frequency (Hz)
ts = 1/fs;                   % sample interval (s)
fmat = 25.0e+03;             % centre frequency of filter (Hz)
bmat = 10.0e+03;             % bandwidth of filter (Hz)
wl=2*ts*(fmat-bmat/2);       % lower band limit
wh=2*ts*(fmat+bmat/2);       % upper band limit
wn=[wl,wh];
[b,a]=butter(3,wn);          % 6th order Butterworth band-pass filter

sig = randn(1,200);          % normally distributed white noise
sig_fil = filter(b,a,sig);   % time domain implementation of the filter

subplot(211),plot(sig),ylabel('Signal (V)');
subplot(212),plot(sig_fil),ylabel('Filtered Signal (V)'), xlabel('Sample (N)')
```

5.5.6.1 Filter impulse response

The impulse response is the time-domain output of the filter if it has been excited by a unit impulse (Dirac function) at its input. Because a true impulse contains equal contributions from all frequencies (as shown in Figure 5.15), it is equivalent to injecting white noise into the filter and hence characterises it over the full frequency band.

Frequency response and impulse response methods are equivalent and are related by the Fourier Transform as shown in Figure 5.61 for the band-pass filter described in the previous section.

In general, because the impulse response dies down quickly, it is a convenient and quick method of determining how effective filters are at attenuating white noise. This is achieved in the time-domain by inputting an impulse with a known variance into the filter and measuring the variance of the output. The ratio of the two is known as the variance reduction ratio (VRR).

5.5.6.2 Filter step response

The step response of filters is often used to quantify their dynamic performance in terms of a rise-time and overshoot. The rise-time is an indication of the bandwidth of the filter and is one way of determining how quickly it will track a changing input, while the overshoot is an indication of the amount of damping. Figure 5.62 shows the responses of a 3rd order Butterworth filter and the same order Chebyshev and Elliptical filters with a 0.1 dB ripple in the passband.

5.5.7 Envelope detection

Two different methods are commonly used to extract the envelopes from amplitude-modulated signals. The conventional method is to rectify the signal

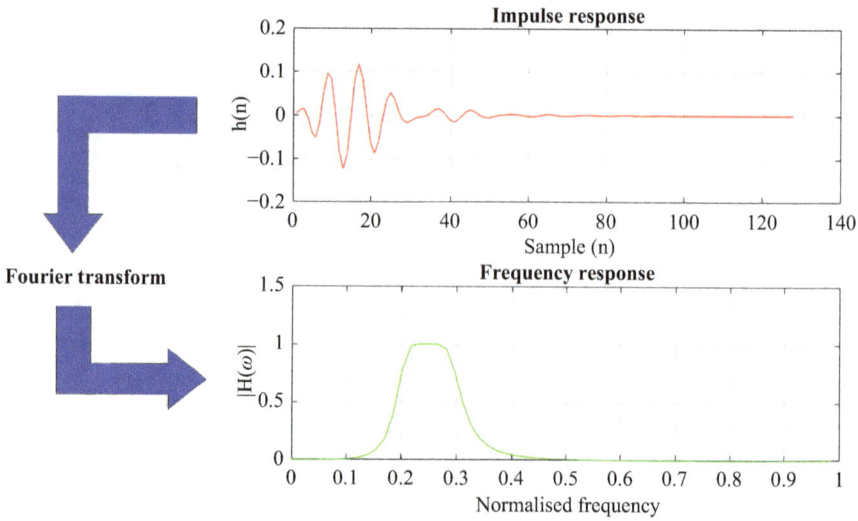

Figure 5.61 Relationship between the frequency response and the impulse response for a band-pass filter (Brooker 2008)

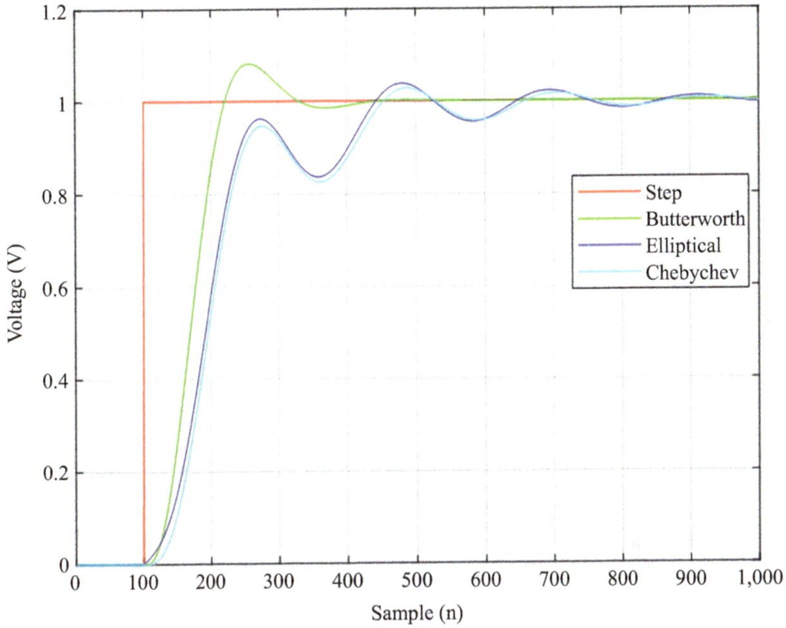

Figure 5.62 Comparison between the step responses of several low-pass filters

followed by a low-pass filter with a cutoff frequency dependent on the frequency content of the modulation. The second method uses the Hilbert transform.

In a digital implementation of the conventional method, the rectification process is achieved by taking the absolute value of the signal before it is processed through a low-pass algorithm such as the Butterworth filter discussed earlier in this chapter. The cutoff frequency of the low-pass filter determines the modulation depth of the envelope, with lower cutoff frequencies resulting in smaller modulation depths.

The Hilbert transform is a mathematical function that can represent a time waveform as the product of a slowly varying envelope and a carrier containing fine-structure information. In mathematical terms, the filtered waveform, $x_i(t)$, can be represented as

$$x_i(t) = a_i(t)\cos f_i(t) \tag{5.55}$$

where $a_i(t)$ represents the envelope and $\cos f_i(t)$ represents the fine structure waveform. Note that $f_i(t)$ is the instantaneous phase of the signal, and its derivative produces the instantaneous frequency (the carrier frequency).

Figure 5.63 shows the outputs of several different envelope detectors used in a cochlear implant. It can be seen that the Hilbert transform method reproduces the envelope more accurately than the conventional envelope detector even with a filter bandwidth of 400 Hz.

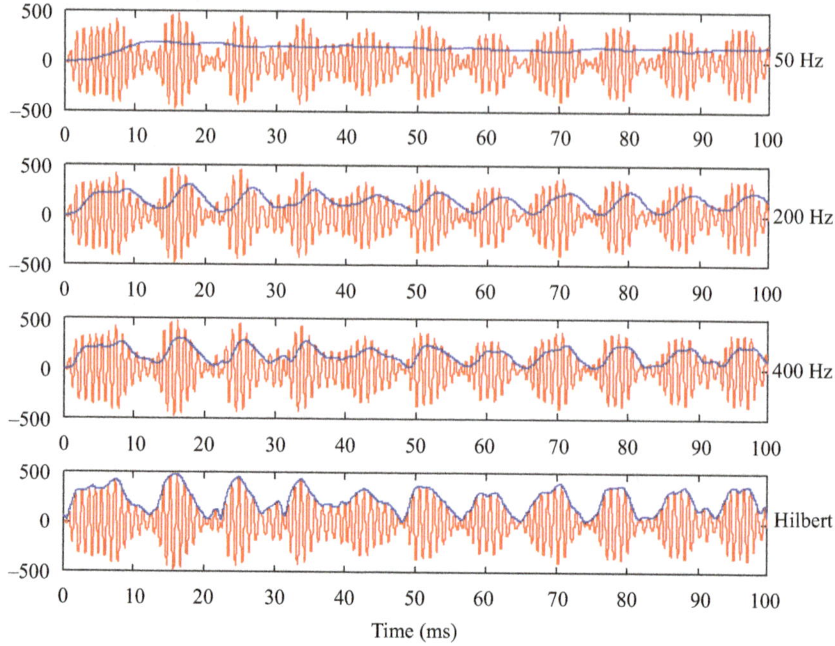

Figure 5.63 Envelope detection with various low-pass filter cutoff frequencies

5.5.8 Spectral estimation

The traditional method of frequency analysis is based on the Fourier transform that is generally evaluated using the discrete Fourier transform (DFT), or the fast Fourier transform (FFT) if the number of points corresponds to $k = 2^N$ where N is an integer.

The expression for the PSD as a function of the frequency, $P(f)$, can be obtained directly from the time series $y(n)$ using the periodogram expression

$$P(f) = \frac{1}{T_s}\left|T_s\sum_{k=0}^{N-1} y(k)e^{-j2\pi fkT_s}\right|^2 = \frac{1}{NT_s}|Y(f)|^2 \tag{5.56}$$

where T_s (sec) is the sample period, N is the number of samples and $Y(f)$ is the DFT of $y(n)$.

The PSD can also be obtained from the FFT of the autocorrelation function $\widehat{R}_{yy}(k)$ of the signal, where $\widehat{R}_{yy}(k)$ is estimated as

$$\widehat{R}_{yy}(k) = \frac{1}{N}\sum_{i=0}^{N-k-1} y(i)y^*(i+k) \tag{5.57}$$

where * denotes the complex conjugate. The PSD is then

$$P(f) = T_s \sum_{k=-N}^{N} \widehat{R}_{yy}(k) e^{-j2\pi f k T_s} \tag{5.58}$$

Both the autocorrelation function and the Fourier transform are theoretically defined over infinite data sequences, so their truncation can lead to errors, particularly in respect to spectral leakage caused by the implicit rectangular windowing of the data.

5.5.8.1 Fourier transform

All continuous periodic signals can be represented by a fundamental frequency sine wave and a collection of sine and/or cosine harmonics of that signal.

The Fourier series for any waveform can be expressed as (Edminster 1972, Carr 1997)

$$v(t) = \frac{a_0}{2} + \int_{n=1}^{\infty} a_n \cos(n\omega t) + \int_{n=1}^{\infty} b_n \sin(n\omega t) \tag{5.59}$$

where a_n b_n are the amplitudes of the harmonics, which can be zero and n is an integer.

The amplitude terms for each spectral component can be calculated by integrating the product of the time domain signal with a sample phasor of the correct frequency

$$\begin{aligned} a_n &= \frac{2}{T}\int_0^t v(t)\cos(n\omega t)dt \\ b_n &= \frac{2}{T}\int_0^t v(t)\sin(n\omega t)dt \\ c_n &= \sqrt{a_n^2 + b_n^2} \\ \theta_n &= \tan^{-1}\left(\frac{a_n}{b_n}\right) \end{aligned} \tag{5.60}$$

The magnitude, c_n, of the nth harmonic can be calculated, along with its phase, θ_n, as shown in (5.60).

If there is a component of the signal at, or near, the selected frequency this phase-sensitive integration process will produce a non-zero amplitude. This process only examines the signal at discrete integer frequencies determined by the value of n and at DC where the term $a_0/2$ is the average value of $v(t)$ over a complete cycle.

Care must be used when interpreting the results of this spectral analysis, as the process assumes that individual harmonics are present with constant amplitude and phase for the duration of the measurement. However, harmonics are seldom constant in amplitude or phase, and so the coefficients, which represent only the averages, may be significantly in error.

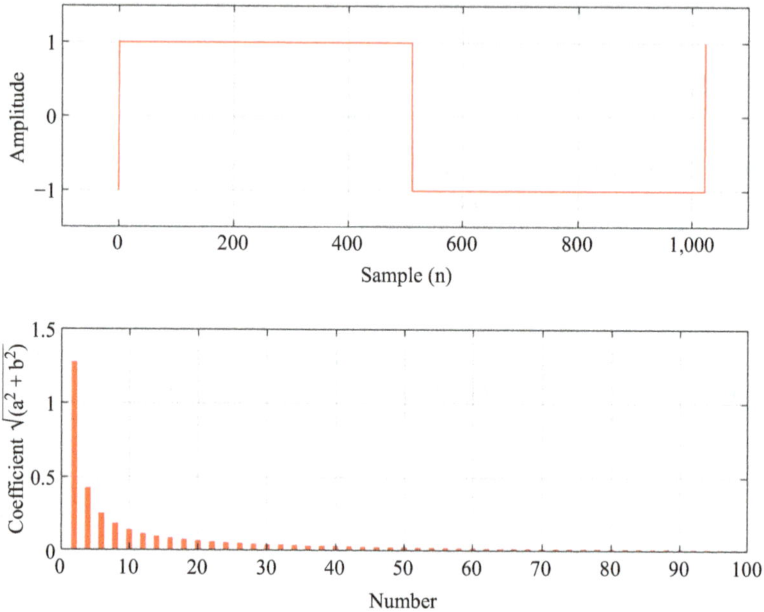

Figure 5.64 Amplitudes of Fourier coefficients to produce a square wave (Brooker 2008)

Depending on the nature of the signal and the information that must be reproduced, the number of frequency components that are required will vary. For example, a square wave may require up to one thousand harmonics to reproduce the sharp transitions that define the switching points (Carr 1997). The harmonic analysis in Figure 5.64 shows that the amplitudes of the coefficients are reduced by progressively smaller amounts with the result that their individual contributions become progressively smaller.

However, as discussed earlier is this chapter, an ECG trace can be reproduced using spectral components up to about 100 Hz.

The effect of truncating the series is demonstrated in Figure 5.65 where the waveform is reconstructed by summing the appropriately scaled sinusoidal components. For example, when five components are used, the signal is reconstructed as follows:

$$v(t) = \sum_{n=1}^{5} a_n \cos(n\omega t) + b_n \sin(n\omega t) \tag{5.61}$$

The figure shows the effectiveness of the reconstruction for five, fifty and five hundred coefficients. It can be seen that the transitions are still not perfect even after the largest number.

The spectrum of a repeated time-domain sequence such as an ECG produces a comb of frequency components at intervals corresponding to the reciprocal of the

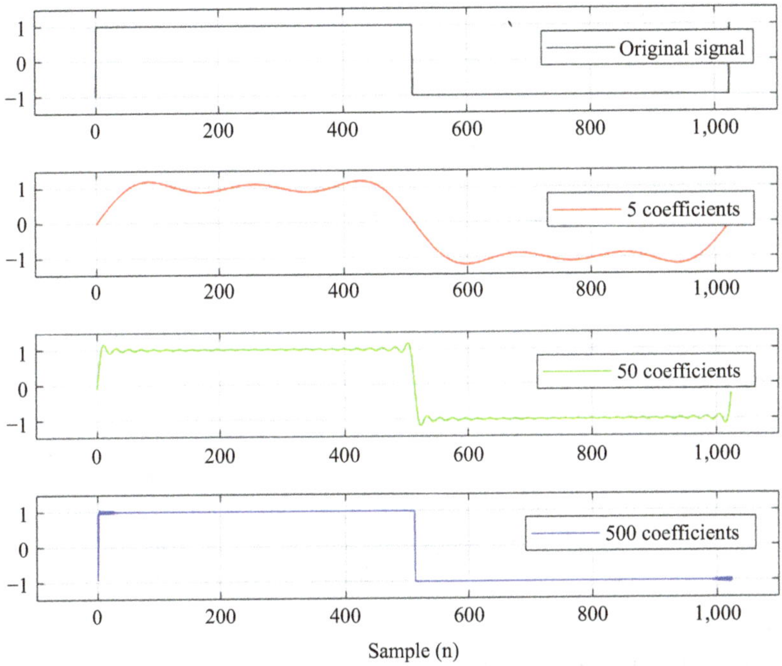

Figure 5.65 Effect on the reconstructed signal of limiting the number of Fourier coefficients

repeat period. As shown in Figure 5.66, these fall within the envelope of the spectrum of a single ECG cycle.

To understand this, examine Figure 5.67 and consider that the repeated time-domain signal can be generated by convolving a single ECG cycle with a sequence of impulses at the correct time intervals, Δt. It can be shown that the spectrum of a sequence of impulses is a regularly spaced comb of frequency components with spacing $\Delta f = 1/\Delta t$. In addition, in the frequency domain, the convolution process is replaced by the product of the two spectra; therefore, the composite spectrum will be the product of the spectrum of a single ECG cycle and the spectrum of a sequence of impulses.

5.6 Statistical techniques, machine learning and AI

Living organisms are so complex that it is often difficult, if not impossible, to use any of the conventional signal processing methods discussed previously to obtain useful outputs. In such cases, automated processes such as statistical methods, pattern recognition, data mining or machine learning are often used to extract useful information from raw measurements. The following points provide an overview of the more common applications of the technology:

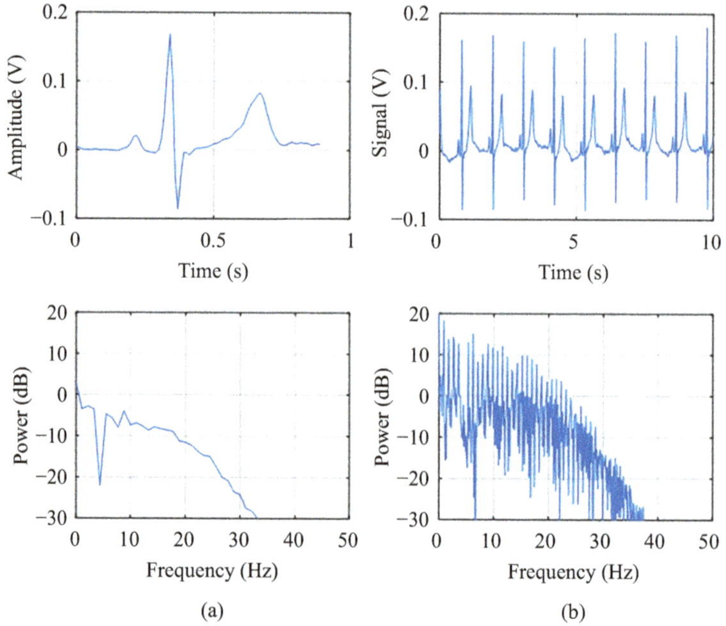

Figure 5.66 Comparison between the spectra of (a) a single ECG cycle and (b) a sequence of cycles

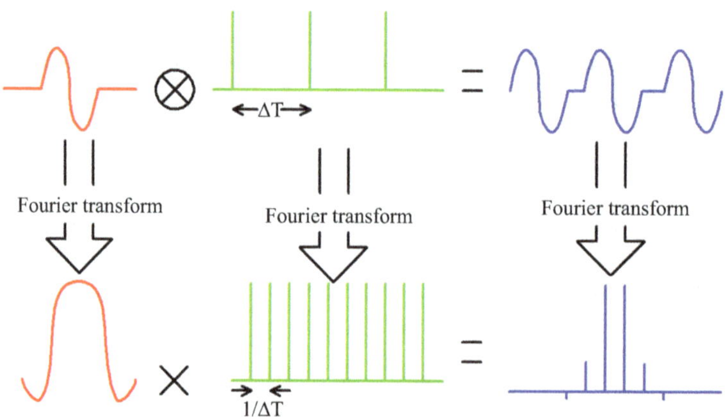

Figure 5.67 Graphical explanation showing the relationship between single and repeated spectra

- **Medical Imaging Analysis:** AI algorithms are used to analyse medical images such as X-rays, including mammograms, MRIs and CT scans. These algorithms can assist in the early detection of diseases like cancer, tumours, and fractures, improving diagnostic accuracy and efficiency.

- **Biomechatronics and Prosthetics:** AI-powered prosthetic limbs and exoskeletons utilise machine learning algorithms to interpret signals from the user's muscles or nervous system, enabling intuitive control of the devices. These advancements improve mobility and functionality for individuals with limb loss or mobility impairments.
- **Rehabilitation and Therapy:** AI-based systems design personalised rehabilitation programs and therapeutic interventions tailored to individual patient needs. These systems monitor progress, adjust treatment plans dynamically, and provide feedback to optimise rehabilitation outcomes.
- **Brain-Computer Interfaces (BCIs):** BCIs enable direct communication between the brain and external devices, such as prosthetics or computer systems, through neural signals. Machine learning algorithms decode these complex signals to interpret the user's intentions, allowing for seamless interaction and control.
- **Remote Patient Monitoring:** AI-driven monitoring systems can track patient health metrics in real-time, detecting anomalies and predicting potential health issues before they escalate. This facilitates early intervention and enables healthcare providers to deliver timely care, particularly for patients with chronic conditions.
- **Virtual Health Assistants and Telemedicine:** AI-powered virtual assistants and "chatbots" provide personalised health advice, answer medical queries, and assist with appointment scheduling. Telemedicine platforms leverage AI algorithms for remote diagnosis, monitoring, and treatment, expanding access to healthcare services.
- **Drug Discovery and Development:** AI models are employed to predict the properties of molecules and their interactions with biological targets. This accelerates the drug discovery process by identifying potential candidates for further study and optimising drug design.
- **Genomics and Personalised Medicine:** Machine-learning techniques analyse genomic data to identify patterns and associations with diseases, enabling personalised treatment plans based on an individual's genetic makeup. This approach leads to more targeted therapies and better patient outcomes.
- **Healthcare Management and Predictive Analytics:** AI systems analyse electronic health records (EHRs), patient data, and other healthcare-related data to identify trends, predict disease progression, and optimise treatment plans. This assists healthcare providers in making informed decisions and improving patient care.
- **Biomedical Research and Discovery:** AI algorithms analyse large-scale biomedical datasets, such as those from clinical trials or research studies, to uncover hidden patterns, biomarkers, and potential therapeutic targets. This accelerates scientific discovery and enhances our understanding of complex diseases.

Broadly speaking, these application domains all have similar goals – modelling for classification or hypothesis testing.

5.6.1 *Statistical techniques*

Statistical methods have traditionally emphasised models that can be solved analytically and that result in some restrictions to the expressive power of the models that can be used. However, these methods have been around for a long time and are well understood.

5.6.1.1 Regression analysis

Regression analysis is probably the oldest and best-known statistical method. It is a technique that provides a mathematical description of the relationship between two or more measured variables. It is particularly useful when dealing with continuous quantitative data like the amplitude of myoelectric signal levels as a function of muscle force for example.

In its simplest form for a given series of paired variables (x_1,y_1), (x_2,y_2), ..., (x_n,y_n) where $n \geq 2$, and following a relationship known to be linear

$$y = a + bx \tag{5.62}$$

The best fitting curve in a least squares sense minimises the residual, ε, to determine the coefficients a and b

$$\varepsilon = \sum_{i=1}^{n} [y_i - f(x_i)]^2 = \sum_{i=1}^{n} [y_i - (a + bx_i)]^2 = \min \tag{5.63}$$

To obtain the least squared error, the partial derivatives of the residual with respect to a and b respectively must both be zero

$$\begin{aligned} \frac{\partial \varepsilon}{\partial a} &= 2\sum_{i=1}^{n} [y_i - (a + bx_i)] = 0 \\ \frac{\partial \varepsilon}{\partial b} &= 2\sum_{i=1}^{n} x_i[y_i - (a + bx_i)] = 0 \end{aligned} \tag{5.64}$$

Expanding and solving for a and b

$$a = \frac{\sum_{i=1}^{n} y_i \sum_{i=1}^{n} x_i^2 - \sum_{i=1}^{n} x_i \sum_{i=1}^{n} x_i y_i}{n\sum_{i=1}^{n} x_i^2 - \left(\sum_{i=1}^{n} x_i\right)^2} \tag{5.65}$$

$$b = \frac{n\sum_{i=1}^{n} x_i y_i - \sum_{i=1}^{n} x_i \sum_{i=1}^{n} y_i}{n\sum_{i=1}^{n} x_i^2 - \left(\sum_{i=1}^{n} x_i\right)^2} \tag{5.66}$$

The correlation coefficient, r, is a measure of the degree of linear association between the two variables and is defined as

$$r = \frac{\sum_{i=1}^{n} x_i y_i - \frac{\sum_{i=1}^{n} x_i \sum_{i=1}^{n} y_i}{n}}{\sqrt{\left[\sum_{i=1}^{n} x_i^2 - \frac{\left(\sum_{i=1}^{n} x_i\right)^2}{n}\right]\left[\sum_{i=1}^{n} y_i^2 - \frac{\left(\sum_{i=1}^{n} y_i\right)^2}{n}\right]}} \tag{5.67}$$

It is seldom necessary to use these equations directly as many scientific calculators and computer software packages are already programmed to perform regression analysis.

Consider the measured EMG signals from the biceps as a function of the applied force shown as the "+" markers in Figure 5.68. The coefficients a and b are obtained in MATLAB and used to plot the red best-fit line $y = -0.1562 + 0.3319x$. The green lines above and below the red line show y +/− *delta* and they will contain at least 50% of future observations at x.

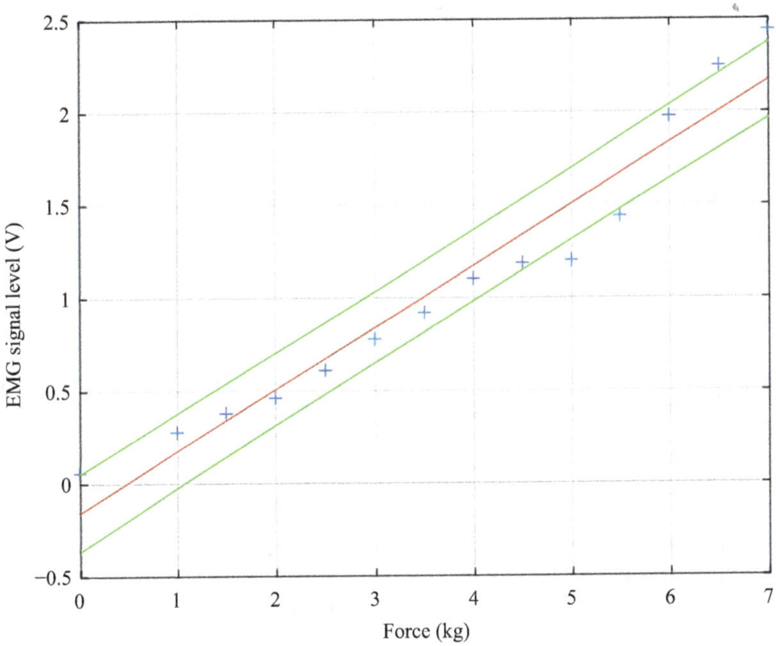

Figure 5.68 Measured relationship between applied force and measured EMG signal

```
% mapping force vs emg
%emg_force.m
force = [0,1,1.5,2,2.5,3,3.5,4,4.5,5,5.5,6,6.5,7];
emg = [0.06,0.278,0.38,0.46,0.61,0.778,0.92,1.103,1.18,1.198,1.436,1.97,2.24,2.44];
% Linear fit to measured force and EMG data
[P,S] = polyfit(force,emg,1);                    % obtain P=(b,a) from the measured data
x = (min(force):1:max(force));                   % generate the x vector
[y,delta] = polyval(P,x,S);                      % generate the y vector using a and b
plot(force,emg,'+',x,y,'r',x,y+delta,'g',x,y-delta,'g');  % plot and label the graph
grid
xlabel('Force (kg)')
ylabel('EMG Signal Level (V)')
```

Multiple regression provides an estimate of the coefficients of a function when more than one control variable is involved. For example, in the case where there are two independent variables, x and y, and one dependent variable, z, the linear equation is

$$z = a + bx + cy \tag{5.68}$$

For a given data set (x_1,y_1,z_1), (x_2,y_2,z_2), …, (x_n,y_n,z_n) where $n \geq 3$, the best fitting curve in a least squares sense minimises the residual, ε, to determine the coefficients a, b and c

$$\varepsilon = \sum_{i=1}^{n} [z_i - f(x_i, y_i)]^2 = \sum_{i=1}^{n} [z_i - (a + bx_i + cy_i)]^2 = \min \tag{5.69}$$

Equating the partial derivatives with respect to the coefficients as before, allows us to construct a set of simultaneous equations that can be solved to find the coefficients

$$\begin{aligned}
\sum_{i=1}^{n} z_i &= a\sum_{i=1}^{n} 1 + b\sum_{i=1}^{n} x_i + c\sum_{i=1}^{n} y_i \\
\sum_{i=1}^{n} x_i z_i &= a\sum_{i=1}^{n} x_i + b\sum_{i=1}^{n} x_i^2 + c\sum_{i=1}^{n} x_i y_i \\
\sum_{i=1}^{n} y_i z_i &= a\sum_{i=1}^{n} y_i + b\sum_{i=1}^{n} x_i y_i + c\sum_{i=1}^{n} y_i^2
\end{aligned} \tag{5.70}$$

When working with categories of outputs rather than numbers then classification algorithms are a better choice than the regression methods so far discussed.

5.6.2 Data mining

Data mining has been defined as the process of extracting patterns from very large data sets by using a combination of statistical and machine learning techniques in conjunction with database management. Typically, the focus of data mining is primarily on data storage, integration and retrieval with the analysis functions being of secondary importance.

5.6.3 Machine learning

Machine learning (ML) uses computationally powerful methods to learn very complex non-parametric (or quasi-parametric) models of the data. The models used are often close to human representations, and the process is one of learning or improved performance with "experience".

Machine Learning techniques in biomechatronic engineering include the following:

- **Decision Trees and Random Forests:** Decision support for movement classification and control.
- **Support Vector Machines (SVM):** Classification tasks such as gesture recognition or gait analysis.
- **K-Nearest Neighbours (KNN):** Pattern recognition and classification.
- **Gaussian Mixture Models (GMM):** Clustering and modelling complex distributions.
- **Principal Component Analysis (PCA):** Dimensionality reduction and feature extraction.
- **Neural Networks (including Deep Learning):** Signal processing, feature extraction, and control in BCIs or EMG systems.
- **Hidden Markov Models (HMM):** Modelling sequential movements and recognising temporal patterns.
- **Reinforcement Learning:** Adaptive control and learning in prosthetics or exoskeletons.

Machine Learning can be divided into two broad categories: supervised and unsupervised learning. In supervised learning, the algorithms are presented with multiple examples. This allows the algorithm to learn to predict the correct output corresponding to previously seen inputs as well as previously unseen ones. In unsupervised learning the algorithm is required to generate its own categories of outputs.

5.6.3.1 Decision/classification trees

Traditional classification methods such as Discriminant Analysis, Cluster Analysis, Nonparametric Statistics and Nonlinear Estimation are commonly used where assumptions regarding the statistical nature of the measurements are met. However, in cases where traditional methods fail, Classification Trees can be used.

Classification trees predict membership of objects into classes of a categorical dependent variable from measurements of predictor variables using a hierarchical process. An example given in (Breiman *et al.* 1984) that was used to identify heart attack victims most at risk of dying was a simple three question binary classification tree shown in Figure 5.69.

Discriminant Analysis of the heart attack data would produce a set of coefficients for a linear combination of blood pressure, patient age and heart rate that best differentiates between low and high-risk patients. The sum of each of the measurements weighted by the respective discriminant coefficient provides a single score based on the simultaneous evaluation of the three measurements.

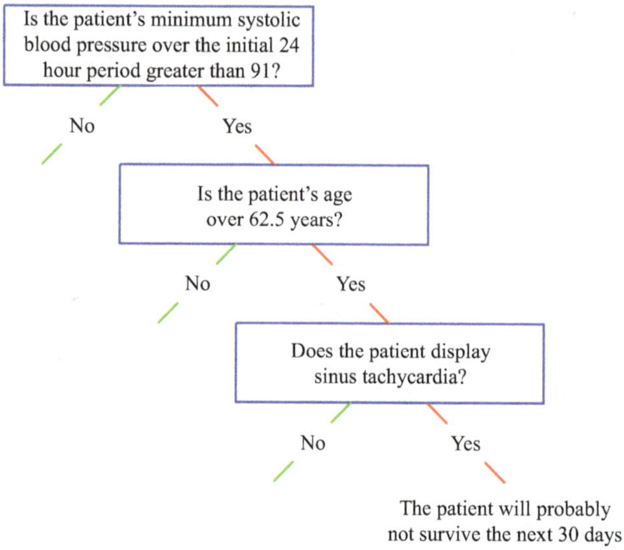

Figure 5.69 Example of a simple classification tree used to determine the likelihood of a heart attack victim dying in the next 30 days

For example, if BP_{min} is the minimum systolic blood pressure measured over a 24 hour period, A is the patient's age and T is a binary value (1,0) representing the presence or absence of tachycardia are the predictor values and b,a,t are the discriminant function coefficients, then a decision equation can be written based on a threshold value, *Thresh*, that is used to decide whether the patient is at high risk or not.

If
$$bBP_{\text{min}} + aA + tT > Thresh$$
Then
The patient is at high risk
End

5.6.3.2 Clustering

In the case where the user has no prior knowledge of the relationships between groups, then k-means clustering is a simple Machine Learning algorithm that is commonly used in Biomechatronics and medical imaging

k-means is an evolutionary unsupervised learning algorithm that clusters observations into *k* groups, where *k* is provided as an input parameter. Each observation is assigned to a specific cluster based upon its proximity to the mean of that cluster. The cluster's mean is then recomputed and the process is repeated as described:

1. The algorithm arbitrarily selects *k* points as the initial cluster means.
2. Each point in the dataset is assigned to a cluster, based upon the Euclidean distance between that point and each cluster mean.

3. Each cluster mean is recomputed as the average of the points in that cluster.
4. Steps 2 and 3 repeat until the clusters converge (typically when no change in the mean is seen when steps 2 and 3 are repeated)

A disadvantage of k-means is the fact that the user specifies the number of clusters as an input to the algorithm. As designed, the algorithm is not capable of determining the appropriate number of clusters and depends upon the user to identify this in advance. For this reason, it is a good idea to experiment with different values of k to identify the value that best suits the data.

5.6.3.3 Support vector machines

Support vector machines (SVMs) are a set of kernel-based supervised learning methods that perform their classification by constructing an N-dimensional hyperplane that separates the measured data into two categories. In most cases the raw data cannot be separated by a hyperplane, so it must be transformed by the kernel into another "feature" space which maximises the separation between the two classes.

In the two-dimensional example shown in Figure 5.70 the two different categories of data are represented by circles and squares. In the original data shown in the "input" space, a complex curve is required to separate the categories. However, the appropriate kernel transforms the data in such a manner that a straight line separates the two categories as shown.

In this example, the support vectors are shown as thin green lines that define the edges of each of the categories. Their orientations are defined to maximise the margin (distance between the two vectors).

5.6.3.4 Transformers

Transformers are a type of deep learning architecture and have revolutionised the field of medicine by enabling advanced data analysis, personalised care, and improved healthcare processes. Examples include natural language processing where Transformers are used to process unstructured text data from clinical notes,

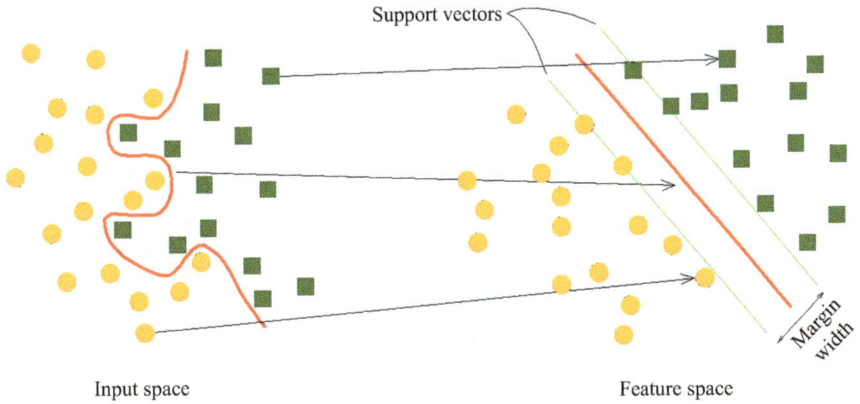

Figure 5.70 Kernel transformation for a 2D Support Vector Machine

Electronic Health Records (EHR), and scientific literature. This includes tasks like summarising patient records, extracting medical entities, and generating patient-friendly explanations of complex medical information. Large Language Models (LLMs) such as BioBERT, PubMedBERT, and GPT variants are tailored for biomedical text mining, enabling tasks like drug discovery, disease mechanism analysis, and clinical trial matching.

Another common application is in medical imaging where vision transformers enhance image analysis tasks such as segmentation, classification, and reconstruction in radiology and pathology. They help improve diagnostics by identifying patterns in X-rays, MRIs and CT scans.

5.7 Isolation barriers

Many biomedical applications require that power be conveyed one way across an isolation barrier and that signals be conveyed in the other direction. These serve two purposes. In the first, isolation is required to ensure that, in the case of electronic failure, high voltages cannot be introduced into the patient through electrodes attached to the body. In the second, electronics are mounted under the skin or within the body cavity, so that a method of charging internal batteries and obtaining telemetry is required.

5.7.1 Implant systems

Most implant systems share a common framework, as shown in Figure 5.71, which consists of an external signal processing unit for biological sensory information

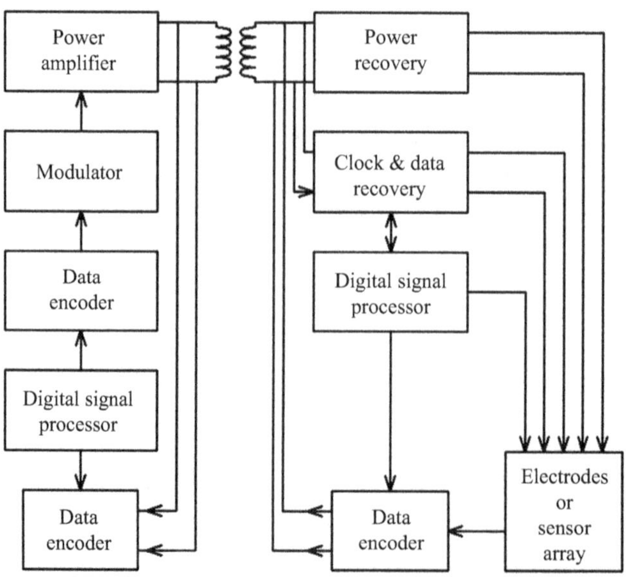

Figure 5.71 Schematic diagram for a generic implantable system [adapted from Finn and LoPresti (2003)]

(sound, images etc.), a bidirectional telemetry unit, an internal signal processing module, a stimulus generator/driver and an electrode array to interface to tissue or nerves (Finn and LoPresti 2003).

It is envisaged that ultimately most prostheses will be internal, with sensors, processing and actuation electrodes all self-contained within a biocompatible module. However, during the development phase, or while the technology is immature, there are myriad advantages of including both internal and external modules, including decreased risk of adverse reaction to implanted materials, lower internal heat dissipation and, of course, ease of refinement to both the hardware and software.

Percutaneous connectors (physical wired links between implanted and exterior components) raise the risk of infection due to perpetual breach of the skin through which the wires must pass. In addition, anchor points can be a problem, and movement between the internal and external components can result in unnecessary stresses to the body and to the connection itself.

Transcutaneous (wireless) connections are therefore justified in most cases for bidirectional telemetry and to supply power to the embedded portion of the prosthesis. Three types of telemetry are considered to be appropriate under these circumstances for providing power and communication. These are optical, magnetic (inductive) and electromagnetic (radio frequency), and are discussed in the following sections (Finn and LoPresti 2003).

Much work is being done in this field to improve the efficiency of RFID tags which work on similar principles, albeit in applications where the tag requires very little power other than to generate a response.

5.7.1.1 Optical telemetry

Optical telemetry appears most commonly amongst research groups developing ocular prostheses as the cornea and lens are usually transparent. This form of telemetry usually uses high-energy light from a laser to power the photoelectronics that form part of the implant. In an MIT/Harvard prosthesis an array of photodiodes was illuminated by a 30 mW 820 nm laser to produce a current of 300 μA at 7 V.

5.7.1.2 Inductive telemetry

Inductive telemetry has been the standard means of wireless connection to implanted devices for many years. It is based on the coupling between a pair of coaxial coplanar coils. The secondary is mounted subcutaneously along with the internal modules, while the primary coil remains exterior. A low-frequency carrier, usually between 1 MHz and 10 MHz, supplies the primary coil and power is coupled into the secondary coil where it is rectified and filtered to supply power. Additionally, the power carrier can be modulated to transmit data to the implant.

Problems with this method of power transfer include the fact that, because the coils have no cores, very little of the energy couples through to the other coil and most of it radiates omnidirectionally. This results in a large drain on the batteries, and the potential for electromagnetic interference.

The coupling between coils is dependent on coil loading, the excitation frequency and number of turns as well as coil alignment and spacing. Typical values for coil coupling vary from 0.01 and 0.1. However, this can be improved if the coils are series or parallel resonant using an appropriate capacitor. If L (H) is the self-inductance of the coil, then the capacitance, C (F), should be

$$C = \frac{1}{(2\pi f)^2 L} \tag{5.71}$$

The resonant coil is driven by a class-E amplifier (usually just a FET switch) that shunts the supply to ground as shown in Figure 5.72. As with conventional switch-mode power supplies, unused energy is not dissipated as heat but is traded back and forth between electric and magnetic fields in the resonant structure.

The low coupling coefficient to the secondary means that shifts in the load impedance have very little effect on the primary circuit. However, coil distortion and the close proximity of any metallic objects do have an effect, so feedback using a sensing coil is often used to maintain performance.

Data modulation usually follows standard digital protocols and this is accomplished using amplitude, frequency or phase-shift keying, though amplitude shift keying (ASK) is the most common. This can be achieved by altering the supply voltage or by shifting the switching frequency slightly to move off resonance. In this case a 10 Hz shift on a carrier frequency of 760 kHz is sufficient. However, because the resonant circuit is high Q, the data rate is limited.

Secondary coils are also resonant to maximise power transfer efficiency and then conventional diodes are used to rectify the alternating current (AC) signal for use by the prosthesis.

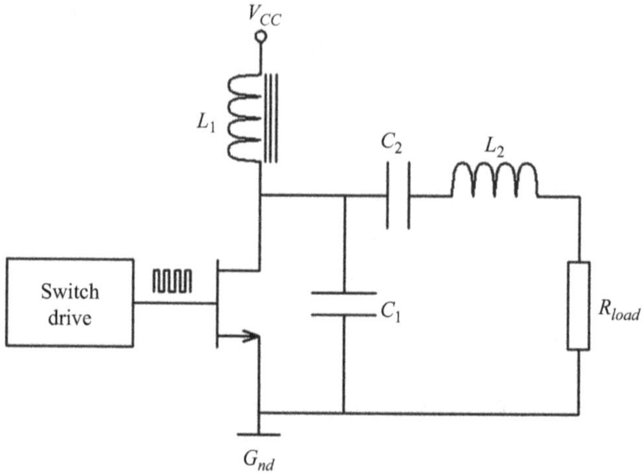

Figure 5.72 Class-E amplifier driving a resonant coil

5.7.1.3 Radiofrequency telemetry

Radiofrequency telemetry is similar to inductive coupling except that a travelling electromagnetic (EM) wave is employed rather than magnetic induction, and antennas are used rather than coils. Problems with this method include the size of the antennas required and the dielectric constant and conductivity of the tissue, which introduces high losses.

5.7.1.4 Back telemetry

Back telemetry refers to the capability of a prosthesis to relay information from the implanted electronics out to the exterior of the body. Optical means can be used in which an LED is modulated to provide the signal, or in magnetic systems a coil-based inductive link. An alternative is to use a passive method with a power transfer link. This is achieved by changing the loaded Q of the inductive link by altering the load impedance as shown in Figure 5.73.

5.7.2 Isolation amplifiers

Signal isolation enables digital or analogue signals to transmit without galvanic connections between the transmitting and receiving side of the barrier. This in turn allows ground or reference levels on the transmitting and receiving side to differ by thousands of volts and prevents circulating currents between differing ground potentials which might contaminate signals. Noise on a signal ground may corrupt the signal; isolation can separate the signal from a clean signal subsystem ground.

In addition, a galvanic connection between reference levels might provide a current path that would present a safety risk to an operator or medical patient. To avoid this, safety regulations in medical care prescribe maximum leakage currents

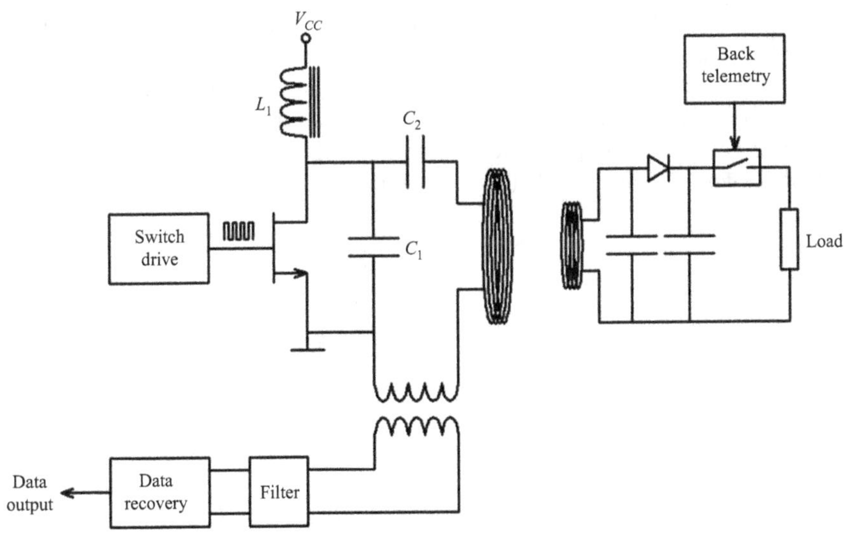

Figure 5.73 Back telemetry implemented by altering the secondary load

in case a patient should touch ground or mains power during a bioelectric recording (Webster 1978). Although the regulations vary between countries, sufficient galvanic isolation and an isolation capacitance less than 800 pF are usually demanded (maximum leakage current is 50 μA_{rms} when the patient touches the line voltage of 220 V_{rms}, 50 Hz).

Isolation can be realised using three different methods: transformer isolation, capacitor isolation and opto-isolation. Because there is no current flow across the barrier, the potential difference across it is determined by the difference between the input and output grounds (common points).

Power transfer is not a problem in instrumentation because magnetic coupling between the primary and secondary coils can be enhanced by using magnetic cores. Conventional DC/DC converters with isolation voltages of many kV are available with efficiencies of 70% to 90%. These are discussed in detail in Chapter 2. In some cases, particularly if a fibre optic link is used for signal isolation batteries, are used to provide an isolated power source.

PCs for display and analysis of medical data pose a major isolation problem. The design of power supplies in PCs permits high leakage currents that are inadequate for medical environments, and even if the patient is separated from the PC by an isolation barrier, care must be taken not to touch them simultaneously because of the danger of electrocution. If PCs are to be used in this role, then medical-rated devices are the right choice, or as a last resort, an additional medical-grade isolation transformer should be installed between the PC and the mains supply (Bronzino 2006).

Galvanic isolation in digital systems is easily achieved using opto-isolators. These devices consist of a light emitting diode physically separated from a phototransistor as shown in Figure 5.74. The digital input on–off modulates the LED output which in turn changes the current flow through the phototransistor in an identical manner. Note that in this configuration, the digital signal is inverted.

Galvanic isolation of analogue signals can be achieved in several ways. The simplest is to convert the analogue signal to a digital one. This can use pulse width modulation, voltage to frequency conversion or even using an ADC to generate a digital word. This digital information is then transmitted across the barrier using the method shown in Figure 5.74 before conversion back to an analogue signal.

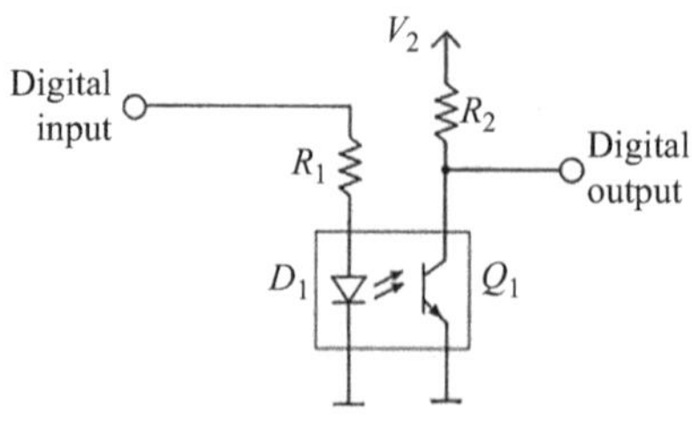

Figure 5.74 Digital opto-isolator schematic

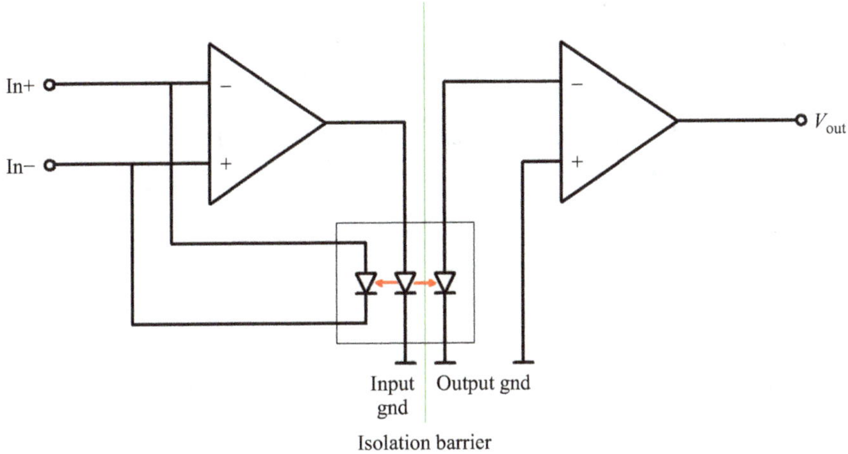

Figure 5.75 Opto-isolated analogue link

For pulse width modulation, the conversion is achieved using a simple low-pass filter, while in the frequency encoded signal a frequency to voltage converter is used. In the case where a digital word is transmitted a conventional DAC converts the word back to its analogue equivalent.

An interesting alternative is a true analogue link, where the intensity of the LED encodes for the analogue voltage, as shown in Figure 5.75. In this configuration, the isolation consists of a pair of matched photodiodes and an LED. The signal from the photodiode is fed back to the input of the amplifier to counteract the nonlinear characteristics of both the LED and the photodiodes.

References

Alciatore, D. and M. Histand (2003). *Introduction to Mechatronics and Measurement Systems*, 2nd ed. Boston: McGraw Hill.

Benedict, T. and G. Bordner (1962). "Synthesis of an optimal set of radar track while scan smoothing equations." *IRE Transactions on Automatic Control* AC 7: 27–32.

Blackman, S. (1986). *Multiple-Target Tracking with Radar Application*. Norwood: MA, Artech House.

Breiman, L., J. Friedman, R. Olshen and C. Stone (1984). *Classification and Regression Trees*. Chapman and Hall/CRC.

Bronzino, J., (Ed.) (2006). *Medical Devices and Systems*. Boca Raton, FL: CRC Press.

Brooker, G. (2008). *Introduction to Sensors for Ranging and Imaging*, SciTech.

Brooker, G. (2021). *Sensors for Ranging and Imaging*, 2nd ed. Stevenage, Herts: The Institution of Engineering and Technology (IET).

Carlson, G. (1998). *Signal and Linear System Analysis*, 2nd ed. New York: John Wiley & Sons, Inc.

Carr, J. (1997). *Electronic CircuitGuidebook: Sensors*. Pennsylvania State University, Prompt.

Cromwell, L., F. Weibell, E. Pfeiffer and L. Usselman (1973). *Biomedical Instrumentation and Measurements*. Englewood Cliffs, New Jersey: Prentice-Hall, Inc.

Edminster, J. (1972). *Shaum's Outline Series: Theory and Problems of Electric Circuits*. New York: McGraw Hill.

Finn, W. and P. LoPresti, (Eds.) (2003). *Handbook of Neuroprosthetic Methods*. London: CRC Press.

Kuphaldt, T. (2003). "Lessons in Electric Circuits: Volume III – Semiconductors." Retrieved August 2008 from http://www.faqs.org/docs/electric/Semi/index.html.

Mahafza, B. (2000). *Radar Systems Analysis and Design Using MATLAB*. Boca Raton, FL: Chapman & Hall/CRC.

Stanley, W. (1975). *Digital Signal Processing*. Reston, VA: Reston Publishing Company, Inc.

Walpole, R. and R. Myers (1978). *Probability and Statistics for Engineers and Scientists*, 2nd ed. New York: Macmillan.

Webster, J., (Ed.) (1978). *Electrical Safety in Medical Instrumentation: Application and Design*. Boston, MA: Houghton Mifflin Co.

Young, P. (1990). *Electronic Communication Techniques*, 3rd ed. New York: Merrill.

Chapter 6

Feedback and control systems

6.1 Introduction

A system can be defined as the artificial boundary surrounding a collection of interacting components which allow those components to be considered as a "black box" consisting of a limited number of inputs and outputs. In the natural world, this could be a complete ecosystem, a single human being or one of the physiological systems, such as the cardiovascular or respiratory system, within that human being. From a mechatronic perspective, systems can be as complex as atomic power stations or as simple as a pneumatic actuator or a single electronic component.

From a biomechatronic perspective, it is often important that a system's characteristics be well enough understood to be controlled. The control can occur in an open-loop manner, which implies that the system is provided with an input, and this will determine the output state. For example, a skin patch that provides a constant supply of insulin to a diabetic would be an open-loop system. Unfortunately, open-loop control does not cater for changes in the system's characteristics, so if the diabetic drank a large glass of fruit juice, their blood sugar level would increase substantially with potentially serious consequences. If, however, the skin patch was replaced with a more sophisticated device that measured the blood sugar level and compensated for changes by adjusting the rate at which insulin was infused, the blood sugar level would remain reasonably stable. This is known as closed-loop control, and it relies on feedback of an output parameter to help regulate the system.

This controlled process may be applied on a macroscopic scale, that is, feedback from the output to the input of a complex system, or on a much smaller scale, such as a simple filter that applies a proportion of the output voltage to the input.

6.2 Biological feedback mechanisms

The human body is, of course, a riot of cross-coupled feedback processes, some of which are understood, but many of which are still beyond our comprehension. Figure 6.1 illustrates the control aspects that govern the interaction between insulin and glucose levels mentioned above. It is one of the simpler internal processes and yet it involves adjusting the levels of both insulin and glucagon to control glucose production and uptake.

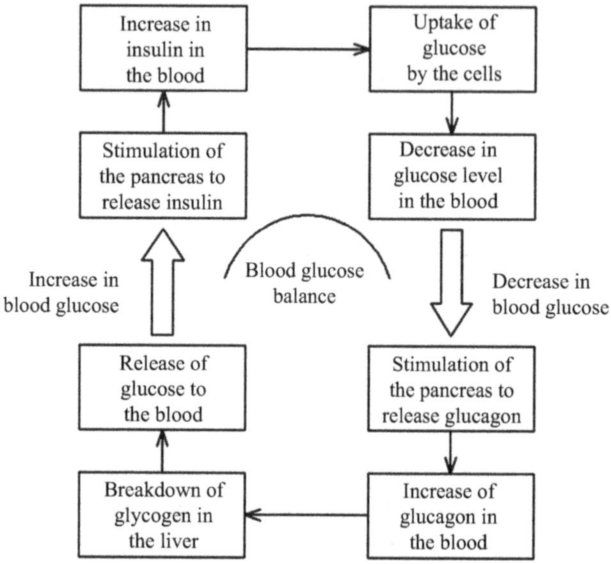

Figure 6.1 Feedback mechanisms used to control blood glucose levels

Insulin is synthesised in the pancreas and its secretion from β-cells into the blood is controlled by the concentration of glucose in the blood. As the level of glucose rises (say, after a meal), the insulin level follows suit. Most cells in the body have receptors to which insulin binds, and when this occurs, the cell is stimulated to absorb glucose from the bloodstream.

When glucose levels fall below a certain threshold, α-cells in the pancreas are stimulated to secrete glucagon that is carried directly to the liver where it stimulates the breakdown of glycogen to glucose, which is released into the bloodstream.

Other biological feedback mechanisms are discussed briefly in later chapters. One example, discussed in Volume 2, Chapter 4, is the short- and long-term control of blood pressure, and others are the less well-understood neural feedback mechanisms in the retina and in the visual cortex that allow us to interpret visual signals and images.

6.3 Biomechatronic feedback mechanisms

Almost all biomechatronic systems incorporate some form of feedback for control purposes. Some examples discussed in later chapters include the control of the output level of a hearing prosthesis (Volume 2, Chapter 1), and the control of blood flow in artificial hearts and ventricular assist devices (Volume 2, Chapter 4). The latter example is interesting because the parameter being controlled is not, in fact, measured and must be estimated from other available data like the current drawn by the pump motor. However, in most cases, it is possible to measure the value of the parameter to be controlled and to apply feedback to control it. This chapter examines the basic principles of modelling systems and applying feedback to improve their control.

6.3.1 Limit switches

Probably the simplest of all of the feedback mechanisms is the limit switch. These devices can be mechanical or electrical, as discussed in Chapter 2, and their function is to produce an output when a mechanism has reached the end of its allowable range of travel. An example of such a system is shown in Figure 6.2.

In this example, an electric motor drives an interrupter plate that is designed to break the optical path of a pair of optical switches at the clockwise and anticlockwise limits of travel. These outputs are fed into a controller that can be used to cut power to the motor, or to reverse the direction of travel. This form of feedback can be applied to a pulsatile ventricular assist device to produce a continuous reciprocating pumping action. In most cases, however, limit switches are only used as warning devices in association with more complex mechanisms to ensure that mechanical damage does not occur by driving a device beyond its allowable limits.

6.3.2 Proportional and higher order controllers

In a proportional control system, the set point is compared to the measured output of the system, and the difference between the two is applied as a correction input to the system. In most cases, the sense of the correction signal drives the output to minimise, and sometimes even eliminate the error completely. Under some conditions, simple proportional controllers are either not sufficiently fast or are not stable and need to be augmented by additional states such as the derivative or the integral of the error signal.

In addition, as discussed, systems often comprise multiple inputs and outputs that are often cross-coupled, which makes efficient control that much more difficult. For example, consider the sophisticated prosthetic arm controller shown in Figure 6.3, which is commanded to pick up a ball.

In this case, both the position of the hand used to reach the ball and the positions of the fingers used to grasp it need to be controlled. In the case of the position

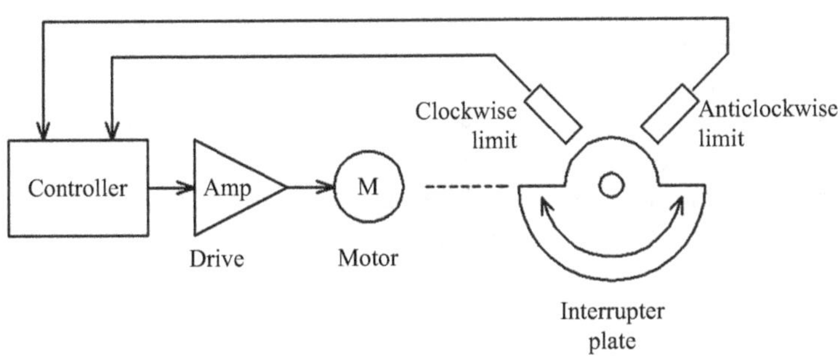

Figure 6.2 Application of limit switches as feedback elements

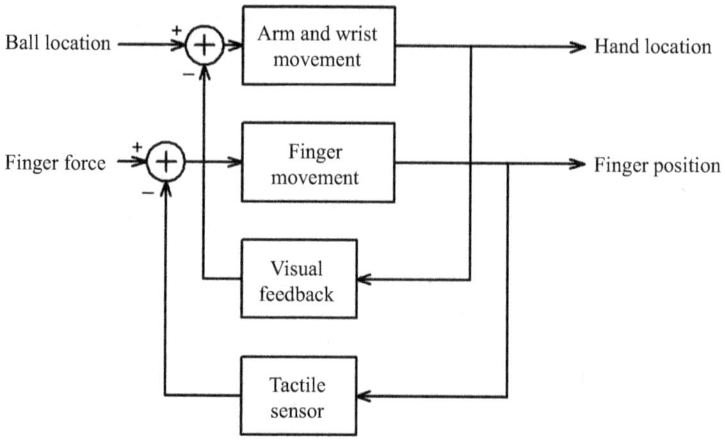

Figure 6.3 Block diagram for multi-input multi-output control

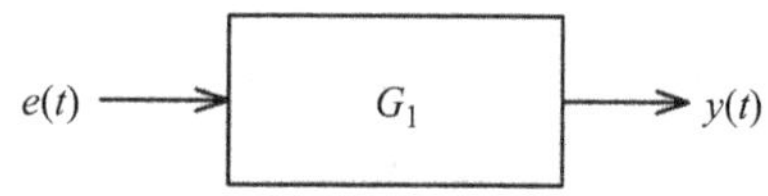

Figure 6.4 Open-loop system

of the hand, it is assumed that visual feedback is used to control this function, and that when it is in place, the hand can be commanded to close and tactile feedback from force sensors in the fingertips is used to control the grip strength. What is not shown in the diagram is the cross-coupling between the actual finger positions and the position of the arm which ensures that the ball is grasped accurately and not bumped out of reach.

6.4 System representation

The classical representation of a system is a rectangular box that represents the system transfer function. A single box may be used, or multiple boxes may be linked in series, in which case, the overall system transfer function is the product of the individual transfer functions. An open-loop system can be represented by a box with an input and an output, as shown in Figure 6.4.

In this case, the input $e(t)$ is referred to as the actuating signal and $y(t)$ the controlled variable, where G_1 is the control element, so

$$y(t) = G_1 e(t) \tag{6.1}$$

With the application of feedback, as shown in Figure 6.5, the block diagram starts to resemble those in the previous examples.

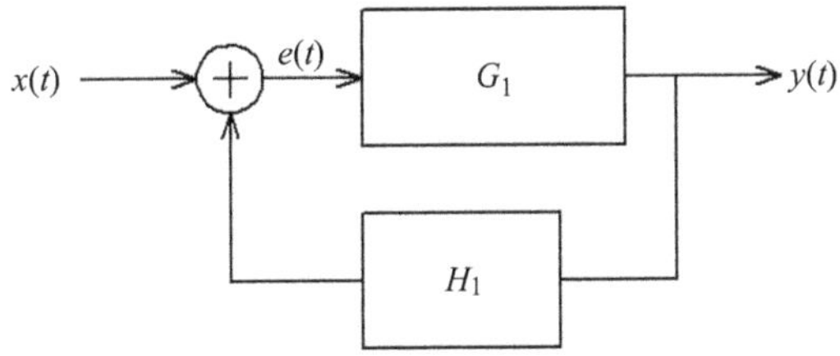

Figure 6.5 Closed-loop system

In this case, $x(t)$ is the set-point and $e(t)$ is the error signal, being the difference between the set-point and the feedback signal

$$e(t) = x(t) - b(t) \tag{6.2}$$

where $b(t)$ is the feedback signal, and it is related to the output signal, or controlled variable $y(t)$ by the feedback element H_1

$$b(t) = H_1 y(t) \tag{6.3}$$

If equation (6.2) is written in terms of (6.1) and (6.3)

$$\frac{y(t)}{G_1} = x(t) - H_1 y(t)$$

$$y(t)\left[\frac{1}{G_1} + H_1\right] = x(t) \tag{6.4}$$

$$y(t)\left[\frac{1 + G_1 H_1}{G_1}\right] = x(t)$$

The system transfer function is the ratio of the output to the input

$$T_1 = \frac{y(t)}{x(t)} = \frac{G_1}{1 + G_1 H_1} \tag{6.5}$$

The main advantages of using closed-loop feedback are that the system becomes less sensitive to disturbances and small changes to the characteristics of the control element G_1. It can also result in faster response to changes in the set-point. However, the overall gain is reduced from G_1 to T_1, and the system can become unstable under some circumstances.

6.4.1 *Worked example: motor control*

In an open-loop controller, the transfer function of a motor is $G = 250$ rpm/V. Determine the steady-state speed if 12 V is applied to the input.

In this case, the open-loop model shown in Figure 6.4 is used, where $G_1 = 250$ and $e(t) = 12$ V

$$\begin{aligned} y(t) &= Ge(t) \\ &= 250 \times 12 \\ &= 3000 \text{ rpm} \end{aligned}$$

In the closed-loop case, consider that a tacho-generator is coupled to the motor shaft and its transfer function is $H_1 = 3$ mV per rpm. What is the new transfer function?

From equation (6.5)

$$\begin{aligned} T_1 &= \frac{G_1}{1 + G_1 H_1} \\ &= \frac{250}{1 + 250 \times 3 \times 10^{-3}} \\ &= 142.86 \text{ rpm/V} \end{aligned}$$

The application of 12 V will produce an output speed of $12 \times 142.86 =$ 1715 rpm.

At this speed, the tacho-generator will produce an output of $1714 \times 3 \times 10^{-3} =$ 5.14 V producing an error voltage of $12-5.14 = 6.86$ V, which, in turn, produces an output speed of

$$\begin{aligned} y(t) &= e(t) G_1 \\ &= 6.86 \times 250 \\ &= 1715 \text{ rpm} \end{aligned}$$

as expected.

The control elements G_1 and H_1 can be simple gains, as shown in this example, but in general, they are more complex and are modelled as differential equations.

6.5 System models

One way of analysing a system is to produce a lumped parameter model of each of the elements and combine them to form a system. Fortunately, there are strong similarities between the behaviours of most types of elements of interest in biomechatronics, including mechanical, electrical, fluidic and thermal systems, and this makes their analysis reasonably straightforward.

6.5.1 Mechanical elements

Mechanical systems consist of springs, dashpots and masses, where the springs represent stiffness, dashpots, the forces opposing motion (damping and friction) and masses the inertial components, or resistance to acceleration.

The stiffness of a spring is described by the linear relationship between the compressive or expansive force, F (Newton), and the displacement x (m)

$$F = kx \tag{6.6}$$

where k (N/m) is the spring constant.

A dashpot consists of a piston moving within a closed fluid-filled cylinder. As the piston moves, fluid must pass through a small hole through the piston from one side of the cylinder to the other. This flow produces a resistive force, F (N), which is proportional to the velocity v (m/s) of the piston

$$F = cv \tag{6.7}$$

where c (N per m/s) is the constant of proportionality.

This can be rewritten in the form of a derivative, because velocity is the rate of change of position

$$F = c\frac{dx}{dt} \tag{6.8}$$

Finally, in the case of the mass, m (kg), for a given acceleration, a (m/s^2), the greater the mass, the higher the required force, F (N)

$$F = ma \tag{6.9}$$

where the constant of proportionality between the acceleration and the force is the mass.

Because acceleration is the rate of change of velocity, which is in turn equal to the rate of change of displacement, equation (6.9) can be rewritten as

$$F = m\frac{dv}{dt} = m\frac{d^2x}{dt^2} \tag{6.10}$$

It is interesting to note that potential energy is stored in the compression of the spring, and kinetic energy in the velocity of the mass, but energy is dissipated in the dashpot. In this case, the potential and kinetic energy (Joules) is

$$\begin{aligned} E_{spring} &= \frac{1}{2}kx^2 \\ E_{mass} &= \frac{1}{2}mv^2 \end{aligned} \tag{6.11}$$

and the power, P_{dash} (Watt), dissipated in the dashpot is proportional to the square of the velocity

$$P_{dash} = cv^2 \tag{6.12}$$

Similar mechanical models can be derived for the rotary equivalents of these linear equations.

For a torsion spring

$$T = k\theta \tag{6.13}$$

where T (Nm) is the torque, θ (rad) is the angle of rotation and k (Nm per rad) is the constant of proportionality.

In the case of a rotary damper, often modelled as a fan in a fluid bath, the equation is

$$T = c\omega = c\frac{d\theta}{dt} \tag{6.14}$$

where c (Nm per rad/s) is the damping constant and ω (rad/s) is the angular rate.

The moment of inertia I (kgm^2) is the constant of proportionality between the torque and the angular acceleration α (rad/s^2)

$$T = I\alpha = I\frac{d\omega}{dt} = I\frac{d^2\theta}{dt^2} \tag{6.15}$$

As before, potential energy is stored in the torsion spring

$$E_{spring} = \frac{1}{2}k\theta^2 \tag{6.16}$$

and kinetic energy in the rotating inertia

$$E_I = \frac{1}{2}I\omega^2 \tag{6.17}$$

but power is dissipated by the rotary damper

$$P_{damp} = c\omega^2 \tag{6.18}$$

These equations are summarised in Table 6.1.

Table 6.1 Description of mechanical blocks

Block	**Describing equation**	**Energy/power**
Spring	$F = kx$	$E_{spring} = \frac{1}{2}kx^2$
Torsion spring	$T = k\theta$	$E_{spring} = \frac{1}{2}k\theta^2$
Mass	$F = m\frac{dv}{dt} = m\frac{d^2x}{dt^2}$	$E_{mass} = \frac{1}{2}mv^2$
Moment of inertia	$T = I\alpha = I\frac{d\omega}{dt} = I\frac{d^2\theta}{dt^2}$	$E_I = \frac{1}{2}I\omega^2$
Dashpot	$F = cv = c\frac{dx}{dt}$	$P_{dash} = cv^2$
Rotational damper	$T = c\omega = c\frac{d\theta}{dt}$	$P_{damp} = c\omega^2$

6.5.2 Mechanical model

Many models used in biomechatronic systems can be modelled using the building blocks described in the previous system. For example, consider the actuator that is driving the incus in a middle-ear implantable hearing device (MEIHD) shown in Figure 6.6.

As a first approximation, the driven mass consists of the sum of the mass of the actuator pin and the ossicles. The spring component results from the springiness of the flexible linkages between the ossicles, the attached ligaments and their connection to the eardrum and the round window, while the damping component is also determined by the energy absorbed in these linkages.

In this model, it is assumed that the applied forces all operate in the direction of movement of the ossicles. Therefore, the net force applied to the mass m (kg) is $F - kx - cv$, and it results in an acceleration a (m/s^2). Written as an equation

$$ma = F - kx - cv \tag{6.19}$$

This can be written as a differential equation

$$m\frac{d^2x}{dt^2} = F - kx - c\frac{dx}{dt} \tag{6.20}$$

which can be rearranged thus

$$m\frac{d^2x}{dt^2} + c\frac{dx}{dt} + kx = F \tag{6.21}$$

This is the classic form for a second order differential equation which can be rewritten in terms of its natural oscillation frequency, ω_n and damping ratio, ξ, where

$$\omega_n = \sqrt{\frac{k}{m}}, \tag{6.22}$$

and

$$\xi = \frac{c}{2\sqrt{mk}} \tag{6.23}$$

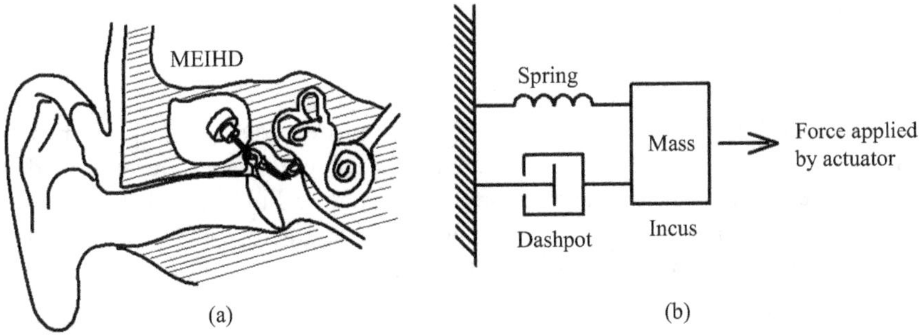

Figure 6.6 Diagram of (a) a middle-ear implantable hearing device and (b) simplified free body diagram of the system

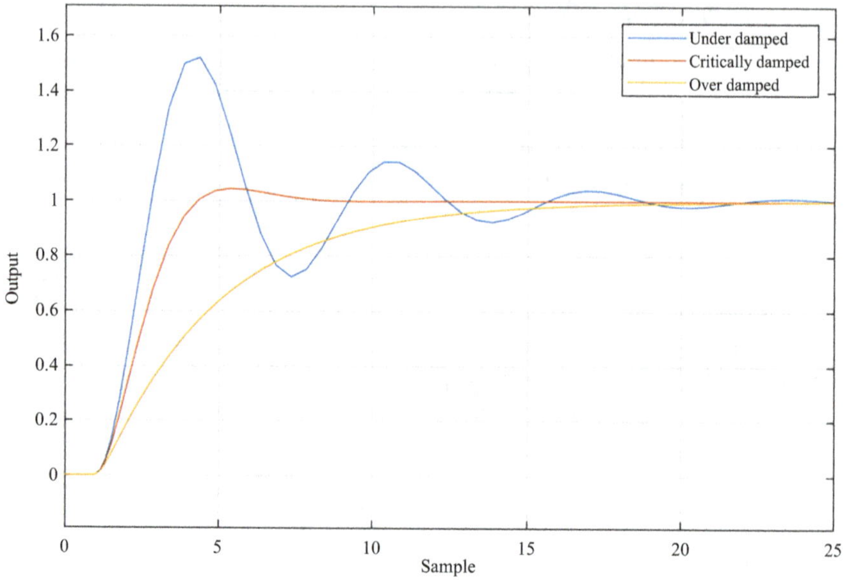

Figure 6.7 Response of a second-order differential equation to a unit step

Rewriting equation (6.21) in terms of these two parameters

$$\frac{1}{\omega_n^2}\frac{d^2x}{dt^2}+\frac{2\xi}{\omega_n}\frac{dx}{dt}+x=\frac{F}{k} \tag{6.24}$$

The response of the system depends on the natural frequency and the damping ratio. In this application, the actuator characteristics would be adjusted so that the natural frequency of the system is outside the normal audio range to ensure that audio signals do not excite any resonances, and the damping ratio would be adjusted to ensure that the response is critically damped, or even over damped so that no ringing occurs. Analysis of the frequency response of the auditory path in conjunction with a MEIHD is considered in Volume 2, Chapter 1.

The step responses (Figure 6.7) are typical for a second-order differential equation.

6.5.3 Electrical elements

The basic building blocks for electrical circuits are inductors, capacitors and resistors, as discussed in Volume 1, Chapter 2. These elements are described in terms of the relationship between the applied voltage and the current flow.

For an inductor, the potential difference, V (volts), across it is proportional to the rate of change of current, i (amps), through it

$$V=L\frac{di}{dt} \tag{6.25}$$

where L (Henry) is the inductance, and the polarity of the induced potential difference is opposite to the polarity of the potential difference used to drive the current.

This relationship can also be described in terms of an integral

$$i = \frac{1}{L}\int Vdt \tag{6.26}$$

For a capacitor, C (farads), the potential difference, v (volts), depends on the charge, q (coulomb)

$$V = \frac{q}{C} \tag{6.27}$$

From the definition, current is the rate of change of charge

$$i = \frac{dq}{dt}, \tag{6.28}$$

which can be rewritten as an integral

$$q = \int idt \tag{6.29}$$

Substituting into equation (6.27)

$$V = \frac{1}{C}\int idt \tag{6.30}$$

or

$$i = C\frac{dV}{dt} \tag{6.31}$$

The final electrical element is the resistor, R (ohms), and in this case the potential difference, V (volts), across it is proportional to the current, i (amps), flowing through it

$$V = Ri \tag{6.32}$$

As with the mechanical model, in the electrical case, both the inductor and the capacitor store energy, where

$$\begin{aligned} E_{ind} &= \frac{1}{2}Li^2 \\ E_{cap} &= \frac{1}{2}CV^2 \end{aligned} \tag{6.33}$$

while power is dissipated by the resistor

$$P_{res} = \frac{1}{R}V^2 \tag{6.34}$$

6.5.4 Electrical model

The equations that describe how electrical building blocks are combined are known as Kirchhoff's Laws, and can be summarised as follows:

- The sum of all the currents flowing into a junction (or a node) is zero. That is, the same amount of current that flows into a junction must flow out again.
- In a circuit, the algebraic sum of the potential differences around the circuit is equal to the applied emf.

Consider a simple circuit that contains a resistor, an inductor and a capacitor as shown in Figure 6.8, and determine the equations that describe the potential difference across the capacitor as a function of time

Applying Kirchhoff's Second Law

$$V = V_R + V_L + V_C \tag{6.35}$$

Because there is a single loop, the same current must flow through each of the components; therefore

$$V_R = iR$$
$$V_L = L\frac{di}{dt}$$

Substituting into equation (6.35)

$$V = iR + L\frac{di}{dt} + V_C \tag{6.36}$$

and from Table 6.2, we know that

$$i = C\frac{dV_C}{dt} \tag{6.37}$$

Therefore, substituting (6.37) into (6.36) gives

$$V_c = V - RC\frac{dV_C}{dt} - LC\frac{d^2V}{dt^2} \tag{6.38}$$

This is a second-order differential equation and can therefore be expressed in the same way as its mechanical counterpart.

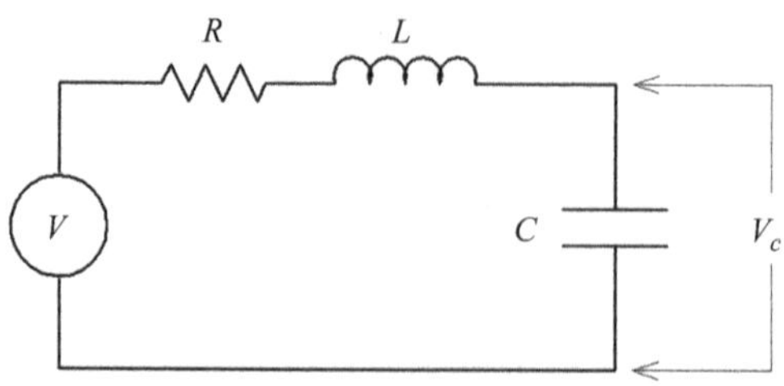

Figure 6.8 Simple RLC circuit

Table 6.2 Description of electrical blocks

Block	Describing equation	Energy/power
Inductor	$V = L\frac{di}{dt} = \frac{1}{L}\int Vdt$	$E_{ind} = \frac{1}{2}Li^2$
Capacitor	$V = \frac{1}{C}\int idt = C\frac{dV}{dt}$	$E_{cap} = \frac{1}{2}CV^2$
Resistor	$V = Ri = \frac{V}{R}$	$P_{res} = \frac{1}{R}V^2$

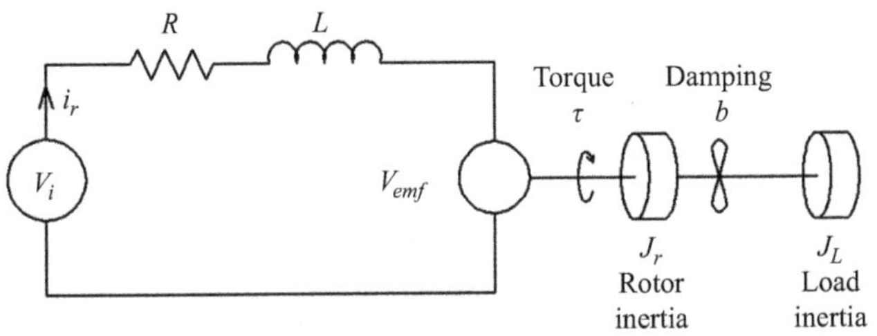

Figure 6.9 Block diagram of elements of an electric motor

Example: DC motor model

A simple model for an electric motor combines electrical and mechanical elements as shown in Figure 6.9 and described in the following section.

The torque, $\tau(t)$ (Nm), generated by the motor is proportional to the current, $i_r(t)$ (A) in the rotor

$$\tau(t) = K_m i_r(t) \tag{6.39}$$

where K_m (Nm/A) is the motor torque constant, and it is related to the physical properties of the motor including the number of turns and the strength of the magnetic field, etc.

The back emf, $V_{emf}(t)$ (volts), is proportional to the motor speed, $\omega(t)$ (rad/s)

$$V_{emf}(t) = K_e\omega(t) = K_e\frac{d\theta(t)}{dt} \tag{6.40}$$

where K_e (V/rpm) is the back emf constant, also known as the speed constant.

The equations that relate the motor output to the input current are described for the electrical and mechanical sections of the circuit. A second-order differential equation describes the mechanical section

$$(J_r + J_L)\frac{d^2\theta(t)}{dt^2} + b\frac{d\theta(t)}{dt} = K_m i_r(t) \tag{6.41}$$

where J_r (kgm^2) and J_L (kgm^2) are the moments of inertia of the rotor and the load, respectively, and b is the damping coefficient.

A first-order differential equation describes the electrical portion of the motor

$$L\frac{di_r(t)}{dt} + Ri_r(t) = V_i(t) - K_e\frac{d\theta(t)}{dt} \tag{6.42}$$

Alternatively, equations (6.41) and (6.42) can be written in terms of the speed of the motor

$$(J_r + J_L)\frac{d\omega(t)}{dt} + b\omega(t) = K_m i_r(t) \tag{6.43}$$

$$L\frac{di_r(t)}{dt} + Ri_r(t) = V_i(t) - K_e\omega(t) \tag{6.44}$$

6.5.5 *Similarities of the two models*

Spice software used to analyse electrical circuits is both accurate and easily available; therefore, mechanical models are often converted to their electrical counterparts to make use of this software. For example, the electrical equivalent of the standard mass, spring, damper block diagram is the parallel capacitor, inductor, resistor circuit shown in Figure 6.10.

Table 6.3 lists the equations that describe the equations for the linear and torsional mechanical elements along with their electrical equivalents.

6.5.6 *Fluid flow elements*

Similar models can be derived for fluid flow components. However, in this instance, there are two different formulations; one for incompressible hydraulics and another for compressible pneumatics, where changes in pressure result in changes in density. This book only examines the incompressible hydraulic model.

The relationship between the volume rate of flow q (m^3/s) and the pressure difference $\Delta p = p_1 - p_2$ is determined by the hydraulic resistance, R. This is the hydraulic equivalent to Ohm's law, where the hydraulic resistance is analogous to electrical resistance, the volume flow rate is equivalent to current, and the pressure difference is equivalent to the potential difference.

$$\Delta p = p_1 - p_2 = qR \tag{6.45}$$

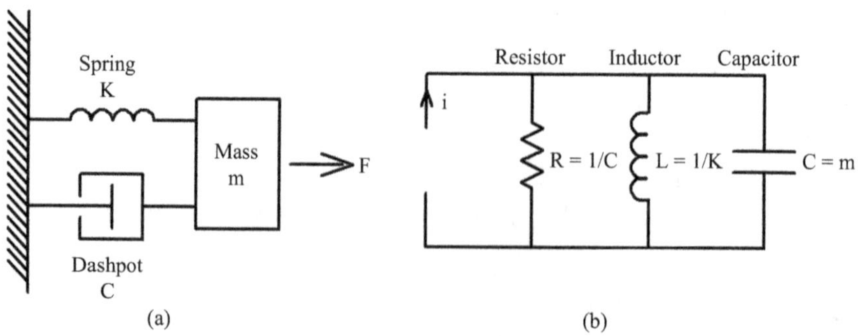

Figure 6.10 Mechanical and electrical equivalent circuits

Table 6.3 Comparison between electrical and mechanical elements

Block	**Describing equation**	**Energy/power**
Inductor	$i = \frac{1}{L}\int Vdt$	$E_{ind} = \frac{1}{2}Li^2$
Spring	$F = kx = k\int vdt$	$E_{spring} = \frac{1}{2}kx^2$
Torsion spring	$T = k\theta = k\int \omega dt$	$E_{spring} = \frac{1}{2}k\theta^2$
Capacitor	$i = C\frac{dV}{dt}$	$E_{cap} = \frac{1}{2}CV^2$
Mass	$F = m\frac{dv}{dt} = m\frac{d^2x}{dt^2}$	$E_{mass} = \frac{1}{2}mv^2$
Moment of inertia	$T = I\alpha = I\frac{d\omega}{dt} = I\frac{d^2\theta}{dt^2}$	$E_I = \frac{1}{2}I\omega^2$
Resistor	$i = \frac{V}{R}$	$P_{res} = \frac{1}{R}V^2$
Dashpot	$F = cv = c\frac{dx}{dt}$	$P_{dash} = cv^2$
Rotational damper	$T = c\omega = c\frac{d\theta}{dt}$	$P_{damp} = c\omega^2$

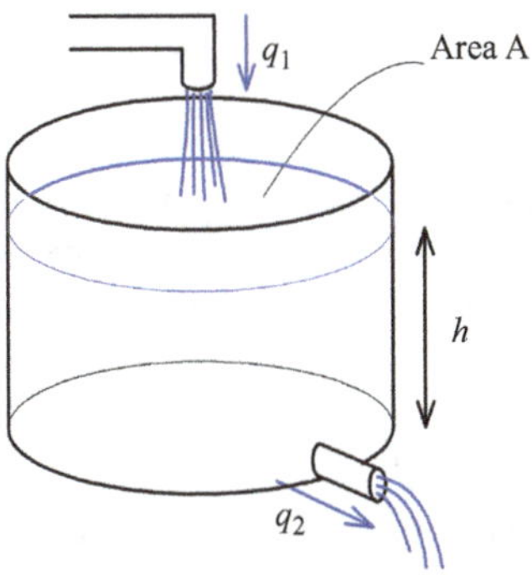

Figure 6.11 Potential energy of fluid in a reservoir

Hydraulic capacitance is the term that describes the potential energy storage by a fluid in terms of its pressure head. Consider the diagram shown in Figure 6.11, where the rate of change of volume V (m^3) in the reservoir is equal to the difference in the volume flow rates into and out of the reservoir

$$q_1 - q_2 = \frac{dV}{dt} \quad (6.46)$$

However, if the volume is written in terms of the cross-sectional area A (m^2) and the height of the fluid, h (m), in the reservoir, then

$$q_1 - q_2 = A\frac{dh}{dt} \tag{6.47}$$

The pressure difference between the input and the output is a function of the density ρ (kg/m^3) and the acceleration due to gravity g (m/s^2)

$$P = \rho g h \tag{6.48}$$

Rewriting equation (6.47) in terms of the pressure

$$q_1 - q_2 = \frac{A}{\rho g}\frac{dP}{dt} \tag{6.49}$$

The hydraulic capacitance C is defined as

$$C = \frac{A}{\rho g} \tag{6.50}$$

Writing equation (6.49) in terms of the hydraulic capacitance

$$q_1 - q_2 = C\frac{dP}{dt} \tag{6.51}$$

or, in terms of integrals

$$P = \frac{1}{C}\int (q_1 - q_2)dt \tag{6.52}$$

Finally, hydraulic inertance is equivalent to electrical inductance. Consider the cylindrical section of fluid shown in Figure 6.12. The difference between the two forces can be described in terms of the difference between the two pressures $(P_1 - P_2)$ across the cylinder

$$F_1 - F_2 = P_1A - P_2A = (P_1 - P_2)A \tag{6.53}$$

The net force, F (N), that causes the mass of fluid to accelerate at a (m/s^2) is

$$F = ma \tag{6.54}$$

Therefore

$$(P_1 - P_2)A = ma \tag{6.55}$$

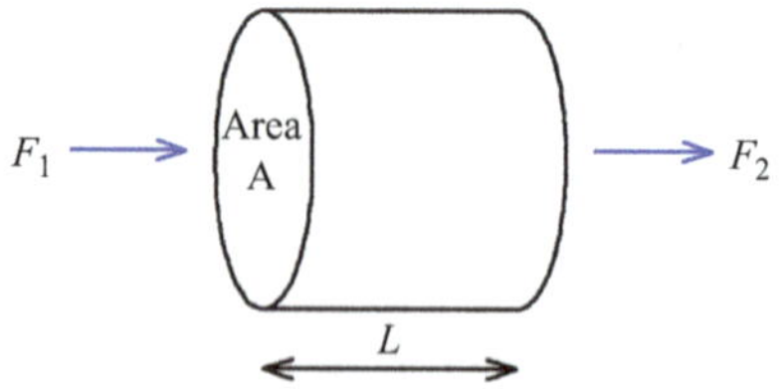

Figure 6.12 Forces applied to a cylinder of fluid

Equation (6.55) can be rewritten in terms of the rate of change of velocity

$$(P_1 - P_2)A = m\frac{dv}{dt} \tag{6.56}$$

The mass of the fluid cylinder is equal to the product of the volume of the fluid and the density

$$m = \rho V = \rho AL \tag{6.57}$$

Therefore

$$(P_1 - P_2)A = \rho AL\frac{dv}{dt} \tag{6.58}$$

But the volume flow rate $q = Av$ therefore

$$(P_1 - P_2)A = \rho L\frac{dq}{dt} \tag{6.59}$$

The pressure difference can be written as

$$(P_1 - P_2) = \frac{\rho L}{A}\frac{dq}{dt} = I\frac{dq}{dt} \tag{6.60}$$

where the hydraulic inertance, I is defined as

$$I = \frac{\rho L}{A} \tag{6.61}$$

As with the mechanical elements discussed earlier, hydraulic elements can be thought of as equivalent to their electrical counterparts which allows spice models to be used to simulate fluid flow. The equivalence is shown in Table 6.4.

Table 6.4 Comparison between electrical and hydraulic elements

Block	**Describing equation**	**Energy/power**
Inductor	$i = \frac{1}{L}\int Vdt$	$E_{ind} = \frac{1}{2}Li^2$
Hydraulic inertance	$q = \frac{1}{I}\int (P_1 - P_2)dt$	$E_{HI} = \frac{1}{2}Iq^2$
Capacitor	$i = C\frac{dV}{dt}$	$E_{cap} = \frac{1}{2}CV^2$
Hydraulic capacitance	$q = C\frac{d(P_1 - P_2)}{dt}$	$E_{mass} = \frac{1}{2}C(P_1 - P_2)^2$
Resistor	$i = \frac{V}{R}$	$P_{res} = \frac{1}{R}V^2$
Hydraulic resistance	$q = \frac{1}{R}(P_1 - P_2)$	$P_{HR} = \frac{1}{R}(P_1 - P_2)^2$

6.6 System response

If the input to a system changes, there will be two different stages of changes to the output. The first will be a transient phase that, in a stable system, will settle into the steady-state response. The solution to the differential equations discussed earlier is used to determine analytically what these responses will be.

In a real system, the input may change in an arbitrary way, but in order to quantify its behaviour, responses are usually determined for impulses, steps, ramps and sinusoidal inputs. If the system is linear and time invariant (LTI), the output will be the algebraic sum of any number of responses to any of these functions.

It is, in fact, possible to construct any of the inputs described from the sum of a sequence of impulses and, therefore, the impulse response provides a common method to characterise a system in the time domain. However, the manipulation of sums of impulse responses is cumbersome, so another method of analysis using the Laplace Transform is generally used as it makes things easier.

Consider the simple example of a float-valve controller that could be used to maintain the level of blood in an open reservoir in a heart–lung machine as shown in Figure 6.13. In this system, the rate at which blood enters the tank is dependent on the difference between the current blood depth, h (m), and the final depth, h_{fin} (m).

A simple first-order differential equation can be written as

$$\frac{du}{dt} = k(h_{fin} - h) \tag{6.62}$$

The solution of this form of differential equation can be determined by inspection

$$h = h_{fin}(1 - e^{-kt}) \tag{6.63}$$

In the case of a second-order differential equation, such as the mass, spring, damper system discussed earlier, the response is described by the natural frequency of the system and the damping ratio. Depending on the magnitudes of these parameters, the response to a step input could be an undamped sinusoid, a damped sinusoid or an exponential, as shown in Figure 6.7.

It is often more difficult than this to fit the response of a system to one of the standard solutions, particularly if the order is high and there are feedback components.

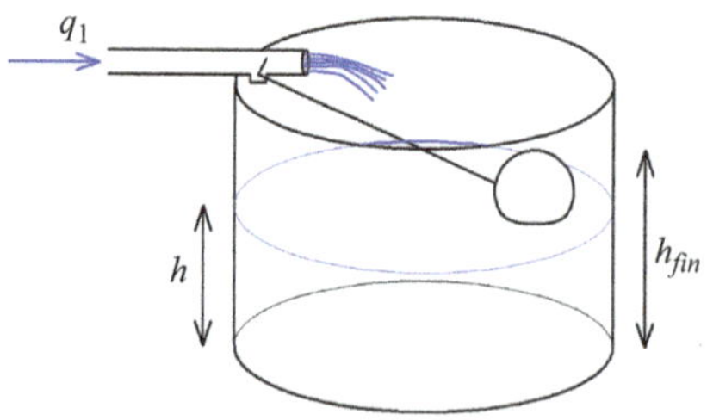

Figure 6.13 Float controller reservoir for a heart–lung machine

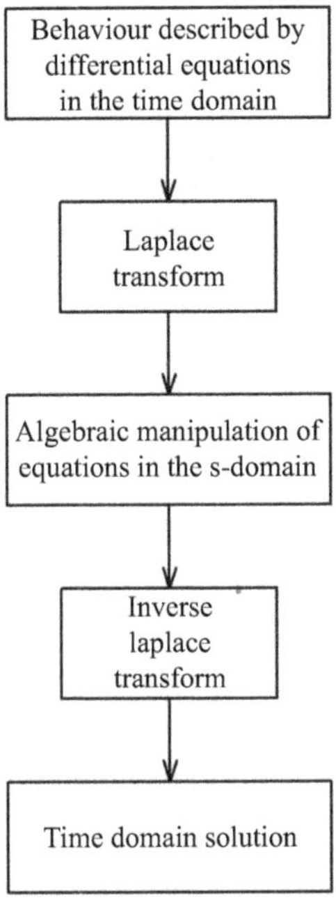

Figure 6.14 Process of using Laplace Transforms to determine the time domain responses of complex systems

This is another reason for describing systems in terms of their Laplace Transforms which makes them far easier to manipulate and solve. The process is shown in Figure 6.14.

For a function $f(t)$ in the time domain, the Laplace Transform is

$$F(s) = \int_0^\infty f(t)e^{-st}dt \tag{6.64}$$

Under normal circumstances, it is not necessary to evaluate this integral because tables have been compiled for most commonly occurring functions, and these combined with a few rules for handling combinations of transforms allow for the solution of most problems encountered in biomechatronics and other disciplines. Some of the rules are listed in Table 6.5.

To solve differential equations using the Laplace Transform method, perform the following steps:

- transform each term in the differential equation into its Laplace Transform,
- perform the algebraic manipulation for a specific input (step, impulse, sine wave, etc.),
- convert the results into the standard form using the partial fraction expansion process, and
- invert the results back into the time domain.

Table 6.5 Rules for handling Laplace Transforms (Bolton 1992)

Time domain	S domain
$f_1(t)+f_2(t)$	$F_1(s)+F_2(s)$
$f_1(t)-f_2(t)$	$F_1(s)-F_2(s)$
$af(t)$	$aF(s)$
$f(t-T)$	$e^{-Ts}F(s)$
$\frac{d}{dt}f(t)$	$sF(s)-f(o)$
$\frac{d^2}{dt^2}f(t)$	$s^2F(s)-sf(0)-\frac{df(0)}{dt}$
$\frac{d^n}{dt^n}f(t)$	$s^nF(s)-s^{n-1}f(0).....-\frac{d^{n-1}f(0)}{dt^{n-1}}$
$\int_0^t f(t)$	$\frac{1}{s}F(s)$

Table 6.6 Laplace Transforms

S domain $F(s)$	Time domain $f(t), t \geq 0$
1	$\delta(t)$
$\frac{1}{s}$	$1(t)$
$\frac{1}{s^2}$	t
$\frac{2!}{s^3}$	t^2
$\frac{3!}{s^4}$	t^3
$\frac{m!}{s^{m+1}}$	t^m
$\frac{1}{s+a}$	e^{-at}
$\frac{1}{(s+a)^2}$	te^{-at}
$\frac{1}{(s+a)^m}$	$\frac{1}{(m-1)!}t^{m-1}e^{-at}$
$\frac{a}{s(s+a)}$	$1-e^{-at}$
$\frac{a}{s^2(s+a)}$	$\frac{1}{a}(at-1+e^{-at})$
$\frac{b-a}{(s+a)(s+b)}$	$e^{-at}-e^{-bt}$
$\frac{s}{(s+a)^2}$	$(1-at)e^{-at}$

(Continues)

Table 6.6 (Continued)

S domain $F(s)$	Time domain $f(t), t \geq 0$
$\dfrac{a^2}{s(s+a)^2}$	$1-(1+at)e^{-at}$
$\dfrac{(b-a)s}{(s+a)(s+b)}$	$be^{-bt} - ae^{-at}$
$\dfrac{a}{s^2+a^2}$	$\sin at$
$\dfrac{s}{s^2+a^2}$	$\cos at$
$\dfrac{s+a}{(s+a)^2+b^2}$	$e^{-at}\cos bt$
$\dfrac{b}{(s+a)^2+b^2}$	$e^{-at}\sin bt$

Example: Response of an RC circuit

Consider the step change to the input voltage of a series RC circuit, and determine the voltage, $V_C(t)$, across the capacitor, as a function of time.

The differential equation can be written as

$$V = RC\frac{dV_C}{dt} + V_C \tag{6.65}$$

with $V_C = 0$ at $t = 0$

Take the Laplace Transform or each of the terms

$$\frac{V}{s} = RCsV_C(s) + V_C(s) \tag{6.66}$$

Rewrite equation (6.66) with $V_C(s)$ the subject of the formula

$$\begin{aligned} V_C(s)[1+RCs] &= \frac{V}{s} \\ V_C(s) &= \frac{V}{s(1+RCs)} \end{aligned} \tag{6.67}$$

Manipulate so that equation (6.67) conforms to one of the standard forms found in tables for which there is a solution. In this case, the obvious choice is

$$\frac{a}{s(s+a)}$$

for which the time domain solution is $1 - e^{-at}$

$$\begin{aligned} V_C(s) &= \frac{V/RC}{s(1/RC+s)} \\ &= V\frac{1/RC}{s(s+1/RC)} \end{aligned}$$

Therefore, $a = 1/RC$
Taking the inverse Laplace Transform

$$V_C(t) = V(1 - e^{-t/RC}) \tag{6.68}$$

In this case, *1/RC* is the time constant for the time domain response.

6.6.1 *Partial fraction expansion*

To simplify more complex algebraic functions into a form that can be solved using tables, it is necessary to use the partial fraction expansion method. To enable this, the denominator must be factorisable and the power of the highest term in the numerator is at least one lower than that of the denominator.

Convert the expression $F(s) = \dfrac{s+3}{s^2+3s+2}$ so that it can be written as a partial fraction expansion.

The denominator can be factorised into two roots $(s+1)(s+2)$, so the expression can be rewritten in the following form

$$\begin{aligned} F(s) &= \frac{A}{s+1} + \frac{B}{s+2} \\ &= \frac{A(s+2) + B(s+1)}{(s+1)(s+2)} \end{aligned}$$

For the two expressions to be equal requires that the numerators must be equal

$$A(s+2) + B(s+1) = s+3$$

This can be solved using simultaneous equations, by selecting the value for s in a pair of equations in which $A = 0$ in one and $B = 0$ in the other

$$\begin{aligned} s &= -2 \\ B(-2+1) &= -2+3 \\ -B &= 1 \\ B &= -1 \\ s &= -1 \\ A(-1+2) &= -1+3 \\ A &= 2 \end{aligned}$$

Therefore

$$F(s) = \frac{2}{s+1} + \frac{-1}{s+2}$$

6.6.2 *Analysing complex models*

The ability to manipulate time domain equations in the s-domain makes it practical to generate and analyse complex models of biomechatronic devices.

In a typical system with two feed-forward blocks $G_1(s)$ and $G_2(s)$, and one feedback block $H_1(s)$ shown in Figure 6.15, the feed-forward blocks can be combined

$$G_3(s) = G_1(s)G_2(s) \tag{6.69}$$

and secondly, the inclusion of the feedback element can be achieved using the relationship described in equation (6.5) for negative feedback

$$T(s) = \frac{G_3(s)}{1 + G_3(s)H_1(s)} \tag{6.70}$$

If the feedback is positive, then the sign changes

$$T(s) = \frac{G_3(s)}{1 - G_3(s)H_1(s)} \tag{6.71}$$

If the controller has multiple inputs, it is best to deal with them individually by setting all the other inputs to zero and performing the analysis for each in turn. Because the systems are LTI, the output is obtained by producing the algebraic sum of all the results. This method is often used to determine the effects of disturbance inputs.

Consider the previous example with the addition of a disturbance between the two feed-forward blocks $G_1(s)$ and $G_2(s)$ as shown in Figure 6.16.

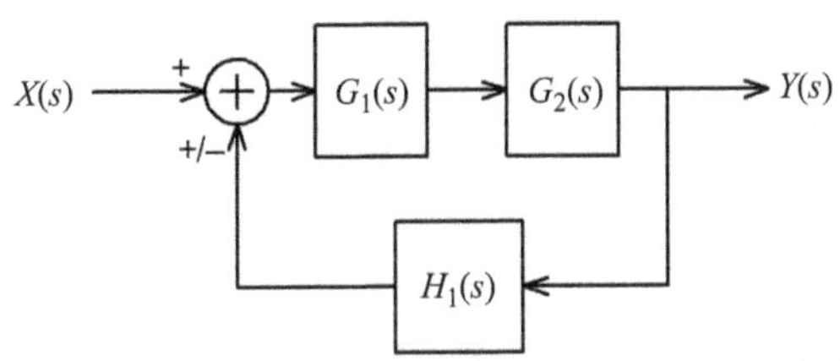

Figure 6.15 System block diagram in the s-domain

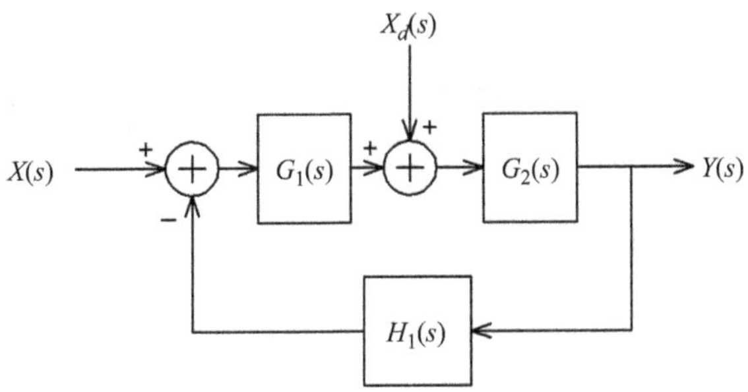

Figure 6.16 System block diagram with a disturbance

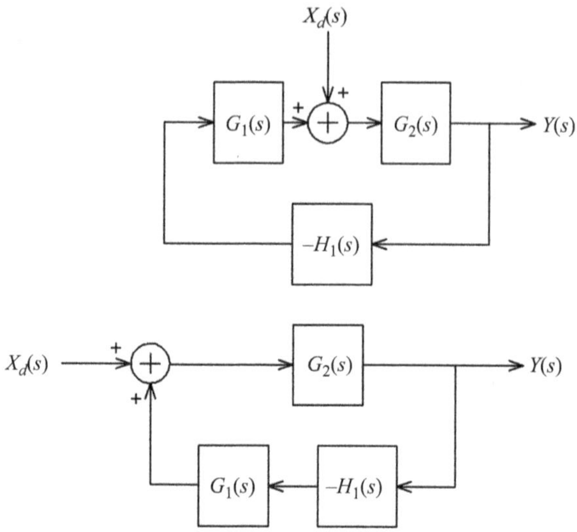

Figure 6.17 System block diagram for the disturbance input

Start by setting $X_d(s) = 0$ to obtain the system transfer function

$$\frac{Y(s)}{X(s)} = \frac{G_1(s)G_2(s)}{1 - G_1(s)G_2(s)H_1(s)} \tag{6.72}$$

Now rearrange the block diagram with $X(s) = 0$, so that $X_d(s)$ is the input as shown in Figure 6.17.

By inspection, the new transfer function can be written as

$$\frac{Y(s)}{X_d(s)} = \frac{G_2(s)}{1 - G_1(s)G_2(s)H_1(s)} \tag{6.73}$$

Combining equations (6.72) and (6.73) to produce the output of the system in terms of the two inputs

$$Y(s) = \frac{G_1(s)G_2(s)X(s) + G_2(s)X_d(s)}{1 - G_1(s)G_2(s)H_1(s)} \tag{6.74}$$

This equation can then be used to determine the relative effect of the disturbance, or interference signal, compared to the effect of the input signal.

6.7 System Stability

A system is considered to be stable if every bounded input produces a bounded output. An alternative definition is that a system is stable if the impulse response dies away to zero.

A transfer function $G(s)$ can be written as the ratio of two polynomials that can be factorised into their respective roots. Where the roots can be real or can occur in complex pairs.

$$G(s) = \frac{K(s + z_1)(s + z_2)....(s + z_m)}{(s + p_1)(s + p_2)....(s + p_n)} \tag{6.75}$$

The roots of the numerator $z_1, z_2, \ldots z_m$, are called zeros, and the roots of the denominator $p_1, p_2, \ldots p_n$, are called poles.

The zeros are the values of s for which the transfer function becomes zero, while the poles are the values of s for which the transfer function becomes infinite.

Consider a simple transfer function

$$G(s) = \frac{s - 1}{s^2 + 4s + 5} \tag{6.76}$$

The denominator can be factorised to produce a complex pole pair and plotted on the s-plane in Figure 6.18.

$$G(s) = \frac{s - 1}{(s - [1 + j2])(s - [1 - j2])} \tag{6.77}$$

The stability can be determined by considering how the output changes with time after an impulse. For example, consider a first-order system with a pole at -3,

$$G(s) = \frac{1}{s + 3}$$

The output and input are related by

$$Y(s) = X(s)G(s)$$

For a unit impulse, $X(s) = 1$, therefore

$$Y(s) = G(s) = \frac{1}{s + 3}$$

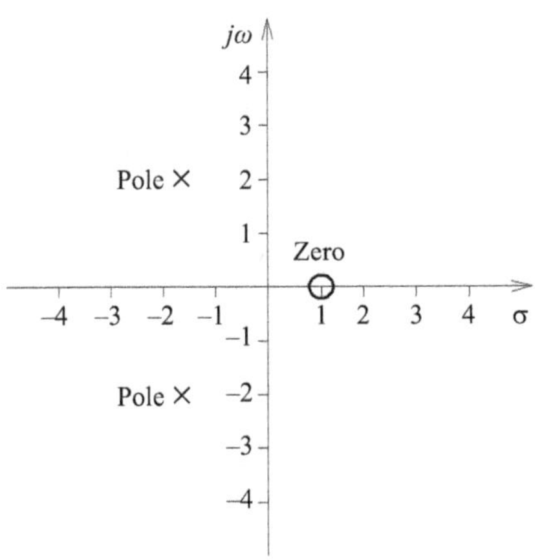

Figure 6.18 S-plane plot showing the poles and zeros

Taking the inverse Laplace Transform gives

$$y(t) = e^{-3t}$$

which decreases to zero as time increases.

Now consider a similar system with the pole at +3

$$G(s) = \frac{1}{s - 3}$$

Repeating the previous analysis results in the following response

$$y(t) = e^{+3t}$$

This increases with time; therefore, the system is unstable.

In general terms, the output of a system excited by an impulse will be the sum of a number of exponential terms. If any of those exponentials increases with time, the system is unstable.

In terms of the poles, this translates into the following, as illustrated in Figure 6.19: If all of the poles are in the left-hand half of the plane (shown shaded), then the system is stable. If any lie on the y-axis, the system is said to be critically stable, and if any are in the right-hand half of the plane, the system will be unstable.

If it is difficult to factorise the denominator, then it is possible to use the Routh–Hurwitz criteria to determine the system's stability (Raven 1978). For the denominator written as a polynomial

$$a_n s^n + a_{n-1} s^{n-1} + \dots + a_1 s + a_0 \tag{6.78}$$

- If any of the coefficients are negative, then the system will be unstable.
- If any coefficient is missing then the system is, at best, critically stable.

Other criteria for stability exist, and it is suggested that the reader investigate these further if they are of concern.

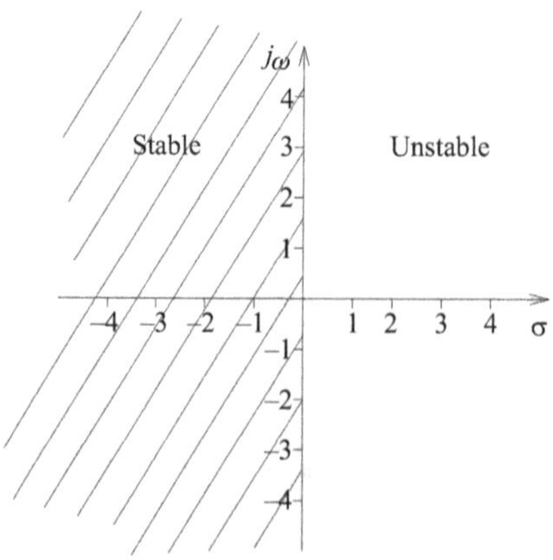

Figure 6.19 S-plane showing the criterion for system stability

As illustrated in Figure 6.20, the actual time-domain response that occurs for an impulse is a function of the position of the poles, and whether they are real or complex.

6.7.1 Root locus

The stability of a system, based on the positions of the poles, as the controller gains are changed, is known as the root locus method. It is one way to design controllers and to determine the limits in the gains that can be applied by a controller before instability results.

The simplest controller that can be implemented is one that includes a gain K in the forward path, as shown in Figure 6.21.

The system transfer function is

$$\begin{aligned} T(s) &= \frac{G(s)}{1+G(s)} \\ &= \frac{K/(s+3)}{1+K/(s+3)} \\ &= \frac{K}{s+3+K} \end{aligned}$$

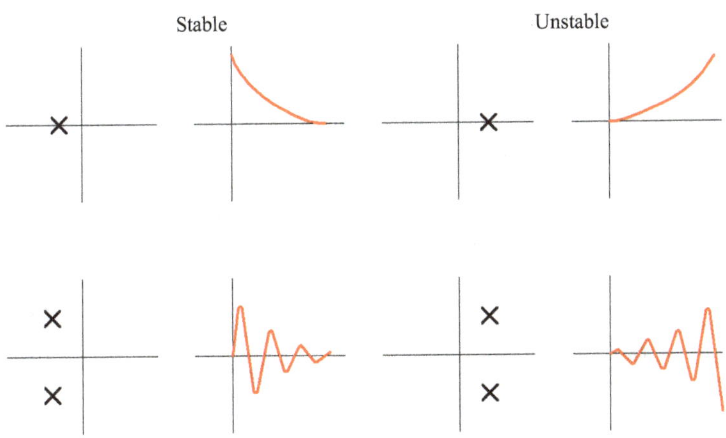

Figure 6.20 Relationship between pole position and system response

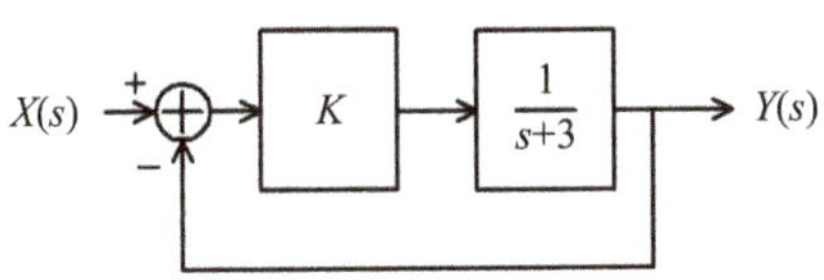

Figure 6.21 Proportional controller for a first-order system with unity feedback

The characteristic equation is $s + 3 + K = 0$, so the open-loop pole occurs when $K = 0$, and this will be at $s = -3$. However, when $K > 0$, then the pole will occur at $s = -(3 + K)$, so as K increases, the pole will become more negative, as shown in Figure 6.22.

The system response can be solved for an impulse or a step response, and it will be found that the speed of response increases as K increases.

In a second-order system, the response is completely different. Consider the proportional controller shown in Figure 6.23, also with gain K.

In this case, the system transfer function is

$$
\begin{aligned}
T(s) &= \frac{K/s(s+3)}{1 + K/s(s+3)} \\
&= \frac{K}{s^2 + 3s + K}
\end{aligned}
$$

The characteristic equation is now a quadratic, $s^2 + 3s + K = 0$, and the roots are

$$
\begin{aligned}
s &= \frac{-b \pm \sqrt{b^2 - 4ac}}{2a} \\
&= -\frac{3}{2} \pm \frac{1}{2}\sqrt{9 - 4K}
\end{aligned}
$$

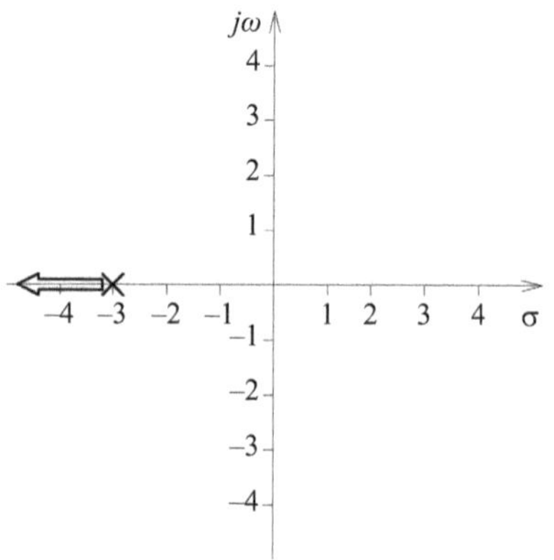

Figure 6.22 Root locus plot for a simple controller

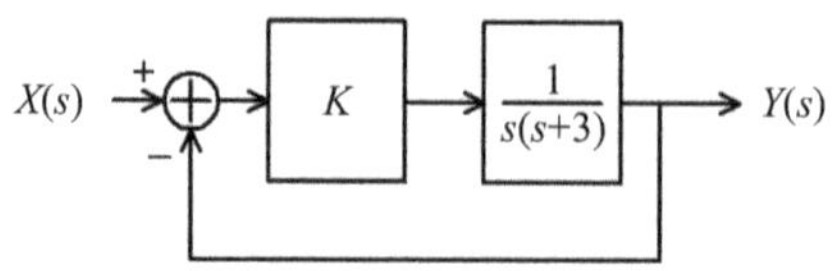

Figure 6.23 Proportional controller for a second-order system with unity feedback

When $K = 0$, the roots are at

$$s = -\frac{3}{2} \pm \frac{3}{2}$$
$$= 0 \text{ or } 3$$

This is as expected.

As K increases from 0 to 9/4, the root at 0 becomes more negative, moving towards $-3/2$, while the root at -3 becomes more positive, also moving towards $-3/2$.

As K continues to increase from 9/4, the roots become complex and split. By the time $K = 3$, the roots are

$$s = -\frac{3}{2} - j\sqrt{\frac{3}{4}}$$

and

$$s = -\frac{3}{2} + j\sqrt{\frac{3}{4}}$$

The root locus plot can easily be plotted as shown in Figure 6.24.

In this case, the response moves from being over-damped, through critical damping to being under-damped for a step input, as shown in Figure 6.25.

As systems become more complicated, it becomes less convenient to plot root loci manually. Fortunately, MATLAB® includes scripts that automate the process, and the previous example is shown below.

Note that the open-loop transfer function is represented by the coefficients of the polynomials describing the numerator and the denominator. The root locus command assumes that the system includes unity gain feedback with a proportional controller gain implicit in the feed-forward path. The root locus plot is shown in Figure 6.26.

```
%rl01.m
% simple root locus
% G = 1/s(s+3)
num = [1];
den = [1 3 0];
rlocus(num,den);
```

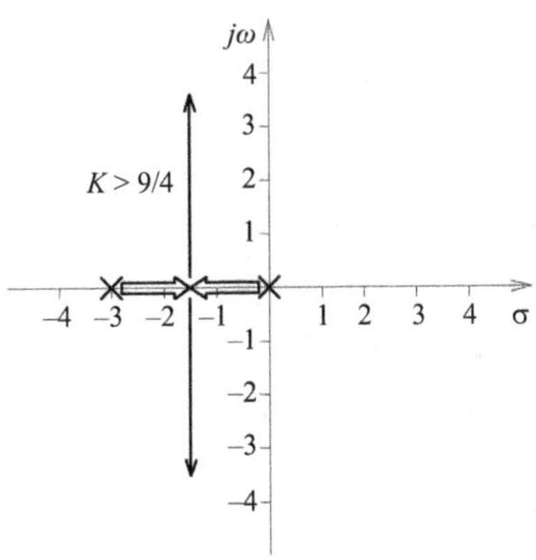

Figure 6.24 Root locus plot for the quadratic system with a proportional controller

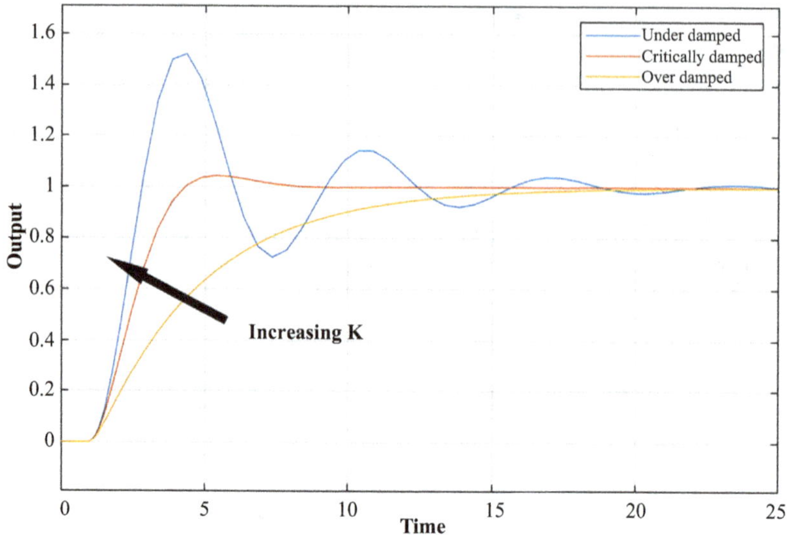

Figure 6.25 Time domain step response for a quadratic system with a proportional controller

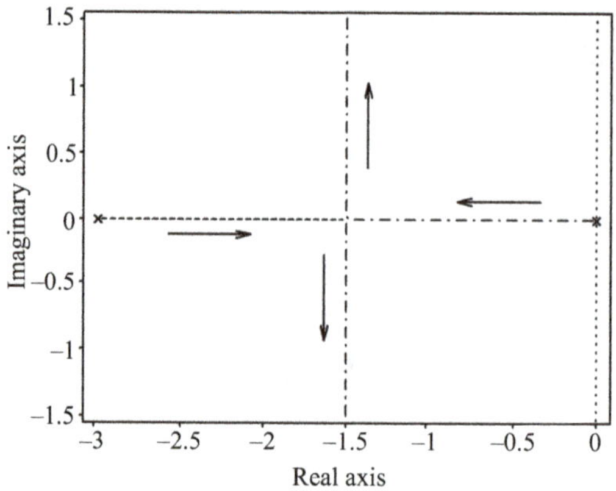

Figure 6.26 MATLAB version of root locus plot for quadratic system with a proportional controller

Consider a more complex function with an open-loop transfer function

$$G_o(s) = \frac{K}{(s+1)(s+2)(s+3)}$$

The block diagram is shown in Figure 6.27.

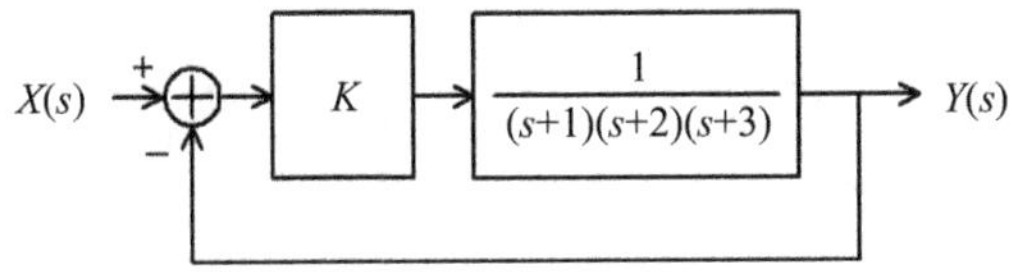

Figure 6.27 Proportional controller for a third-order system with unity gain feedback

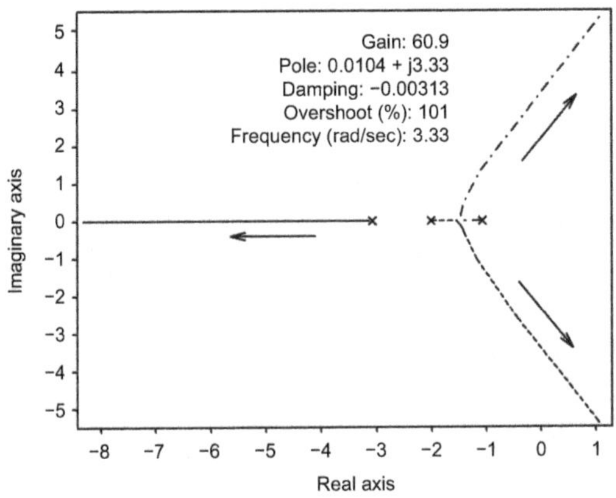

Figure 6.28 MATLAB version of root locus plot for third-order system with proportional controller

The numerator and denominator are expanded into their polynomial forms for representation by MATLAB, with the root locus plot shown in Figure 6.28.

$$G_o(s) = \frac{K}{s^3 + 6s^2 + 11s + 6}$$

```
%r102.m
% simple root locus
% G = 1/(s+1)(s+2)(s+3)
num = [1];
den = [1 6 11 6];
rlocus(num,den);
```

In this case, it can be seen that two of the roots become complex and then move towards the right where, for $K > 60$, they cross into the right-hand half plane and the system will become unstable.

6.7.2 Steady-state error

The steady-state error is the difference between the commanded output and the actual output

$$E(s) = X(s) - Y(s) \tag{6.79}$$

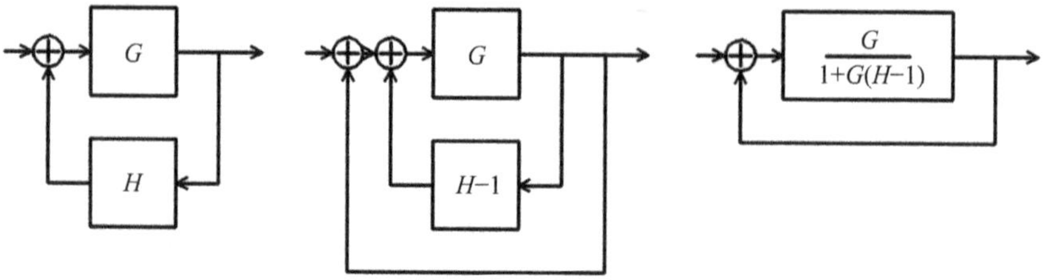

Figure 6.29 Conversion to unity gain feedback

The error depends not only on the system but also on the form of the input $X(s)$. To determine the magnitude of this error, it is necessary to convert a system with feedback to one with unity gain feedback as shown in Figure 6.29.

The forward path transfer function for the unity gain feedback system

$$G_o(s) = \frac{1}{1 + G(H + 1)} \tag{6.80}$$

is then used in the following limit

$$e_{ss} = \frac{\lim}{s \to 0}\left[s\frac{1}{1 + G_o(s)}X(s)\right] \tag{6.81}$$

These limits are considered further in the section on system control.

6.8 Controllers

6.8.1 Proportional controller

The controllers discussed in the previous sections comprise a simple gain stage that results in the output being directly proportional to the input, with K_a being the proportional gain. In reality, these controllers are only proportional over a limited band until the output saturates, as illustrated in Figure 6.30.

In general terms, for a unity gain feedback system and a plant with a transfer function $G_P(s)$, shown in Figure 6.31.

$$T(s) = \frac{K_p G_p(s)}{1 + K_p G_p(s)} \tag{6.82}$$

There are a number of considerations that determine how well the controller works. The first is the ultimate stability of the system, the second is the steady state error for a specific input and the third is the response time until the error is within specific bounds. One other consideration that can be applied is the ability of the controller to reject disturbances within the loop. Methods to determine all of these have been discussed earlier in this chapter.

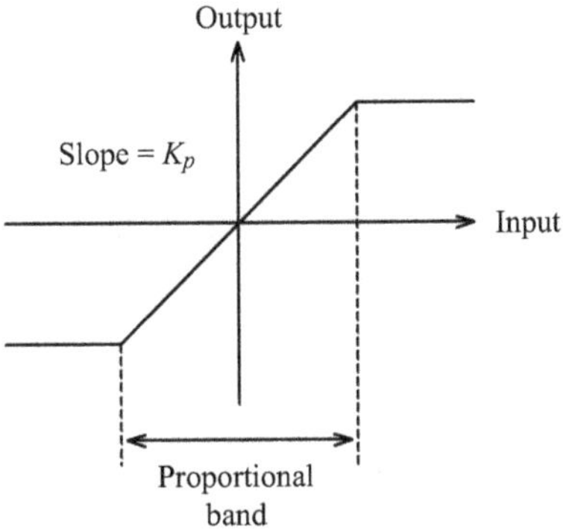

Figure 6.30 Proportional controller with saturation

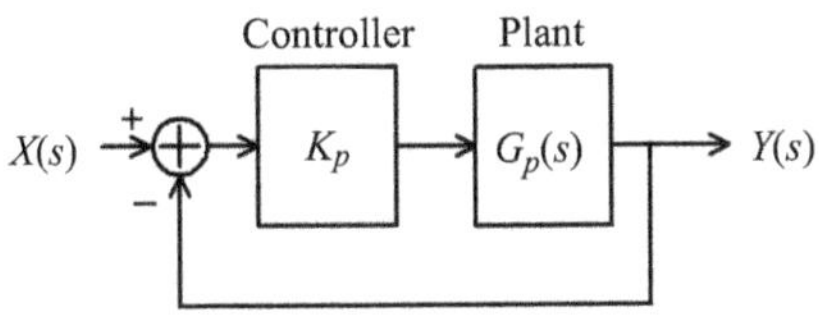

Figure 6.31 Classic configuration of a proportional controller

---

Example: Motor speed and position control
Consider the DC motor model developed earlier, and described by equations (6.43) and (6.44), reproduced here

$$(J_r + J_L)\frac{d\omega(t)}{dt} + b\omega(t) = K_m i_r(t) \tag{6.43}$$

$$L\frac{di_r(t)}{dt} + Ri_r(t) = V_i(t) - K_e\omega(t) \tag{6.44}$$

To develop a controller, it is convenient to convert these to their Laplace Transform form, with the moments of inertia lumped together so $J = J_r + J_L$

$$Js\Omega(s) + b\Omega(s) = K_m I_r(s) \tag{6.83}$$

$$LsI_r(s) + RI_r(s) + K_e\Omega(s) = V_i(s) \tag{6.84}$$

From equation (6.83), the current can be written as

$$I_r(s) = \frac{\Omega(s)}{K_m}(Js + b) \tag{6.85}$$

Substituting (6.85) into (6.84)

$$\frac{\Omega(s)}{K_m}(Js + b)(Ls + R) + K_e\Omega(s) = V_i(s)$$

Simplifying

$$\frac{\Omega(s)}{K_m}[(Js+b)(Ls+R)+K_eK_m]=V_i(s)$$

The transfer function that relates the input voltage to the output speed can be written as

$$T(s)=\frac{\Omega(s)}{V_i(s)}=\frac{K_m}{(Js+b)(Ls+R)+K_eK_m} \tag{6.86}$$

An alternative is to consider the two governing equations separately, with the back emf applied as feedback.

The current drawn by the motor can be obtained by rewriting equation (6.84)

$$I_r(s)(Ls+R)=V_i(s)-K_e\Omega(s)$$

$$I_r(s)=\frac{1}{Ls+R}(V_i(s)-K_e\Omega(s)) \tag{6.87}$$

The equation for the mechanical portion can be obtained by rewriting equation (6.85) in terms of the input current

$$\Omega(s)(Js+b)=K_mI_r(s)$$

$$\Omega(s)=\frac{K_m}{Js+b}I_r(s) \tag{6.88}$$

A block diagram can be constructed, which is equivalent to that shown in Figure 6.32, but which includes the two individual transfer functions as well as the back emf feedback. This is shown in Figure 6.33.

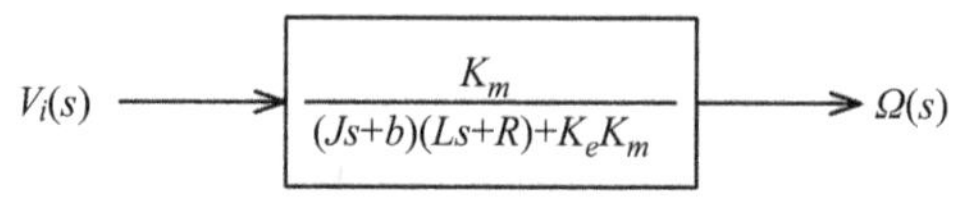

Figure 6.32 DC motor block diagram

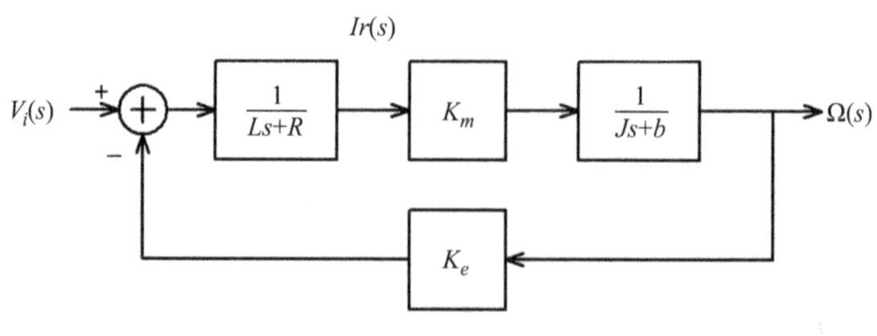

Figure 6.33 Alternative form for DC motor block diagram

The advantage of using this form to describe the open-loop motor model is that each of the first-order systems can be rewritten in a form that is easy to interpret in the time domain. In addition, it gives access to the torque node (just after the K_m block) which then allows for the introduction of an external torque requirement or disturbance.

From the table of Laplace Transforms, it would be convenient to write the first-order blocks in the form

$$F(s) = \frac{1}{\tau s + 1}$$

which has the tabled solution $f(t) = e^{-t/\tau}$. Therefore, the electrical model of the motor can be modified to conform as follows:

$$\frac{1}{Ls + R} = \frac{1/R}{\frac{L}{R}s + 1} = \frac{1/R}{\tau_1 + 1}$$

so $\tau_1 = L/R$ is the electrical time constant of the motor.

Similarly, $\tau_2 = J/b$ is the mechanical time constant for the motor and load.

The block diagram can be redrawn with the time constants for the electrical and mechanical portions of the motor as shown in Figure 6.34.

Consider the Maxon motor selected to drive the turbine of a CPAP air pump discussed in Chapter 3. It has specifications listed in Table 6.7.

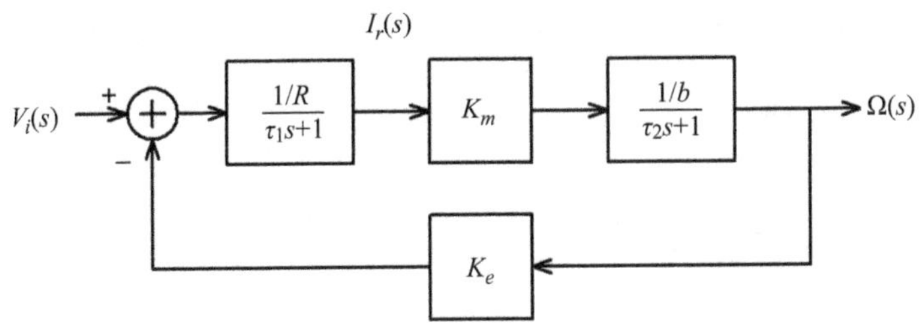

Figure 6.34 DC motor block diagram in time-constant form

Table 6.7 Maxon RE 16 motor specifications

No load speed @ 24V n_o (rpm)	7250
No load current I_o (mA)	3.11
Terminal resistance R (Ω)	42.8
Terminal inductance L (mH)	1.75
Torque constant K_m (mNm/A)	31.4
Speed constant K_e (rpm/V)	304
Output power P_o (W)	3.2
Mechanical time constant τ_2 (ms)	5.28
Rotor inertia J_r (gcm^2)	1.22

Several factors must be determined or scaled so that the model shown in Figure 6.34 can be used. Firstly, the electrical time constant $\tau_1 = L/R = 40.9$ μs. The mechanical time constant is given as $\tau_2 = 5.28$ ms; therefore, the damping, b, is

$$\begin{aligned} b &= \frac{J_r}{\tau_2} \\ &= \frac{1.22 \times 10^{-7}}{5.28 \times 10^{-3}} \\ &= 2.31 \times 10^{-5} \end{aligned}$$

Note that J_r has been converted from $g \cdot cm^2$ to $kg \cdot m^2$ for this calculation.

K_m is converted to 31.4×10^{-3} Nm/A, and K_e is converted to 2903 rad/s per V. It is also important to remember that in the feedback path, the conversion is from speed to voltage; therefore, the reciprocal of the constant is used in the model.

The Simulink® model for the motor is shown in Figure 6.35. In this case, the step input is the motor drive voltage.

The step response shown in Figure 6.36 is determined by a combination of the electrical and mechanical time constants of the motor. The mechanical time constant is by far the slowest and therefore has the dominant effect on the settling time. The open-loop speed is 754 rad/s (7200 rpm) which is close to the expected no-load speed of 7250 rpm, as expected.

The nominal load for the CPAP turbine is 3.5 mNm at 5000 rpm. At a first approximation, the flow rate is directly proportional to the turbine speed and, therefore, the load torque will be directly proportional to the rotation speed.

$$T_L(s) = K_K\Omega(s) \tag{6.89}$$

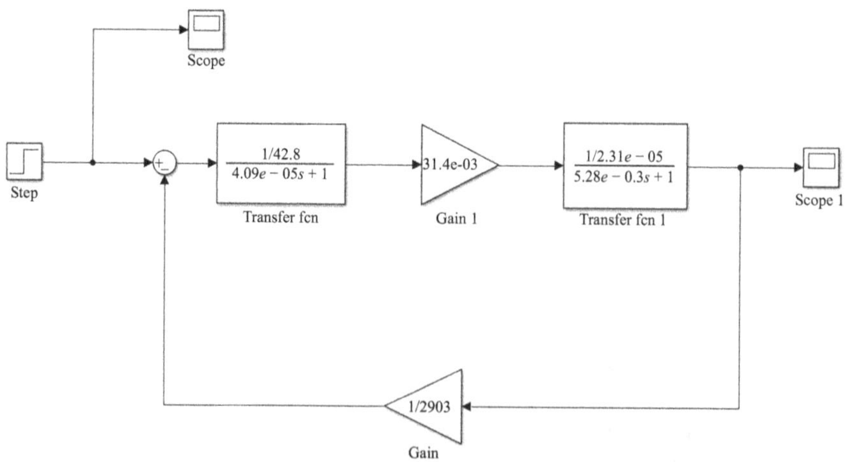

Figure 6.35 Simulink model of the Maxon RE 16 motor

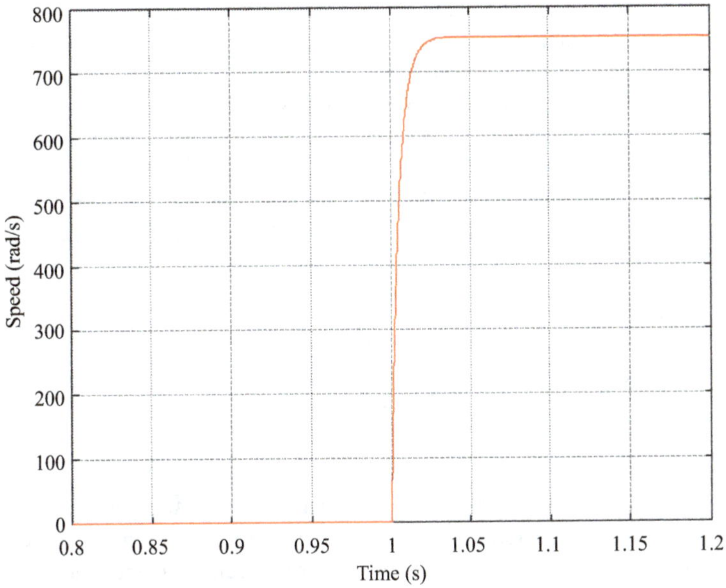

Figure 6.36 Open-loop step response of the motor for a 24 V step

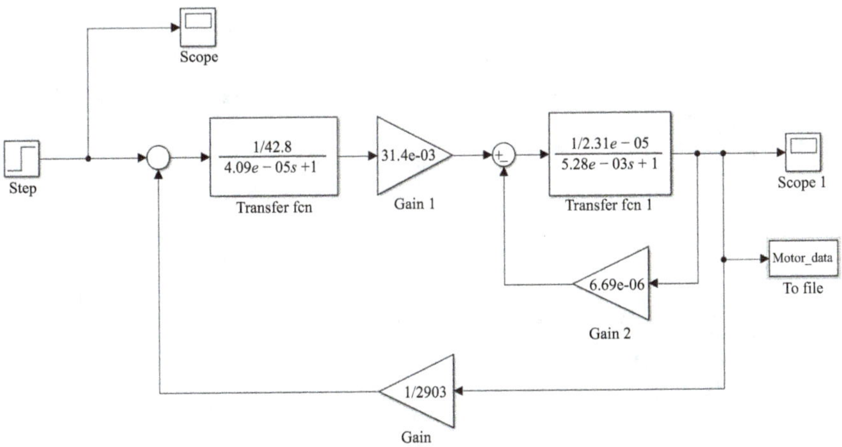

Figure 6.37 Simulink model of the Maxon RE 16 motor driving a load

The constant of proportionality K_L (Nm per rad/s) is

$$K_L = \frac{3.5 \times 10^{-3}}{5000 \times 2\pi/60} = 6.69 \times 10^{-5}$$

This load torque is introduced into the loop as shown in Figure 6.37.

The step response (Figure 6.38) in this case exhibits a similar time constant, but because of the linear relationship between the torque and speed discussed in Chapter 3, the speed is reduced proportionally.

Obviously, the inertia of the turbine (load) must also be accommodated in the model, and this is achieved by adding J_L to the rotor inertia. It may also alter the mechanical time constant of the system.

Assuming that the turbine inertia $J_L = 100$ gcm^2 (about one hundred times that of the motor) and that the damping, b, remains unchanged, then the new mechanical time constant is

$$\begin{aligned}\tau_2 &= \frac{(J_r + J_L)}{b} \\ &= \frac{(1.22 \times 10^{-7} + 100 \times 10^{-7})}{2.31 \times 10^{-5}} \\ &= 0.44 \text{ s}\end{aligned}$$

The result of these changes is that the final speed remains the same as before, but the motor takes much longer to reach steady state, as shown in Figure 6.39.

In Chapter 3, it was shown that motor speed decreases with increasing load torque. Therefore, if a speed controller is required for a motor, an additional loop and a controller must be implemented. A proportional controller to perform this function is shown in Figure 6.40. In this case, the input command is an angular rate.

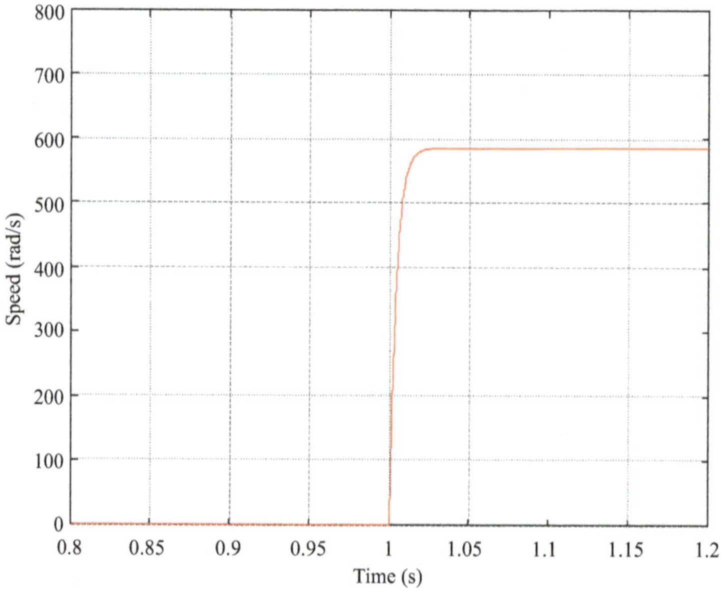

Figure 6.38 Open-loop step response for the motor driving a load requiring some torque

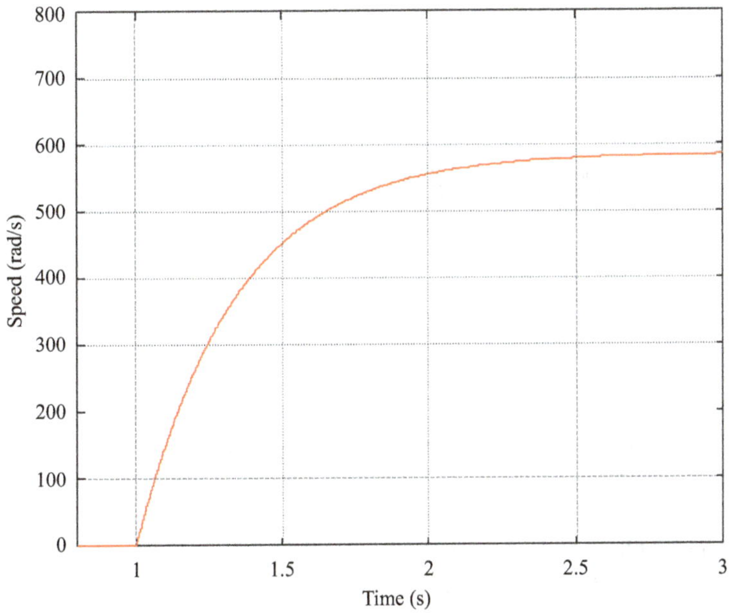

Figure 6.39 Open-loop step response for the motor driving a load with 100 times the inertia

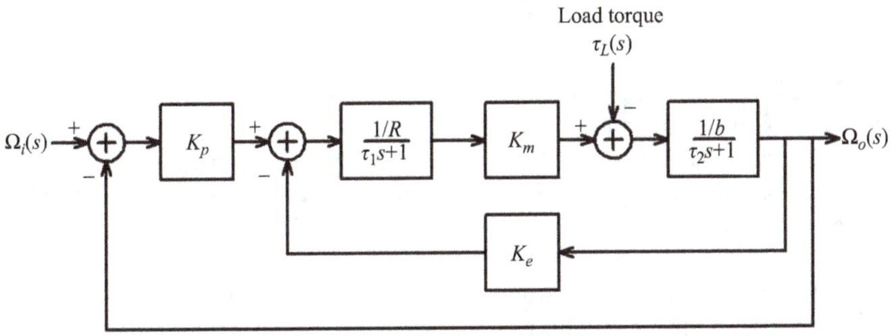

Figure 6.40 Proportional controller of motor speed

It is important to remember that the output of the controller will be limited to the maximum allowable voltage for the motor, so the complete Simulink model is shown in Figure 6.41 where a saturation block limits the maximum voltage to +/−24 V.

The step response to a commanded angular rate of 600 rad/s in Figure 6.42(a) shows a residual speed error of 15 rad/s because the commanded speed is beyond the capability of the motor, given the allowable voltage (24 V) and the load torque. However, if the commanded speed is reduced to 500 rad/s, the controller starts by

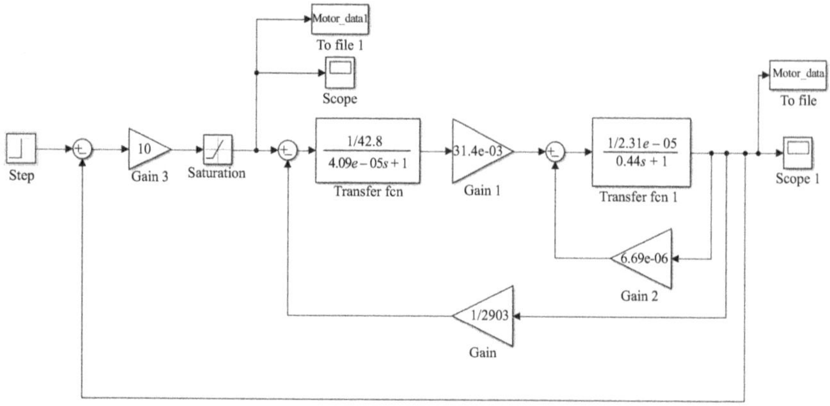

Figure 6.41 Simulink model of a proportional speed controller for a motor driving a load

providing the full 24 V drive until the required speed is reached, when the voltage drops to about 20 V, as seen in Figure 6.42(b).

In many biomechatronic applications, it is not the motor speed that needs to be controlled, but the angle. This requires that an additional block be incorporated into the output of the system to integrate the angular rate and give the angular position.

$$\theta(s) = \frac{1}{s}\Omega(s) \tag{6.90}$$

Therefore, the closed-loop position controller is as shown in Figure 6.43 with the Simulink model shown in Figure 6.44.

In this case, the proportional controller is highly under-damped as can be seen in Figure 6.45. Increasing the gain doesn't improve the response time at all, but it does increase the natural frequency of the system with the result that the oscillation frequency increases.

In general, a gearbox is incorporated at the output of the motor to better match the motor torque to the load. This can be represented by a gain that is representative of the gear ratio.

6.8.1.1 Disadvantages of proportional controllers

The main disadvantage of this form of controller is that the transfer function of the plant with the controller is still the same order, so if the plant exhibits a steady-state error, the controller does not eliminate it completely.

6.8.2 Integral controller

With integral control, the output of the controller is equal to the integral of the error. This is shown in a unity gain configuration in Figure 6.46.

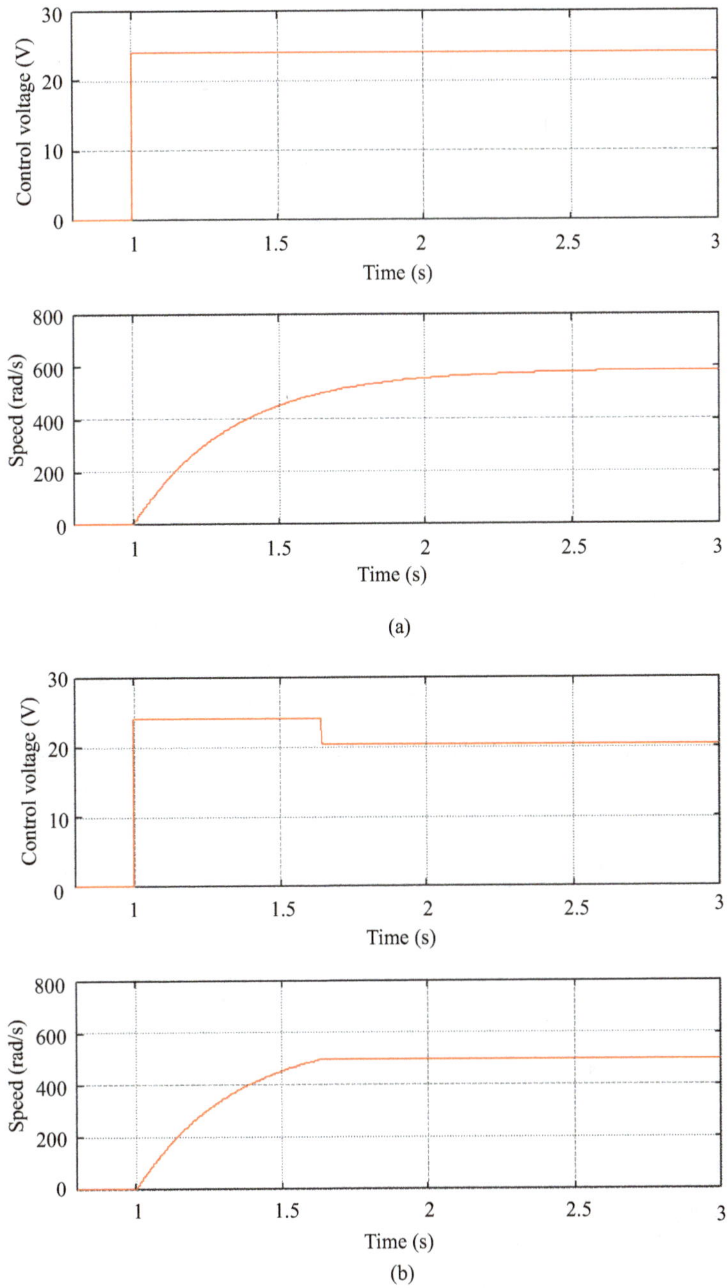

Figure 6.42 Response to a commanded speed of 600 rad/s and 500 rad/s

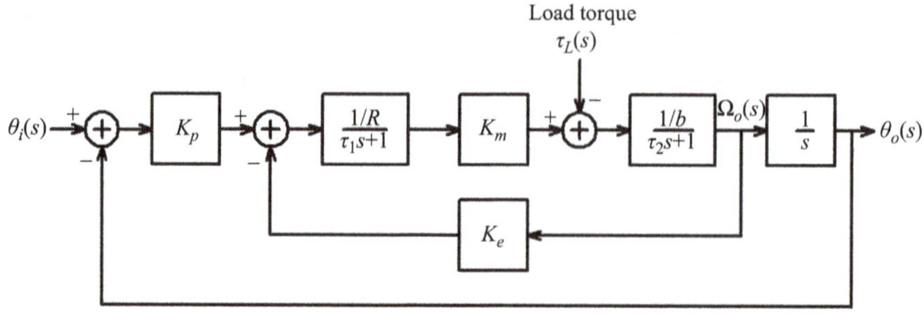

Figure 6.43 Proportional controller for motor angular position

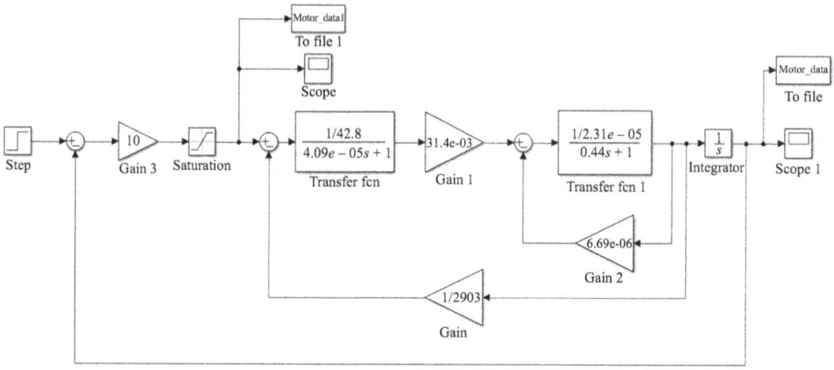

Figure 6.44 Simulink model of a proportional position controller for a motor driving a load

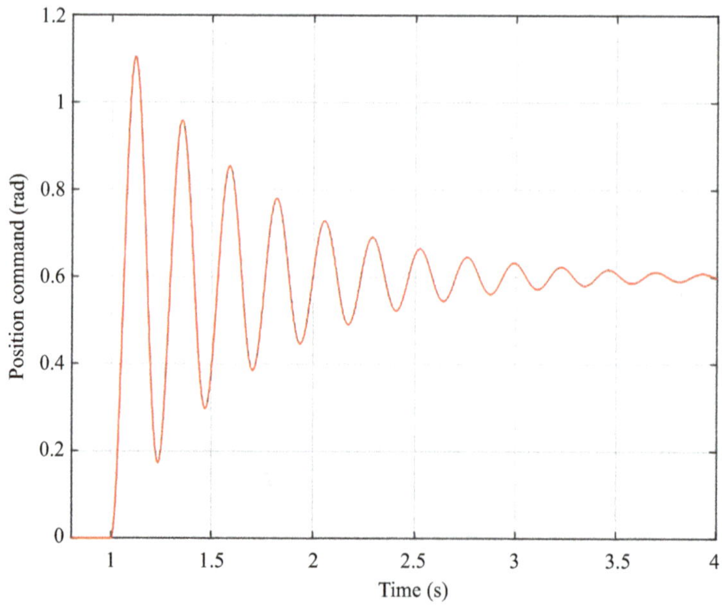

Figure 6.45 Response to a commanded position of 0.6 rad

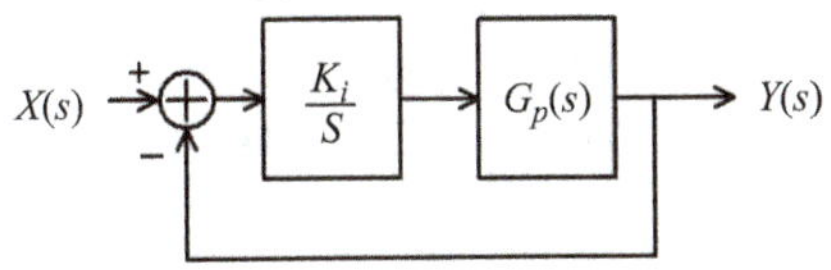

Figure 6.46 Classic configuration of an integral controller

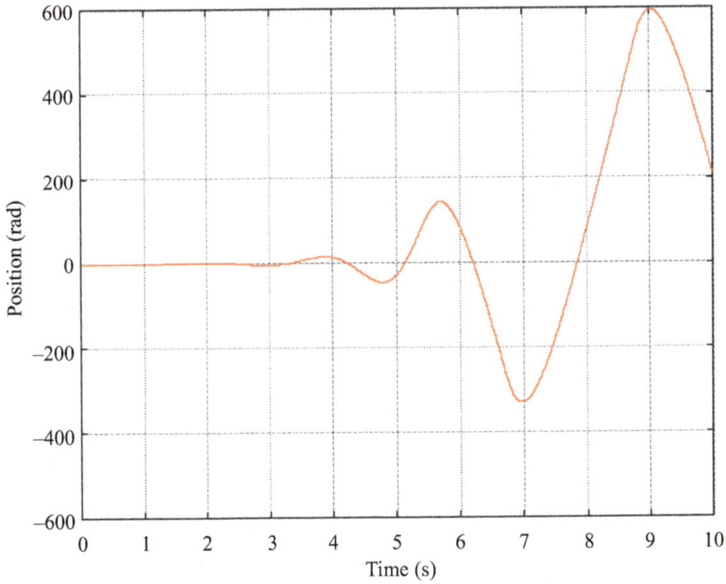

Figure 6.47 Response to a step input for an integral controller of the motor position

The forward path transfer function for this control system is

$$G_o(s) = \frac{K_i}{s} G_p(s) \tag{6.91}$$

It can be seen that the controller introduces an additional s into the denominator, with the result that the order of the system increases. A system that exhibited a steady state error for a step input will now converge with no error. The main disadvantage of the system is that it introduces a pole at the origin ($s - 0$), and that can result in a less stable system.

If an integral controller is used to control the motor position, the system becomes unstable for any gain, as is shown in the step response in Figure 6.47.

6.8.3 Proportional plus integral controller

The reduction of the relative controllability of a plant as a result of using integral control can sometimes be overcome by introducing a proportional term to make the control function

$$G_c(s) = K_p + \frac{K_i}{s} \tag{6.92}$$

Implementation of this controller results in a system block diagram, with unity gain feedback, as shown in Figure 6.48.

The forward path transfer function for this control system is

$$G_o(s) = \left[K_p + \frac{K_i}{s}\right] G_p(s) \tag{6.93}$$

This can be rewritten as follows

$$\begin{aligned} G_o(s) &= \left[K_p + \frac{K_i}{s}\right] G_p(s) \\ &= \frac{sK_p + Ki}{s} G_p(s) \\ &= \frac{K_p(s + K_i/K_p)}{s} G_p(s) \\ &= \frac{K_p(s + 1/\tau_i)}{s} G_p(s) \end{aligned}$$

where $K_p/K_i = \tau_i$ is called the integral time constant.

It is obvious that a zero at $-1/\tau_i$ and a pole at the origin have been added to the system. This maintains the same difference in the number of poles and zeros as the plant had, and it increases the system order. This serves to maintain the improvement in steady state error as well as improve the relative stability somewhat.

In terms of motor control, the proportional plus integral controller returns the system to stability, but it is still highly under-damped.

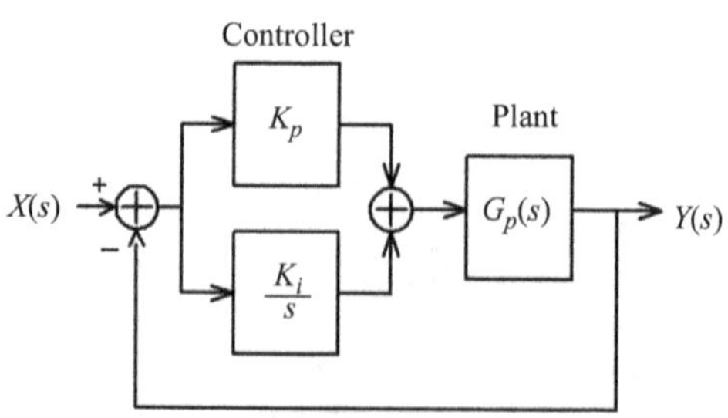

Figure 6.48 Classic configuration of a proportional plus integral controller

6.8.4 Proportional, integral, differential (PID) controller

These are also known as three terminal controllers or just PID controllers. The controller has the transfer function

$$G_c(s) = K_p + \frac{K_i}{s} + K_d s \tag{6.94}$$

The system block diagram is shown in Figure 6.49.

As with the PI controller, equation (6.94) can be rewritten in terms of the integral time constant $\tau_i = K_p/K_i$ and a derivative time constant $\tau_d = K_d/K_p$

$$G_c(s) = K_p\left(1 + \frac{1}{\tau_i s} + \tau_d s\right) \tag{6.95}$$

In this case, the forward transfer function is

$$G_o(s) = \frac{K_p\left(\tau_i s + 1 + \tau_i \tau_p s^2\right)}{\tau_i s} G_p(s) \tag{6.96}$$

It can be seen that this has increased the number of zeros by two and the number of poles by one as well as increasing the overall system order by one.

In the motor control example, a PID controller block replaces the proportional gain block as shown in Figure 6.50.

In this case, the proportional gain $K_p = 10$, the integral gain $K_i = 1$, and the derivative gain K_d is set to a range of values. The step response for the PID controlled position controller is shown in Figure 6.51.

It can be seen that the response is very similar to that for a proportional controller when $K_d = 0$. However, as the derivative gain increases, damping is increased and the ringing decreases. The critically damped response occurs for $K_d = 0.75$, and above that, the response becomes over-damped.

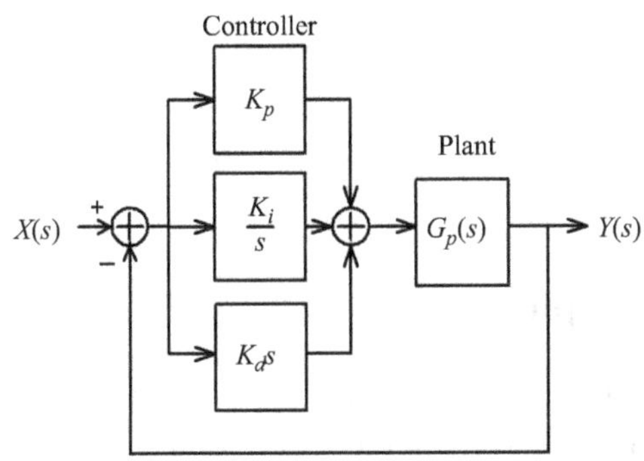

Figure 6.49 Classic configuration of a PID controller

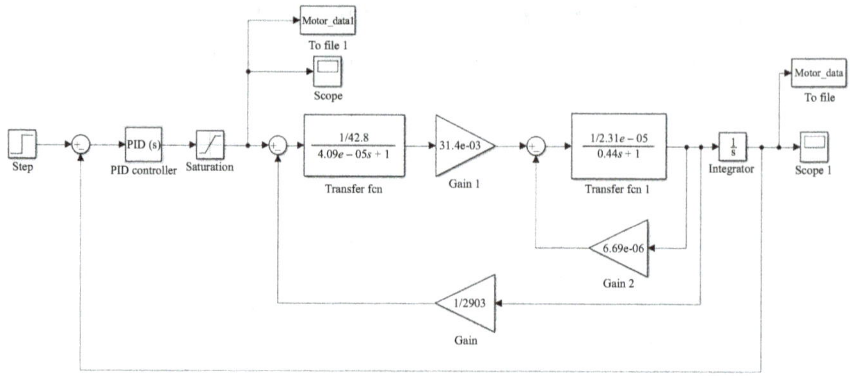

Figure 6.50 Simulink model of a PID position controller for a motor driving a load

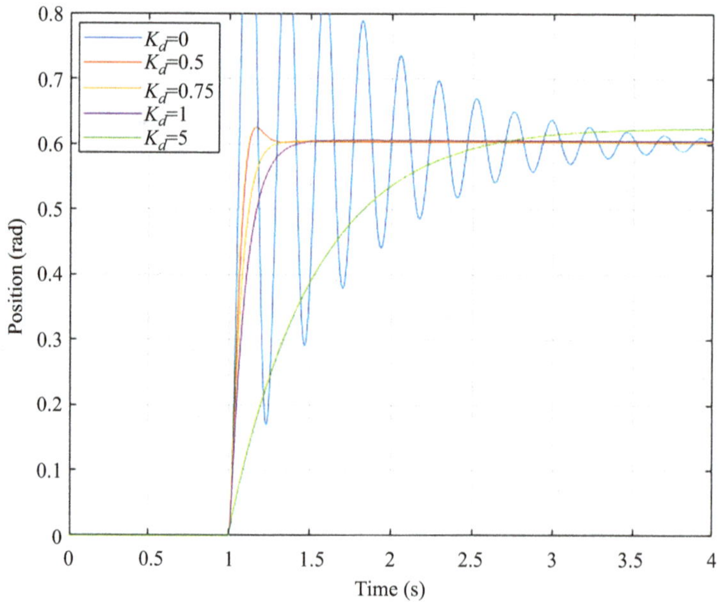

Figure 6.51 Response to a step input for a PID controller of the motor position

6.9 Controller implementation

6.9.1 Selection of controller gains

Selection of the gains for the controllers can be done by trial and error, but a number of techniques have been developed to help with the process. One of these is called the process reaction curve method (Bolton 1992), and it is performed as follows:

- open the control loop so that no control action occurs (break between the controller and the correction unit),
- apply a small test signal step P, expressed as a percentage change in the correction,
- measure the error signal as a function of time, and
- plot as shown in Figure 6.52.
- draw a tangent to the maximum gradient of the error signal curve. The gradient $R = M/T$,
- measure the lag, L, which is defined as the period from the start of the test to the intersection of the tangent with the x-axis,
- Zeigler and Nichols (Franklin *et al.* 1991) suggest adjusting the controller gains based on the measured values of P, R and L, as shown in Table 6.8.

Simulink will optimise the controller automatically based on its analysis of the plant. For this application, the optimum gains are K_p = 0.2503, K_i = 0.173. K_d = 0.0637, and the step response is as shown in Figure 6.53.

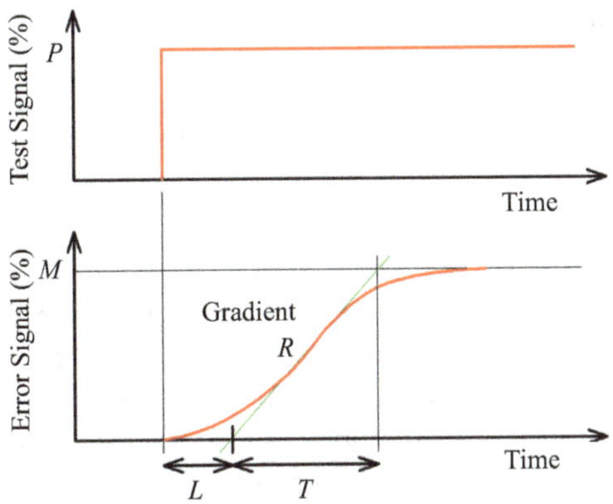

Figure 6.52 Response of the open-loop controller

Table 6.8 Gain relationship for optimum controllers as per Zeigler and Nichols

Control mode	K_p	K_i	K_d
Proportional	P/RL		
PI	$0.9P/RL$	$1/1.33L$	
PID	$1.2P/RL$	$1/2L$	$L/2$

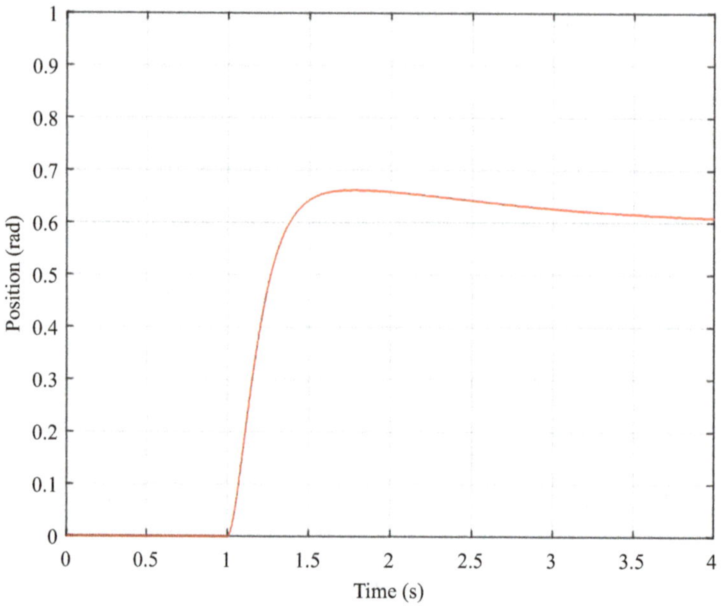

Figure 6.53 Response to a step input for a PID controller with gains optimised in Simulink of the motor position

Other optimisation techniques based on state-space representations are also available (Bryson and Ho 1975), but are beyond the scope of this book.

6.9.2 Controller hardware

Most modern control systems are implemented digitally, as this gives the most scope for optimisation of the controller structure and the gains. However, for simple applications, it is easy to implement proportional, integral, PI and PID controllers using the op amp building blocks discussed in Chapter 5.

Figure 6.54 shows the electronic circuits used to implement proportional and integral controllers.

In these examples, the purpose of the first op amp is to perform the required function and that of the second op amp is to operate as an inverter to restore the correct sign to the output; therefore, $R_3 = R_4$ and $R_6 = R_7$.

The controller transfer function for the proportional case is

$$G_c(s) = K_p = \frac{R_2}{R_1} \tag{6.97}$$

and, for the integral case

$$G_c(s) = \frac{K_i}{s} = \frac{1/R_5C_1}{s} \tag{6.98}$$

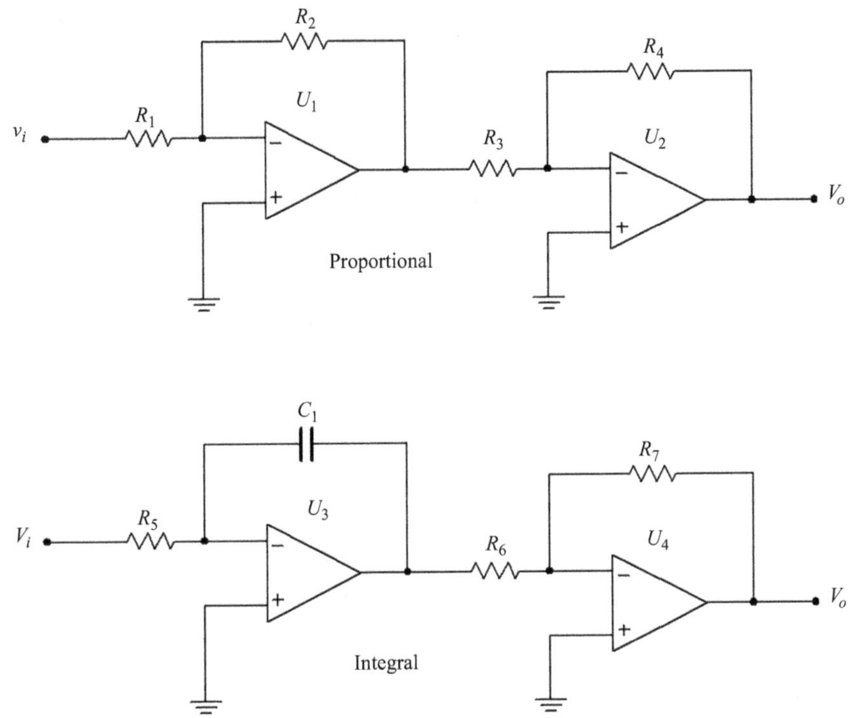

Figure 6.54 Op amp circuits used to implement proportional and integral control strategies

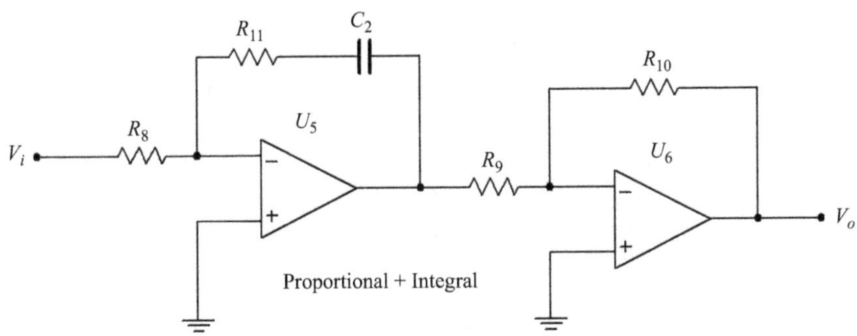

Figure 6.55 Op amp circuit used to implement proportional plus integral (PI) control strategy

The proportional plus integral (PI) controller can be implemented by combining the feedback elements of the previous circuits, as shown in Figure 6.55. In this case, the transfer function is

$$G_c(s) = \frac{R_{11}}{R_8} + \frac{1/R_8C_2}{s} \tag{6.99}$$

The final circuit to implement a PID controller requires an extra op amp to implement the derivative function. However, the pure derivative is seldom used because it adds excessive amounts of high-frequency noise, so a lead compensator is used.

This complete circuit shown in Figure 6.56 has the following transfer function

$$G_c(s) = \frac{R_{15}}{R_{12}} + \frac{1/R_{12}C_3}{s} + \frac{R_{18}C_4 s}{R_{13}C_4 s + 1} \tag{6.100}$$

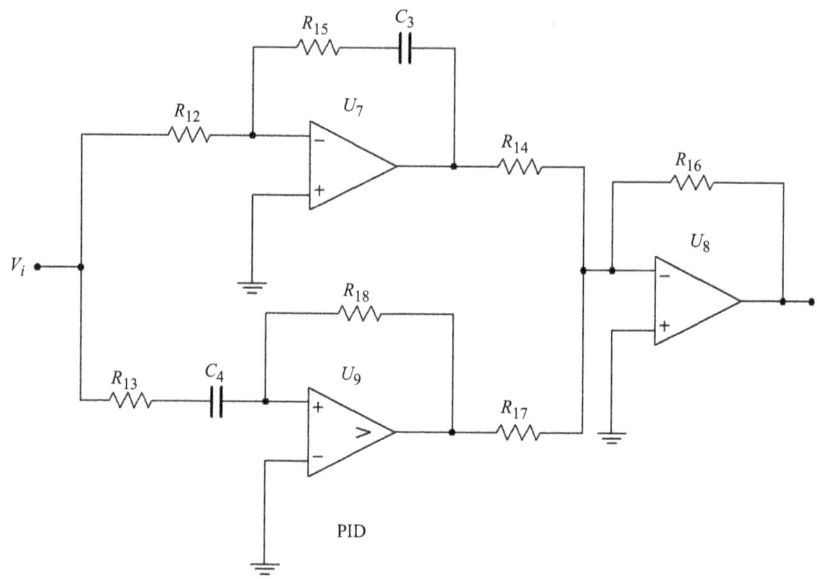

Figure 6.56 Op amp circuit used to implement PID control strategy

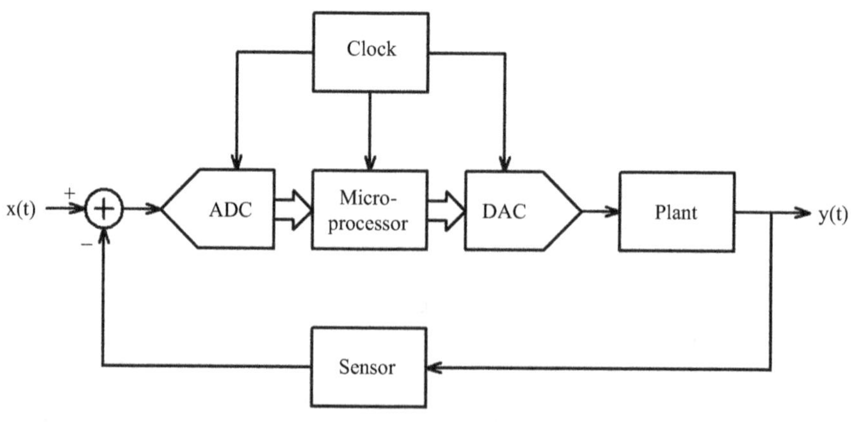

Figure 6.57 Digital controller implementation

In a digital controller, the error signal is digitised and then run through the appropriate control algorithms in a microprocessor before being converted back to analogue and introduced into the plant, as shown in Figure 6.57.

The main difference between analogue and digital controllers is that the latter operate in discrete time steps. Conversion of an analogue signal to its digital representation takes a finite amount of time, as does the reverse process. Additionally, implementing the controller algorithm in a microprocessor also takes time. A clock is generally used to synchronise the various processes. Some of these are discussed in Chapter 5.

References

Bolton, W. (1992). *Control Engineering*. Harlow: Longman Scientific & Technical.

Bryson, A. and Ho, Y.-C. (1975). *Applied Optimal Control – Optimization, Estimation and Control*. New York: Hemisphere Publishing Corporation.

Franklin, G., Powell, J. and Emami-Naeini, A. (1991). *Feedback Control of Dynamic Systems*. Reading, MA: Addison-Wesley Publishing Company.

Raven, F. (1978). *Automatic Control Engineering*, 3rd edn. New York: McGraw Hill.

Index

www.ingramcontent.com/pod-product-compliance
Lightning Source LLC
LaVergne TN
LVHW011833200926
848484LV00002B/352